Vision and Information Processing for Automation

Vision and Information Processing for Automation

Arthur Browne
Philips Research Laboratories
Redhill, Surrey, England

and

Leonard Norton-Wayne
Leicester Polytechnic
Leicester, England

Springer Science+Business Media, LLC

Library of Congress Cataloging in Publication Data

Browne, Arthur.
Vision and information processing for automation.

Includes bibliographical references and index.
1. Computer vision. 2. Automation. I. Norton-Wayne, Leonard. II. Title.
TA1632.B77 1986 006.3'7 86-18679

ISBN 978-1-4899-2030-0 ISBN 978-1-4899-2028-7 (eBook)
DOI 10.1007/978-1-4899-2028-7

Originally published by Plenum Press, New York in 1986
Softcover reprint of the hardcover 1st edition 1986

Preface

Developments in electronic hardware, particularly microprocessors and solid-state cameras, have resulted in a vast explosion in the range and variety of applications to which intelligent processing may be applied to yield cost-effective automation. Typical examples include automated visual inspection and repetitive assembly. The technology required is recent and specialized, and is thus not widely known.

VISION AND INFORMATION PROCESSING FOR AUTOMATION has arisen from a short course given by the authors to introduce potential users to the technology. Its content is a development and extension of material presented in the course. The objective of the book is to introduce readers to modern concepts and techniques basic to intelligent automation, and explain how these are applied to practical problems. Its emphasis is on machine vision.

Intelligent instrumentation is concerned with processing information, and an appreciation of the nature of information is essential in configuring instrumentation to handle it efficiently. An understanding of the fundamental principles of efficient computation and of the way in which machines make decisions is vital for the same reasons. Selection of appropriate sensing (e.g., camera type and configuration), of illumination, of hardware for processing (microchip or parallel processor?) to give most effective information flow, and of the most appropriate processing algorithms is critical in obtaining an optimal solution. Analysis of performance, to demonstrate that requirements have been met, and to identify the causes if they have not, is also important. All of these topics are covered in this volume.

The book assumes no previous knowledge of its specialty, but does require a general background of engineering expertise, including some familiarity with electronics and computers. It is intended to be useful to all involved in the engineering of automation systems using intelligent processing. This includes activities ranging from research, design and development to managerial decision making in which assessment of risk and cost effectiveness and the critical comparison of alternatives are involved. The book is not, on the one hand, an academic textbook on machine vision, of which many excellent examples are available already. Nor, on the other hand, is it a handbook listing data and specifications. Instead, the principles of the subject are developed, implementation is explained, and finally the whole process of engineering a system to meet an operational specification is expounded, illustrated by case studies from the authors' wide experience. The origins and development of intelligent automation, its social and economic implications, and its limitations are given brief mention to provide a context.

The book will be of value both to those developing their own systems and for evaluation of instrumentation offered by outside manufacturers. Its material is designed to have a useful life of at least ten years; thus discussion of the details of specific systems offered commercially has been avoided, since in the current climate of rapid development these are soon superseded. Instead, emphasis has been given to parameters on which systems should be assessed, such as throughput and resolution. The book is novel in discussing benchmark testing for comparative evaluation of automation systems. Sufficient coverage of robotics is included so that the requirement of vision for this important application may be readily appreciated.

The book concludes by considering shortcomings in instrumentation currently available, and anticipates likely developments, as well as those that will be needed to overcome known weaknesses and ensure continued advances in the state of the art.

In producing this book, our thanks are due to many people and organizations. We would like to thank Peter Saraga of the Philips Research Laboratories for writing Sections 7.5–7.7 on the programming of robots. We would like to thank the Philips Research Laboratories in the U.K. for permission to publish the information and figures related to work performed at the Laboratories, which forms part of Sections 6.2.3, 7.7, 9.3.1, 9.3.4, 11.6, 11.7, and 11.8.1. Special thanks are due to the BUSM Company for permission to include material appearing in Sections 3.5.9, 11.2, and 11.3, and to SIRA Ltd. for permission to reproduce Fig. 11.22. Sid Jenkinson of Philips and David Reedman of the BUSM Company have read and criticized drafts of the book, and their advice and that of others who have commented on particular sections is greatly appreciated. Much of the material appearing in Sections 11.1–11.5 is based on work performed at the City University during the period 1972–1981, and the contributions of several research students, in particular that of Geoff West in printed circuit inspection, are gratefully acknowledged; Figures 11.16, 11.17, and 11.18 are adapted from his work. A final note of thanks is due to Airmatic Engineering (UK) Ltd. for allowing publication of the material on tack inspection in Chapter 11.

Arthur Browne
Leonard Norton-Wayne

Redhill and Leicester

Contents

Introduction

1.1. Overview

Much has been written about intelligent automation systems; however, anyone wishing to construct such systems will further need a vast amount of background information on sensors, signal processing methods, electronic hardware, lighting, optics, and much more. This information is available, but is so widely dispersed through so many specialist reference books and scientific papers, often not directed towards automation applications, that it is hard to uncover, access, and hence exploit. We have brought this information together in a single volume.

This book has four principal objectives. The first of these is to present at an introductory level the basic principles which are applied in intelligent automation systems, and to explain the terminology. More advanced concepts are dealt with in specialist monographs, and particularly in papers published in technical journals; references to the most useful of these are given at the ends of chapters. The second objective is to describe the specialist hardware used in intelligent automation. Acquisition of information is of fundamental importance, thus sensors such as electronic camera systems are given particular emphasis. Selection of efficient electronic processing requires appreciation of architectures and methodology, and this is duly covered. The third objective is to discuss systems aspects, particularly the configuration of a system to meet specific performance requirements in a reliable and cost-effective manner. The measurement of performance and the assessment of its significance are also considered. Finally, a number of case studies are examined to illustrate how the principles and methodologies are applied. These are based on the authors' own experience. VISION AND INFORMATION PROCESSING FOR AUTOMATION focuses on intelligent automation applied to production in industry since this is where its use is currently greatest. Though in the future intelligent automation will be used in many other fields, the same technology will apply.

1.2. Introduction

Let us begin by discussing what is meant by "intelligence" in automation. A formal and precise definition of intelligence will not be attempted; its nature is understood too imperfectly, and is in any case manifested in too diverse a selection of forms in nature and in machines. However, it can safely be stated that intelligence is concerned with describing and quantifying an ability to process information. The more sophisticated the processing, then, arguably, the higher the intelligence.

Information exists in a variety of forms that can be arranged in a hierarchy of levels, from lowest to highest, as follows: symbolic, syntactic, semantic, pragmatic, and aesthetic. These are explained more fully in Section 2.2.1. It seems that intelligence should correspondingly be grouped into a hierarchy of levels, so that the higher the level of information being processed, the higher the level of intelligence exhibited. To process the most basic symbolic information—e.g., is the field of vision light or dark?; still or moving?—only the simplest intelligence is required. Conversely, for processing high-level (semantic and pragmatic) information, for example in deducing the strategic aim being pursued by a chess opponent from analysis of his moves, the intelligence required is at the highest level.

Mere speed and accuracy in processing data (as in the arithmetic performed by a computer) cannot be regarded as intelligence at anything but a rudimentary level. As a minimum, the following attributes are required of a system for it to be regarded as being intelligent:

1. An ability to process information.
2. An ability to accept information from the world outside the system.
3. An ability to output information to the outside world.
4. An ability to vary, in a systematic and purposeful manner, what is output as a function of what has been received as input.

However, useful intelligent systems can do rather more; their capabilities include

1. The ability to accept information from outside the system as a symbol string, often in a complex or garbled form, and to selectively filter out and decode semantic, pragmatic, and aesthetic information from redundant and noisy data. Section 2.2 explains how this may be done.
2. The ability to store information in memory, and to recall this information selectively when needed.
3. The ability to perform rational operations on data, including the making of decisions, the perception of relationships and the quantification of similarity and difference, and the prediction of the future by extrapolation from the past (given a model for system behavior).
4. The ability to output information in a form that may be understood and exploited by other intelligent systems, including human beings.

The above list is far from exhaustive.

When applied to automation, intelligent activity is often described as constituting robotics. There are two distinct classes of robotic activity. In *passive* robotics (for example, automated visual inspection), samples are sensed and the information thus obtained is used to characterize and/or classify the samples, but the physical actuation involved is minimal, and normally amounts to no more than sorting of material into batches. In *active* robotics, on the other hand, complex manipulations are performed, and an ability to perceive and anticipate position and orientation in three dimensions is important. Professor Thring[1.1] introduced an interesting distinction between *telechiric* machines, which perform remote manipulations (e.g., for handling objects which are highly radioactive) but are controlled by a human operator at a safe distance, and *robotic* machines, which can think for themselves.

Automation involving "intelligent" processing is used in industry to improve both quality and productivity. It can be applied profitably to many operations in manufacturing, including:

1. *Verification* of items, for example to detect the presence of incorrect product in batched material such as nails or tablets.
2. *Counting* of discrete objects; this is particularly important when dangerous objects such as drugs or valuable objects such as gemstones are being handled.
3. *Gauging,* i.e., checking that critical dimensions on components are within tolerance.
4. *Control,* e.g., for sensing the position of an object during assembly, providing feedback to control a robot.
5. *Recognition* of items from a selection of known alternatives, to permit batching, making-up into kits, etc.
6. *Orientation* and *positioning of objects, e.g.,* as an intermediate stage of an assembly process.
7. *Grading* discrete objects (such as diamonds) according to quality.
8. *Inspection* of discrete items (e.g., engineering components, pc boards) for flaws such as missing and extra holes.
9. *Detection, sizing,* and *identification* of flaws on web material such as steel strip, paper, glass, and textiles.
10. Determination of *surface quality,* such as roughness.
11. *Reading* of labels and other identifying marks, for instance in automated warehousing.

Adding intelligence to automation generally improves upon the performance of existing arrangements; sometimes it enables tasks to be performed which otherwise would be impracticable or uneconomical. The very long production runs on a single design of product that enabled inflexible and usually expensive automation to be used have given way to runs on a smaller number of products. The market now requires many variants of each product, and new designs are introduced frequently; dumb automation is inadequate. However, the flexibility introduced when intelligence is added to the line can make production even of short runs, possibly even of single items, economical. Robots without sensory feedback can perform useful functions, but they require the world about them to be highly ordered; all parts must be in the right place at the right time, and properties of the parts such as dimensions (consider, for example, the welding of a car body) must be tightly controlled. The addition of a sensory system can allow these requirements to be relaxed, making practicable and economical an operation which otherwise would not be.

Thus, adding intelligence to automation generally brings increased speed, accuracy, reliability, consistency, and cost-effectiveness. Increased flexibility is often an additional benefit; when the product is altered or substituted, an automated line may be adapted by being reprogrammed.

Perhaps the most common activity performed by intelligent instrumentation is inspection. *Inspection* involves the detection and classification of defects on materials or objects. A *defect* is some property of a material or object which is not normally present (i.e., represents departure from the "design specification") and which renders the product less acceptable to a purchaser or a user. We must distinguish here *cosmetic* defects which change only appearance without other degradation, and may not require that a product be rejected. Such marks are readily sensed by vision systems and so constitute a nuisance; considerable intelligence may be required if processing is to distinguish them. With increasing pressure to improve product quality, the size of faults considered significant is falling beyond the limits of human perception in reasonable working conditions. Even when defects are detectable by human perception, the variation of performance exhibited by human inspectors enables some faulty material to be passed. Using intelligent processing, inspection can divide the product into batches according to quality. Each may then be sent (at an appropriate price) to the appropriate customer.

1.3. Origins and Development

The origins of intelligent automation are broadly as follows. For almost 200 years, there has been a steady development in the efficiency of the production process, using whatever technology (from the capstan lathe and the moving line to the numerically-controlled machine tool) and methodology (such as work study and computer-aided design) have been available. The chief objective is of course to increase productivity and hence profitability. Other aspects such as a need to make production less boring or safer, or to eliminate a dependence on scarce skills, have also been important considerations.

Recent years have seen vast development in the field of Information Technology, in which electronic hardware (mostly digital) and programs to operate it are used to store, communicate, and process information in a way imitative of animal and even human brainpower. Logical operations and hence arithmetic are easily implemented. Senses such as sight, hearing, and touch can all be imitated using transducers which convert appropriate physical signals into a common electronic format: the digital signal. This is a sequence of integer numbers lying within a fixed range, and is suitable for processing by general purpose computing hardware. Visual displays and speakers are used to transmit information in the opposite direction, from machine to man, and so close the communication loop. The logical power of the processing that is already possible exceeds in its accuracy, speed, consistency, and reliability that of man by many orders of magnitude.

This development was initiated chiefly by human curiosity, reinforced by military needs during World War II, but progress soon revealed a potential for useful commercial application. The obvious requirement in data handling for management and administration, for keeping records, paying bills, and so on, led to commercial production of computing systems far more advanced than the primitive machines of the pioneer universities. These advanced computing systems were much easier to use, having the benefit of high-level languages and friendly operating systems, and were eagerly accepted by industry, commerce, and government. Some were acquired by organizations such as universities for scientific research; it was a natural step to couple a TV camera to such a machine, and to store, process, and display the images thus obtained. This progress might eventually lead to imitation (or at least improved understanding) of human vision. The idea of producing anthropoid machines is intriguing, as is evidenced by the writings of science fiction. The endeavors of these early investigators (i.e., those before 1970) were crowned with only limited success in terms of what they were able to do, but generated a wealth of ideas which were disseminated widely in the scientific literature.

An initial objective of intelligent processing was to read printed characters, for purposes such as directing mail. This has now largely been fulfilled, though early investigators found that problems such as poor print quality and the wide variety of fonts used made the task difficult. In fact, systems are now commercially available which read 1000 characters of good quality print per second, with a rejection rate of 10^{-6} and a substitution rate of 10^{-7}. Reading of handwritten characters has also been attempted, but with very limited success unless the characters are printed along guidelines. The difficulty can be understood by comparing the machine to human reading; in the latter, complete words are recognized rather than individual letters, and the need for text to make sense enables doubtful words to be identified. These aspects have not yet been fully realized or exploited.

Although the incorporation of intelligence into automation has been of interest since the middle of this century, hardware initially available (such as valves and relays) has been too slow, bulky, unreliable, and expensive for anything other than the simplest processing. In the 1960s, however, integrated circuits based on silicon chips became available. These are compact, robust, and economical in power consumption, besides having an almost infinite lifespan. By 1970, the appearance of the minicomputer had made processing using stored programs sufficiently inexpensive for its application to be contemplated outside the laboratories or the computing departments of large industrial firms. Initial application was to on-line control, in which the algorithms of control theory were applied to large scale plant in-stream production, such as chemical plants and rolling mills. Parameters such as speed, temperature, etc. were sensed continuously, and fed back, to implement optimal control in ensuring maximum throughput, minimum waste during a change in product specification, safe operation, and so on. But the cleverer activities of the human operative, those involving vision, hearing, touch, and positioning of objects, were not implemented industrially; they remained a curiosity pursued (sometimes rather expensively) only in academic organizations.

In the early 1970s the microprocessor appeared, yielding another decrease in the cost of computing by at least an order of magnitude. Almost simultaneously there appeared the solid-state linescan camera with its ability to sense optically with great dimensional accuracy, at high speed, at very modest cost, and with computer compatibility. The stepper motor provided an economical means of accurate mechanical actuation compatible with digital processing. Horrific inflation at about this time was causing manufacturing costs to rocket, and cheap labor in the developing countries attracted labor-intensive aspects of manufacturing (such as packaging integrated circuit chips) abroad. Developed nations were quickly realizing that to remain ahead economically, they had to push forward in the development and application of new technology. The opportunity to apply machine intelligence to automation had now arrived.

In the meantime, the military had not been slow to exploit information technology. An original (and still dominant) use of computing power has been for deciphering intercepted communications. The most interesting military application has probably been in the analysis of the images captured by earth satellites. The amount of data thus gathered has been so great that it is quite impractical to analyze it visually, and some effective though expensive machines have been developed for this task. Automation of the battlefield using machines to direct missiles onto targets and to perform dangerous tasks like defusing bombs, has been a natural development. Tasks such as the guidance of an aircraft close to rough terrain require such quick and infallible reactions that they would be impracticable without automation. Use of simulators for training provides realism combined with safety and economy. The ultimate objective would of course be to automate everything, and so eliminate the chief hazard in warfare—the human casualty (on your side!). This has not yet happened—eventually it may!

The exploration of space generated a need for the transmission of images across vast distances with modest power, and for the correction of these images for noise, distortions, and so on. National survival, and to a lesser extent the prestige of a nation's technology, are a preeminent concern of all governments, and cost is generally a minor consideration in ensuring that one's defense systems are better than those of potential enemies. The success of the USSR in preempting the USA in space technology in the late 1950s initiated the Defense Advanced Research Projects Agency (DARPA), which has initiated and supported some excellent work in fields such as computer networking, large-scale integration, artificial intelligence, and so on. DARPA extends even to support of automation in manufacture[2] if this is seen to ultimately support defense needs.

In law enforcement, there exists a need for processing images to eliminate degradation due to poor focus, movement, and uneven lighting. Some very effective instrumentation has been produced to identify fingerprints automatically. Research in pure science, in astronomy to improve images corrupted by atmospheric turbulence, in nuclear physics to examine bubble chamber photographs by the million to detect extremely rare events in collisions which may indicate new fundamental particles, and in electron microscopy to infer the phase and amplitude transmission characteristics (or the three-dimensional distribution of matter) of a specimen from multiple photographs showing amplitude only, has generated a myriad of ideas and methodologies.

In broadcasting for entertainment and information, there is a need to optimize the efficiency of the worldwide transmission system (often dependent on satellite links) for communicating images in real time without degradation perceptible to an average observer. This has led to a wealth of developments such as efficient methods of coding based on orthogonal transforms.

It is noteworthy that pure science has been interested chiefly in the processing of images. Analysis of speech has been more purposeful and is of great antiquity, having been initiated around 1940 in the hope of making more efficient use of the bandwidth available in the telephone network. Intriguingly, no progress has been made toward this initial aim; fortunately, technical developments have made bandwidth cheaper and easier to obtain. But the understanding derived from this work has proved extremely valuable; machine intelligence has been applied to recognize at least isolated words of speech, to identify the source which has generated various nuisance sounds, and to detect malfunctions from sounds emitted by machinery. Generating intelligible speech electronically is easy, though making this artificial speech sound natural is much more difficult. Work on speech processing has been fundamental to extending understanding of the digital specification and processing of signals.

In medicine, novel methods of imaging (x-ray tomography, gamma ray imaging, nuclear magnetic resonance, ultrasonics)[3] have been developed to provide noninvasive, nontraumatic visualization of the internal structure of the human body, often in real time. The effect on medical diagnosis has been revolutionary. Well-engineered instruments are available commercially, that are simple enough to operate to be of routine use. Vicious competition has resulted in hurricane progress in the application (and the development) of processing methods based on very sophisticated mathematics. Life is a precious commodity, and high costs (often of £1,000,000 per machine) are accepted.

Progress toward application of intelligent instrumentation in industry has been much slower. On the one hand, the need for cost-effectiveness to be scrupulously maintained means that costs acceptable per installation rarely exceed a few tens of thousands of pounds. On the other hand, the need for processing to take place in real time (the time available to inspect an assembly produced at one per second is less than one second) means that many excellent methods of processing simply cannot be considered because they are too slow. Instrumentation that works well when handled by skilled technicians in a laboratory environment does not always continue to perform when handled by semiskilled operatives in the grime of a factory. The prestigious and well-funded organizations of gifted research workers who feed innovation in military, medical, and scientific research are not (with a few noteworthy exceptions) available for supporting industrial automation. For commercial reasons, much progress in industrially applicable intelligent automation is not published; there is no sense in giving it away to one's competitors. Secrecy is often a better form of protection than patenting, so the results of industrial research tend to be inaccessible. In contrast, those of the academic world are often hard to understand; articles published in prestigious technical journals are presented so as to impress other academics and referees, and rigor often takes precedence over lucidity.

To remedy this situation, attempts are at last being made to feed government funding into intelligent automation for industrial application; if a nation fails industrially, it will inevitably fail economically. This failure may be less abrupt and dramatic than defeat on the battlefield, but it is equally effective in promoting national humiliation and misery. It can bring about change in government and the displacement of an establishment. Hence the appearance of the Alvey, Esprit (this also has the function of aiding the unification of Europe), ACME, Teaching Company, and other programs in the United Kingdom to make government money available to support industrial automation.

Thus, the engineer in intelligent automation must adopt a different approach from those in medicine, defense, aerospace, or pure science. His objective must be the optimization of efficiency, and in particular of value for money. One factor in his favor is the likelihood that the instrumentation will be produced in quantity, leading to economy through scale. His processes must be ingenious (to conserve resources) rather than complex. For example, he must exploit inescapable factors (such as the movement of items along a belt) instead of being frightened by them.

1.4. Current State of the Art

The market in intelligent automation is potentially very large, and is growing by about 50% annually. Firms supplying instrumentation are often not yet operating profitably; they are kept going chiefly by the promise of eventual bonanza profits as the scale of application, and consumer confidence, develop. Many small firms are involved (with a few large ones, like General Electric); there will probably eventually be a "shake out" in which weaker competitors go to the wall. This is not altogether in the interest of users, since maintenance and support for instrumentation may become unreliable if a vendor disappears. The alternative, whereby small firms possessing narrow but unique expertize are taken over by larger firms that can use this expertise, is more desirable.

1.4.1. Achievements

Intelligent automation is already being applied extensively in industry, with due profit.[4] Offline operations such as particle sizing in soil analysis and region sizing in metallurgy have been possible using standard commercially available instrumentation for many years; simple image processing operations are utilized, with high quality TV cameras such as plumbicons as sensors. A degree of human participation is retained; an operative sits in front of the machine watching a visual display, and possibly interacting via a lightpen. On-line, passive tasks were the first to be tackled, particularly automated visual inspection (for the detection, sizing, and classification of defects) on material produced as a continuous web or strip, such as paper; this is an easy material, since most defects can be sensed by transmission, and it is easy to segment defects from their background. Similar inspection has been performed on magnetic tape, photosensitive film, glass, plastic, and textiles. The last of these represents the most difficult problem, since it is particularly difficult to distinguish defects from the normal structure of the textile. Printed patterns share this problem, and their inspection is currently the subject of much research.

Inspection of isolated discrete objects then followed, particularly of items that can be represented by their silhouette outline. Applications include tasks such as examination of biscuits for correct shape and thickness and analysis of the chocolate coating on sweets. Critical dimensions are measured on engineering components, and holes are counted. Packaged items (such as chocolate bars) are examined to ensure that the wrapper is positioned correctly.

There has also been useful progress in the inspection of isolated objects for which the silhouette representation is not adequate. Corn-cobs are now graded for size and color using machine vision (with three low-resolution solid-state linescan cameras at 120-degree intervals to cover a cylinder completely), and apples are similarly graded. Another successful application has been the trimming of carrots prior to canning. The food industry seems particularly suitable for the introduction of intelligent automation, probably due to the requirement for high throughput of nominally identical articles of low unit value. Instrumentation is also more hygienic than human operatives, prone to infectious diseases.

Machine vision is also being used to orient and position components in the intermediate stages of automatic assembly, to read dials and keyboards, to inspect printed wiring boards and integrated circuit wafers, to inspect assemblies, to guide robots for simple placement tasks, and to read addresses and sort codes.

Rather surprisingly, little instrumentation is as yet commercially available to solve some problems which appear straightforward, and for which a demand clearly exists. An outstanding example here is the automated visual inspection of bare printed circuit boards; certain large firms have made instruments for use "in house" (at great cost), but these are not generally available. However, instruments are available commercially for the closely related problem of inspecting integrated circuit masks (at ca. £250,000 each). It is worthwhile to inspect automatically pc boards with components mounted, but this requires processing having considerable intelligence, since many differences between a sample board and a notionally perfect prototype are not defect indicators.

Some instrumentation capable of solving more complex problems for which silhouette representations do not suffice has been demonstrated publicly and is believed ready for installation. For example, a machine for inspecting the brake drum assemblies of motor cars was shown at RoViSec in 1984.[5] Machine vision has been applied even to reading gas meters and inspecting keyboards; these are characterized by a need to read simple, well-defined patterns. Solution of problems requiring the sensing of color or fine detail (particularly when these requirements occur together) tends to be less advanced.

Acoustic as well as optical systems are good providers of information. The transient noise generated when a component is manufactured by stamping can be analyzed to detect malfunction in the press; the continuous noise generated by a car engine can be processed to detect and identify problems such as broken gear teeth. Physical damage to components may be detected by analyzing the sound they emit when struck; the skill of the wheeltapper may be emulated using instrumentation, for example, for detecting cracked cathode ray tubes. Analysis of the complex signals generated by a battery of sensors responsive to variables such as pressure, temperature, or gas composition can monitor engine condition very thoroughly, and detect and identify malfunctions, without a requirement to dismantle the engine.

Application in active robotics has been much slower. A few activities (such as the fettling of castings, and spray painting) which human operatives find particularly distasteful have already been automated, but the actions required here are essentially repetitive; they can be learned by the machine by imitating the motions of a human, and the sensing required is minimal. Much activity is being devoted to automating arc and spot welding as required in the motor industry; tasks which are often performed poorly by human operatives. Although the visual feedback required to guide the welding head is very complex, several systems are already being marketed.[6] Also in the motor industry, vision is used so that a spray painting robot can select the correct color although the bodies to be painted appear on a line in random order. Force sensors are used to provide feedback during fettling. Another task just being applied "in plant" is the automated handling during assembly of the cathode ray tubes used in color televisions. The advantage here is in avoiding manhandling for the accurate positioning of heavy objects.

The use of Computer Aided Design (CAD) techniques is now widespread in industry; maximum benefit is obtained if CAD systems are used to guide automated assembly and inspection. An instrument to inspect PC boards could, for example, be linked directly to the CAD system used to design the boards.

The general problem of applying active robots to industrial automation remains unsolved, despite much research activity. An important objective here is flexibility in manufacturing, i.e., the provision of production lines which are reprogrammed rather than retooled when the product being manufactured is modified. A critical requirement is the capability to recognize objects mixed and randomly positioned, to grasp and remove selected objects, and to position and orient them, before incorporation into an assembly. The randomness inherent in the presentation demands a highly intelligent robot.

Vision sensors currently available commercially for robots are limited to primitive processors capable of identifying and homing on simple features such as holes. Lightweight, low-resolution (32 × 32 pixels) sensors which are small enough to be carried easily on the arm of a robot are often used. Novel processors such as CLIP and WISARD seem highly appropriate in robot vision, but are (so far as we know) as yet untried in this application.

1.4.2. Factors Limiting Progress

Despite its benefits and almost limitless potential, the development and application of intelligent automation has been surprisingly slow. This is due to several factors; the most important of which is probably the large investment that must be made before the benefits of automating an activity can be enjoyed. A long leadtime generally elapses between initiation and completion of a project. Since the technology involved is novel, considerable risk is involved. Moreover, intelligent automation uses methodologies and concepts (such as machine pattern analysis) generally unfamiliar to those more senior members of an organization, who must guide the project and take responsibility for critical decisions. Skills must be used which are not available within an organization, and they have to be obtained from outside. This means using consultants, hiring new staff, or retraining old ones. All these factors are costly and involve yet more risk. Further, firms may not possess the expertize to manage research performed on their behalf by outside organizations if they have never been involved in this kind of activity before.

1.4.3. Social Aspects

There are also social aspects to automation; though this book is technical rather than sociological or political, the economic and human consequences of automation are dramatic and cannot be ignored. The general effect of automation is to eliminate large numbers of jobs which are essentially unskilled (such as operatives on assembly lines) and smaller numbers of skilled jobs (such as typesetters). This creates unemployment, and hence passionate opposition. But automation creates jobs of its own, in research and development, design and production, and installation and maintenance, and these enjoy a higher status and remuneration than the jobs eliminated. They require a higher level of technical competence and hence of education; they are jobs people like to do. Most important, automation creates wealth by improving efficiency; it is up to the politicians to ensure that this wealth is distributed equitably, to be enjoyed as widely as possible. Although intelligent processing is very new, automation has been evolving steadily since about 1800; machine intelligence is merely the latest phase. The growth of automation is an inescapable part of the human condition, following from general progress in the development of technology, and is stopped only by tyranny. It is imperative that the benefits automation can bring be realized and enjoyed, and that its harmful consequences be anticipated and avoided.

1.5. Intelligent Processing—Comparison of Man vs. Machine

Intelligence is the ability to process information, to perceive relationships, make decisions, and so on. In the next chapter, we will define information and explain the significance of this definition in terms of the desiderata for an efficient automation system. Almost invariably, electronic signals are used to store and convey information; however, optical systems (explained in Section 10.4.2) are used occasionally where the environment is good enough, and their use will certainly grow. Fluidic systems, in which a binary signal is encoded by the presence or absence of a jet of air, are now obsolete, since they are slow, and cannot actuate anything directly. Mechanical systems are extremely slow, expensive, and cumbersome; they consume excessive power in overcoming friction and inertia and are prone to wear and breakage; they are now obsolete even for watches!

The structure of an intelligent system for industrial automation follows the schematic form shown in Figure 1.1. A SENSOR acquires information from the outside world; this is most frequently in the form of an image, hence the most common sensor in intelligent automation is a camera having an electronic signal as output. However, mechanical sensors (which generate a signal indicating force, position, or accel-

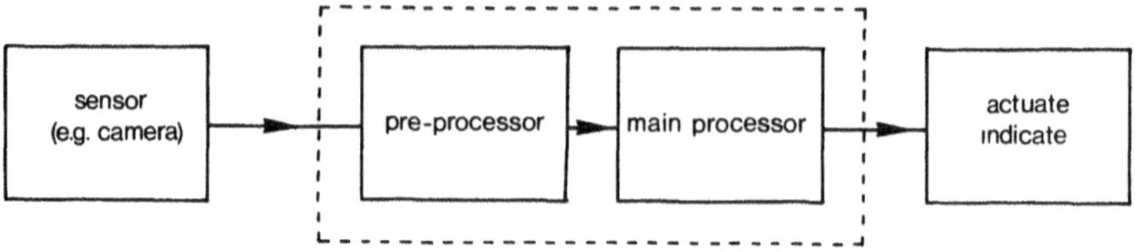

Figure 1.1. Structure of an intelligent system.

eration) are also becoming common. Sensing of acoustic, magnetic, and thermal and other physical quantities is used occasionally. Such natural signals are generally *analogue,* i.e., comprise a continuous variable whose amplitude corresponds to the quantity sensed.

The signal is conveyed to a *processor* that is almost invariably some form of a digital computer. Thus, analog/digital conversion is involved. The form of the processor depends on the application; for investigation off-line to determine a configuration and processing sequence for eventual implementation on-line, a standardized package based on a standard minicomputer is normally used. On-line, however, one or more microprocessors are usual, together with a preprocessor to minimize the load the former must handle. Special purpose processors which are much more powerful than standard micros are beginning to be available commercially, though at the moment they are rather expensive. These aspects are dealt with in Chapter 10.

Usually, some form of actuation is required; in a simple system, this may well be merely a jet of air to blow a defective item into a reject bin. A black art is developing involving bowl feeders, raceways, rotating brushes, and magnetized drums for moving small parts past sensors without jamming, at rates approaching 1000 items per minute; the systems are often designed empirically by craftsmen, without underlying theory. At the other end of the scale, complex robots use vision and other forms of sensing to guide arms to grasp objects and position them. Considerable precision may be needed, and operation must be in real time. The processing may be required to anticipate the position of a moving object at some future time, and hence will need to perceive the world in four dimensions.

Even where actuation is not required, a system will probably produce some kind of output indication that can be used by its human operator. This may comprise merely a statistical listing of the occurrence frequencies of different defect types, or it may involve the marking of regions where defects are present, or the generation of a directive to stop a line because a fault condition is causing bad product to emerge continuously.

Intelligent automation systems replace or supplement human operatives; their value lies to a considerable extent in superior performance. Industrial organizations are generally shy about disclosing the capability of their operatives. However, the performance of human operatives as inspectors has been investigated thoroughly by psychologists, for example, Moraal for the steel strip industry. He studied their efficiency for detection and identification of defects visible on the surface of fast moving strip,[7] and reports that although several years of training are necessary to recognize the 50 or so kinds of defects present on cold rolled steel strip, performance varies considerably between individuals. Different inspectors give widely differing designations of the type and severity of defects present on the same sheet; they are highly mutually inconsistent. Generally, visual inspection is boring to perform (thus, nonproductive rest periods are required), and fatigue leads to fall-off in efficiency as the day proceeds. Besides being slow, expensive, inconsistent, and subjective, in certain circumstances it is also dangerous, for example, underwater, underground, or where there is penetrating radiation.

The main justification for intelligent automation is, however, economic; in the medium to long term, machines are much more cost-effective than people in many applications. Thus, automation of visual activity offers many advantages. The chief shortcoming is an inability (as yet) to mechanize high-level activity such as the interpretation of unconstrained scenes. For example, although machines can easily identify silhouette shapes provided they are isolated, they are generally hopelessly confused by even a small overlap. Humans deal with this problem easily. Whereas even a mentally handicapped human being can perform the simple task of inserting a screw into a threaded hole, a machine finds this very difficult. In addition to locating the hole and distinguishing it from others possibly present, a subtle process of mechanical feedback is involved in lining up the screw coaxially with the hole, and maintaining its position while it is being rotated.

It is interesting to compare the properties of man and machine for processing information; the more important features are indicated in Table 1.1.

Table 1.1
Comparison of Man and Machine

Man best	*Machine best*
High memory capacity	Fast
Versatile	Reliable
Easy input/output	Accurate
Highly miniaturized	Good environmental tolerance
Easy manufacture	Consistent
	Servile

Man is most clearly superior to machine in sheer versatility—in the range of tasks he can tackle. This is due partially to his ease of communication with the world outside, chiefly through speech (with hearing) and vision, though touch is also important. The high degree of miniaturization in animal information processing—which reaches its peak in man—has resulted in a computer (the brain) capable of storing upwards of 10^{11} bits, weighing only 2.5 lb, and consuming merely 25 watts. Redundancy protects against destruction of individual processing elements. Parallel processing via hard-wired paths contributes strongly to the efficiency of the human brain, yet does not destroy the capacity of this processor for modification, which for example allows the system to develop its own software via self-learning.

But the superiority of man ends there; in all other respects the computer reigns supreme. Computers have been undergoing development for only about 40 years, and are still being improved very rapidly, whereas animal systems have developed over hundreds of millions of years and advance very slowly. Thus, it appears certain that machines will exceed man in almost all capacities even in the medium-term future.

Computers are fast; they can, for example, add two 60-bit numbers in 10^{-7} second; a human would probably take 20 seconds, and would get the answer right only after several attempts. This tendency for errors to occur when humans process information is probably due to random noise; while this is present in electronic processing systems also, it is easy by good design to restrict to an acceptable level the corruption caused by noise. This relative absence of noise also means that computers are highly consistent; if the same data is fed repeatedly into a given computer program, an identical result will always be obtained. Human processing is notoriously variable.

Perhaps the greatest advantage of the computer in automation applications is its servility. Human beings are not designed solely to perform work; they have emotions, hopes, fears and ambitions, political and social motivations, sexual drives and domestic stresses, and they get tired. They have a built-in potential for dissatisfaction, and are hence often uncooperative, subjective, and inconsistent. Machines, on the other hand, are purpose–built to perform specific tasks; they are completely servile, have neither wili nor soul, and never crave to do anything else. The ability to tolerate hostile environments, for example, to perform effectively within the temperature range −55°C–+125°C whereas humans are not very good outside 15°C–25°C renders computers even more effective as workers.

And although machines need occasional maintenance and repair, their propensity to illness is nothing compared to that of a human being. Planned maintenance, aided by steady improvement in reliability engineering often arising as fallout from defense and space programs, is making computing hardware even healthier. When worn out or obsolete, machines can be scrapped, destroyed, and replaced without causing the misery and humiliation that would be involved with human operatives.

Thus, it is clear that although man remains superior to machine for most industrial operations, particularly when hand–eye coordination is required or where work is nonrepetitive, machines have vastly more potential for future development. Instrumentation can sense physical quantities that man cannot sense, such as magnetic fields, and covers a wider frequency and dynamic range than man where the capabilities overlap. Man has probably reached the limit of his performance capability; present day intellectual achievements are generally no better than those of the ancient Greeks, and humans breed for appearance, not intelligence. Computers on the other hand, are less than 50 years old, and development is explosive. The rate of fall in the cost of computer memory alone indicates dramatically how fast things are moving.

Human beings receive information via the five senses: sight, hearing, taste, smell, and touch; intelligent instrumentation must be able to accept information in these ways also. Sight is by far the most important (due to its high data rate); fortunately, good progress has been made in producing electronic camera systems which substitute for the human eye. Progress in simulating the actions of the brain in interpreting the information acquired optically has been much slower, but is developing well. Fortunately, hearing is less important than sight to intelligent processing; the overall data rate is several orders of magnitude less than for vision, and communication via speech is much improved upon by use of electronic data channels. An ability to identify and interpret incidental sounds (e.g., indicating that something is being broken) would be useful in active robotics, and is probably easily provided by extending current work in sound analysis. Smell and taste involve microchemical analysis; although the ability of these senses to detect and identify minute amounts of specific chemicals is remarkable, their usefulness in automation is probably confined to safety aspects, such as recognizing smells due to burning and escaping gas, since the information rate of these senses is very low. It is of note, however, that British Leyland have used helium sensing[1.8] to detect leaks in sealing cars.

Touch, on the other hand, is vitally important, and is more sophisticated than the mere measurement of pressure or force.The human hand can measure multicomponent forces, and can respond to variation of the distribution of pressure (at the fingertips) with a resolution quite unmatched by an available instrumentation. These measurements are then fed back to control the forces with which a fragile object (such as an egg) is grasped, or to guide a screw into a threaded hole, when visual sensing of thread registration and coaxiality are impossible. This area is currently the most significant for research; approaches such as measuring the local displacement of a shaped rubber membrane against a glass surface providing total internal reflection except at regions of contact are being investigated.

1.6. Steps to Provision of Intelligent Systems

Let us consider next how the provision of an intelligent system for automation should be approached. The best solution obviously is to purchase existing instrumentation and install it, but this is rarely possible—suitable instrumentation rarely exists! More usually, an intending user has to develop his own. This involves a number of distinct stages, consecutive and mutually indispensible, as summarized in Table 1.2 (and Figure 1.2).

Table 1.2
Stages in the Provision of an Automated System

1. Define problem, e.g., throughput and error rate
2. Demonstrate cost benefit
3. Assess technology
4. Establish feasibility
5. Engineer system
6. Use on-line; install, evaluate, operate, maintain, test.

1.6.1. *Preliminary Work-Problem Specification with Cost–Benefit*

Preliminary work comprises two stages. The first is to define the problem; the more precisely and quantitatively this can be done, the better the project is likely to proceed. Problem definition must include specification of critical parameters such as the throughput rate required and the error rate which can be tolerated. Provision of a specification, even when partly arbitrary because the performance of existing nonautomated methodology is not accurately known, gives a definite objective, which enables progress to be assessed, and guides in the comparison of alternative technologies.

1.6.2. *Cost–Benefit Determination*

The second part of this stage is determination of the cost–benefit that will ensue if the task is successfully automated. It is a truism in technology that virtually any objective, however difficult (like putting a man on the moon), so long as it falls short of the physically impossible may be achieved simply by spending money. However, the expenditure justified in introducing automation is generally limited by the benefits the automation will bring. It is important to confirm at the outset that the expenditure is sufficient to support provision of the automation. It is necessary also to ensure that no more money will be spent (or risk taken) than is absolutely necessary; it may be worth relaxing the specification to keep the cost within acceptable bounds. Automation will bring benefits in increased throughput, thereby increasing utilization of manufacturing facilities, improved quality of product, thereby reducing waste (and increasing customer goodwill), reduced cost of manufacture, and so on. These benefits must be traded against the cost of providing the automation, which must include the

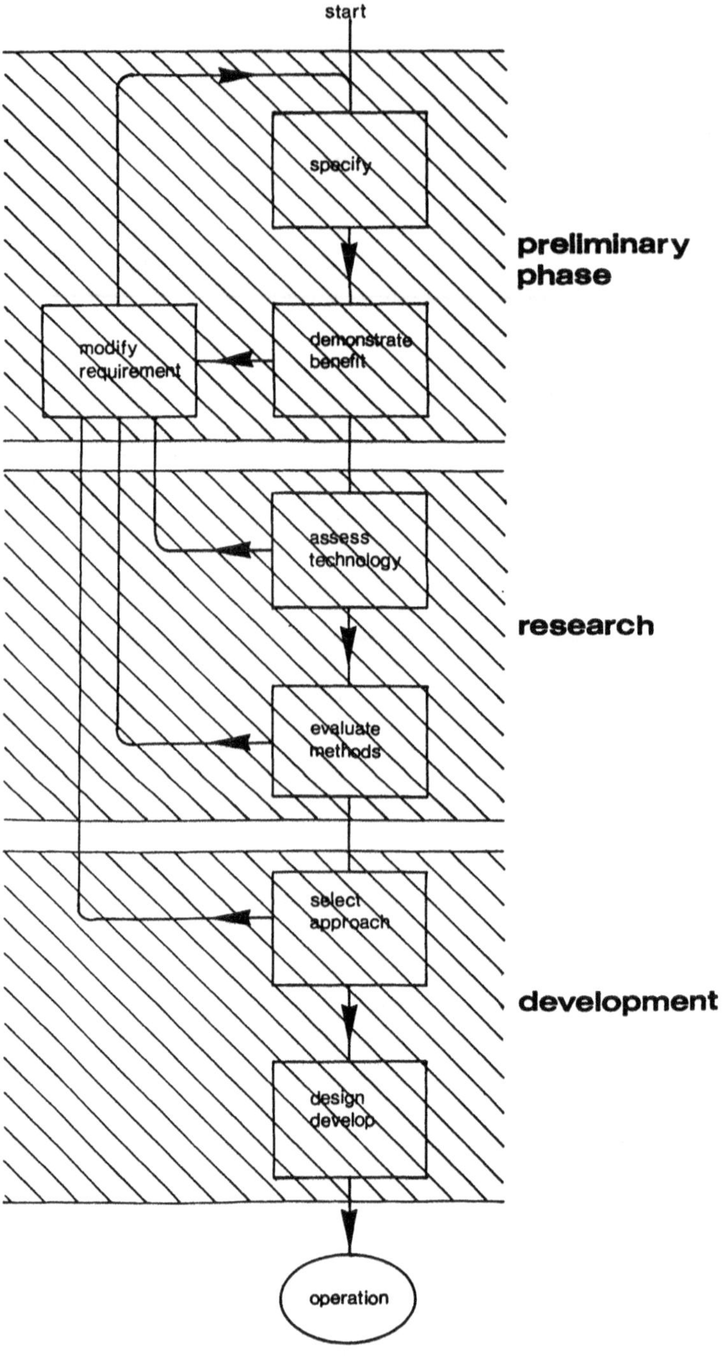

Figure 1.2. Stages in the provision of an intelligent automation system.

cost of research and development as well as the cost of the instrumentation installed on the line and its maintenance and operation. Only when it can be shown that the expected cost of provision is less than the savings to be expected over a reasonable period which is not greater than the expected lifetime of the instrumentation (say, four years) has potential viability been demonstrated, and it is worth proceeding with the project. Thus, quantification of the cost benefit yields a specification of what may reasonably be spent.

The introduction of automation at a particular stage in a process may well have repercussions at other stages. For example, if components are fed or inserted automatically rather than by hand, it may be necessary to introduce additional inspection to ensure that faulty components do not reach the automation stage and cause expensive machinery to jam. When parts are handled by human operatives, they often undergo an informal visual inspection whose existance and importance may not be realized until the automation has been introduced!

In high technology, the cost of providing a minor increase in system performance is often excessive when the law of diminishing returns sets in. For example, the cost of improving the detectability of defects from 90% to 95% may well be tens of times greater than the cost of increasing detectability from 80% to 90%. The benefits available may then justify only the lower performance.

The end product of the preliminary work is a specification of task and performance, and an indication of how much money may be spent in producing the system. It may be worth seeking government funding using one of several schemes according to the amount of innovation involved and the need to maintain confidentiality or to exploit the methodology developed beyond the original application. In the latter case, royalties may have to be paid. The schemes in operation change considerably from year to year.

1.6.3. Tackling the Project

At this point, a decision must be made as to how the project will be tackled. Potential users of automation may not wish to undertake the work themselves. In industries such as food, textiles, apparel, printing, etc., individual firms may have neither the technical resources nor the experience. There are really two possibilities here; either buy a complete (turnkey) system from an outside supplier, or conduct a self-run project in which selected outside organizations undertake portions according to their expertize. Purchase of turnkey systems is attractive in that the risk involved to the purchaser is relatively small; if the equipment does not perform, it is taken out of the plant and no payment is made. But in turn, the supplier will expect a fair profit, and his product may be costly.

If turnkey instrumentation is too expensive, or if a system suitable for the task in hand is not available commercially, there is no alternative to developing one's system. In both cases, at least some of the work may be contracted to outside organizations; it is necessary to decide how much. For major and innovative projects, it may be necessary to break the task into a series of subprojects, each undertaken by a separate organization competent to do the work required, firmly directed and coordinated by the sponsoring organization. This approach was used in the US space program with outstanding success. The outcome for each subproject (is it to produce a report containing information on methodology or hardware, computer software, a laboratory demonstration, or on-line hardware?) is specified by the directing organization, which sets time scales and targets, and ensures these are met. It also sees that participating organizations concentrate on their own tasks, and are provided with the resources they require (including information from the other participants) to do their work. Coordination involves ensuring that participants cooperate, and do not compete with one another or gang up against the sponsor to run the project in their own interest. A long program should be divided into phases each having a definite objective. A regular program of meetings and reports will ensure that progress is monitored and communication is maintained.

Selection of appropriate contributors requires some skill to ensure value for money is obtained. Academic organizations are worth considering the early phases, since they possess state-of-the-art expertize in principles, they receive excellent support from facilities such as libraries and computer centers, and they are good value since the overhead they charge on labor costs is usually closer to 50% than to the 250% normal in industry. Detailed work is normally performed by research assistants who are well-qualified, recent graduates hired for the project in hand. Although the research assistants are generally temporary staff, they anticipate good career prospects (and a higher degree) from a job well done, and are thus well-motivated and they are guided by a permanent academic staff who are often experts of worldwide repute. It may be necessary to provide some special equipment; frequently this is purchased by the sponsor, and loaned for the duration of the project. But academic organizations may know little about specific application areas, and may well need a thorough briefing involving visits to factories and so on. The formal contractual relationship between sponsor and research organization will need to ensure confidentiality and the commercial protection of invention; it is normal in academe to publish as much as possible, and the research assistants will normally wish to submit their work for a higher degree, involving generation of a thesis.

Advice may also be obtained from consultants[1.9] who charge per diem for their services. These are often from academic organizations, though private firms with the necessary expertize are beginning to appear. They are valuable for providing technical advice, often in considerable detail within a particular subject area. They may be less valuable if asked to advise on nontechnical aspects, such as *whether* a task should be automated rather than *how*. The views of consultants are necessarily subjective; a shrewd project manager will seek the advice of more than one, independently, so that he can assess their recommendations by comparison.

1.6.4. Assessment of Technology

The next phase, research for selection and validation of the best approach, comprises two stages. The first is an assessment of technology already available. Technical papers, textbooks, catalogs, theses, and reports are searched, and anything of potential applicability is noted. Consultants may need to be approached. It is generally possible to improve on standard methodology when specific applications are considered, and it may be desirable to do this to obtain patent protection (or to avoid infringing upon someone else's patents), in addition to improving performance or cost–effectiveness.

All candidate methodologies are considered initially; many are easily seen to be unsuitable and are discarded from further consideration. The choice narrows as study and comparison proceed. Mathematical analysis and computer simulation can be used in the comparison to resolve doubts. The former is inexpensive and provides results which are absolutely (logically) true, but most problems are too complex to be tractable completely by analytical techniques, and personnel gifted enough to perform such analysis may be hard to find. Simulation analysis is tedious and often less generally applicable, but is also less demanding intellectually and can tackle problems too complex to be solved by purely mathematical techniques.

The critical problems in past automation projects have often been practical, for example in configuring illumination to highlight important detail, and in handling awkward objects at high speed without jamming. These problems are generally solved by persistent application of elementary techniques (i.e., by "rule of thumb"), often by repeated experimentation. At the research stage there is merely a need to appreciate that such problems will have to be overcome.

It is important always to apply the engineering concept of optimality, which seeks a "best" way for performing each task, according to criteria such as minimizing cost or error rate, or maximizing throughput. Generally, the configuration or parameter set optimal for one criterion will not be optimal for the remainder, and the engineer is required to exercise judgment and decide which is most significant for the task in hand. Typical areas in which optimization brings particular benefit include data comparison and sorting, selective extraction of useful data, and positioning of decision thresholds.

It is worthy of note that partial automation of a task is often worthwhile pending the development of methodology capable of automating it fully. The objective of the partial automation is to eliminate those parts of the task which are routine and boring (though often expensive), while retaining those which are interesting through requiring the exercise of decision and judgment. Generally, the greater the intelligence required to perform a task, the more difficult it is to automate.

1.6.5. Validation of Methodology

The final phase in the research stage is experimental validation, which introduces real–world problems (such as dust and variable ambient lighting) as factors affecting performance. Real sample material is used; processing may well use general purpose instrumentation programmed in a high-level language. This is slow but allows different methodologies to be implemented and investigated very easily. Performance factors such as accuracy of positioning and orientation, rejection, and substitution rates are measured directly and can be assessed quantitatively using statistical techniques where beneficial. This phase is indispensible and conclusive, despite being expensive and time consuming. Its successful outcome clinches establishment of feasibility; only when the validation has run successfully can one confidently commit the resources to engineer a system to go on–line. The methodology to be used is now specified; how it is to be implemented remains to be determined.

1.6.6. Development and Implementation

The final (development) stage involves the design and manufacture of instrumentation that is engineered well enough to operate online in possibly adverse environmental conditions, and that is robust, reliable, safe, and convenient to operate. The configuration and processing to be used are established, and the hardware to be implemented then must be selected. It is now necessary to examine and compare alternative processors and sensors (having established that a 2048-element linescan camera shall be used, whose shall we buy?). Factors such as safety (how should a laser be packaged to ensure no hazard to operatives?) and security (how should a valuable camera lens be protected from being stolen?) must be considered. The project at this point has really reached the stage of conventional development engineering.

Finding someone to construct the on-line instrumentation should not present a problem. There exists in every locality a multitude of small firms able to design and manufacture electromechanical instrumentation once configuration and processing have been established. A tendency is in fact emerging for suppliers of instrumentation to form two groups. In the first are manufacturers of hardware, including processors, sensors, and so on. The second comprise systems houses, who configure and program the hardware supplied by the first group into systems applicable to specific problems.

Some applications of this approach to real tasks are discussed as case studies in Chapter 11.

References

1. M.W. Thring, *Robots and Telechirs,* Ellis Horwood, Chichester, England (1983).
2. Techniques and applications of image understanding, *SPIE Proceedings 281* (1981).
3. Deconinck, F., Quantitation of medical imaging, *Proc. 3rd BPRA Intl. Conf. St. Andrews, Pattern Recognition Letters* (1985).
4. Chin, R.T., Automated visual inspection techniques and applications–a bibliography, *Pattern Recognition 11* (4), 343–357 (1982).
5. P.J. Gregory and C. J. Taylor, Knowledge based models for computer vision, *Proc. 4th Intl. Conf. on Robot Vision and Sensory Controls* (1984), 325–330.
6. P. Davey, Leading the way on robot weld guidance, *The Industrial Robot,* pp. 104–107 (June 1983).
7. J. Moraal, Visual inspection of sheet steel–an ergonomic study in the dutch steel industry, *Le Travail Humain 37* (1), 35–52 (1974).
8. J. Hollingum, *Machine Vision,* IFS Publications, Bedford, UK (1984).

2

Signal Processing for Intelligent Automation

2.1. Overview

In this chapter, we discuss basic principles governing signal processing in intelligent automation. These may appear to be somewhat academic, but their practical utility is considerable in ensuring effective and economical processing. The first principle concerns the nature of information and its definition; how it is provided in signals of various kinds (particularly images), and how processing may be configured to economize on the quantity of information to be processed and stored, with the ultimate benefit of minimizing hardware costs and processing times. The second principle concerns the nature of signals, their representation in various alternative forms, transforms to move reversibly from one form to another, and ways of exploiting the various forms to facilitate particular tasks. The third principle involves noise, a form of signal containing unwanted (and generally random) information, which masks the *message* information within a signal which is wanted and useful. We describe the origins and properties of noise, and show how its deleterious effects may be minimized. Finally, we consider decision making for instrumentation systems, and particularly the design of the decision process to minimize the harmful consequences of errors which are inevitable in a noisy process. Detection of the presence of a message in noise is covered as a prime application. This forms in turn an introduction to the statistical pattern recognition considered in Chapter 3.

2.2. Information—Nature and Quantification

2.2.1. Definitions

Information is the commodity handled in an intelligent processing system. Concisely, information is "that which reduces uncertainty." A theory for dealing quantitatively with information was developed by C. E. Shannon between 1940 and 1950. This theory considers information to be generated by *sources* that emit sequences of messages that are random so far as a recipient is concerned. The messages are generally symbols, selected from some alphabet of finite size. Some examples of practical sources are:

1. *Written text,* in which the message symbols are the 27 alphabetic characters (26 letters plus the space) used in the English language.
2. *Speech,* in which the message symbols are the distinct sounds (called phonemes) of which speech is composed. There are 39 phonemes in English.
3. *Images,* in which the message symbols are the pixels (samples of intensity) which specify the image. If (as is usual) 256 levels of pixel amplitude are possible, the symbol alphabet has 256 members.

Thus, the text in a book communicates information from the author to his reader, and speech likewise conveys information from one person to another. Information can also be conveyed from computers to people and vice-versa or even between machines; in the latter case, the binary alphabet, comprising only the two symbols "0" and "1," is used. The same theory applies to storage (communication of information in time, from "then" to "now") as to normal communication in space (from "here" to "there").

There exist several types of information. *Symbolic* information concerns sequences of symbols. *Syntactic* information concerns relationships that can exist between symbols. Rules of grammar, for example, limit the kinds of words which may be included within a sentence, and ensure that each sentence includes at least one verb. Similar rules occur, e.g., in images; only certain relationships are permitted between successive pixels in the boundary of a silhouette (see

Section 2.2.2). *Semantic* information comprises the ideas (concepts) which a message communicates; as for symbols, rules of syntax are often present that specify certain relationships between concepts as being valid and others as being invalid. The laws of logic are examples of syntactic constraints governing semantic concepts. *Pragmatic* information concerns the use which a particular recipient may make of a message. *Aesthetic* information concerns the ability of a message to affect the recipient's feelings, and is conveyed by poetry and music particularly. But television images (used for human entertainment) must convey aesthetic information as well as the other kinds; engineering drawings do not need to do so. It is thus more complex and expensive to communicate TV images than engineering drawings, given equal resolution. Somewhat surprisingly, the latter convey less information for given complexity of the object described. The various forms of information may be arranged in a hierarchy, with symbolic information (the most basic form) at the bottom, and aesthetic information (the most complex, i.e., the highest form) at the top[(1)] as follows:

- Aesthetic
- Pragmatic
- Semantic
- Syntactic
- Symbolic

In all practical systems which store, process, or convey information, the information is actually contained in sequences of symbols. Also, symbolic information is the only kind for which a universally accepted quantification exists. Thus, our discussion of information will concentrate on symbolic information.

The arrival of a symbol reduces uncertainty in the recipient; the more uncertainty is reduced; i.e., the less probable a symbol is, the more information its appearance will provide. For example, the message "it is raining" (the only alternative being that it is not) conveys little information (it often rains in the UK); the message "men have landed from Mars" conveys a lot of information (assuming it is true!), since it is so unlikely. Thus, the information I provided by receipt of a symbol k whose probability of reception is $p(k)$ is defined as:

$$I = -\log p(k)$$

The units in which information is measured are determined by the base of the logarithms used; if logs are to base 2, the units are bits; since most information transfer involves binary symbols, this unit is most commonly used. Sometimes logs to base 10 are used (giving units called Hartleys), or to base e, giving units called nats.

Another way of interpreting this definition of information is to note that one bit of information provides a yes/no answer (two mutually exclusive alternatives) which are equally probable. Note that one bit is *not* the same as one binary digit, and the whole utility of information theory in designing efficient systems for intelligent automation stems from this distinction. For, while one binary digit *can* carry one bit of information, this is the absolute maximum it can hold, and it holds this amount only when the probabilities of the "1" and "0" symbols are exactly equal at 0.5. If the probabilities of the symbols are unequal, or if the symbols in a sequence are not mutually independent (i.e., if a symbol provides "clues" as to which symbol will follow), then each digit will provide much less than the maximum it can hold. This inefficiency (termed "redundancy") must be minimized by good system design.

For a particular source, the average information provided by each symbol is termed its entropy, H. This must be a maximum for efficient operation. The entropy of a source which emits an alphabet containing N possible symbols, for which the kth appears with probability $p(k)$, and for which the symbols are mutually independent, is determined as follows. The information provided by the symbol k each time it appears is $I(k) = -\log p(k)$ bits. But, only a fraction, $p(k)$, of the symbols received are of type k. Thus, on the average, symbol k provides $-p(k) \log p(k)$ bits. Summing these contributions over all N symbols yields for H the expression:

$$H = \sum_{k=1}^{N} -p(k) \log_2 p(k)$$

The units of H are bits per symbol.

It is easily shown[2] that H is maximum for all N, and for logarithms to any base, when the symbols are equiprobable, i.e., when $p(1) = p(2) = p(3) = p(\mathrm{k}) = \cdots 1/N$. Thus, $H_{\max}$ is equal to log(no.

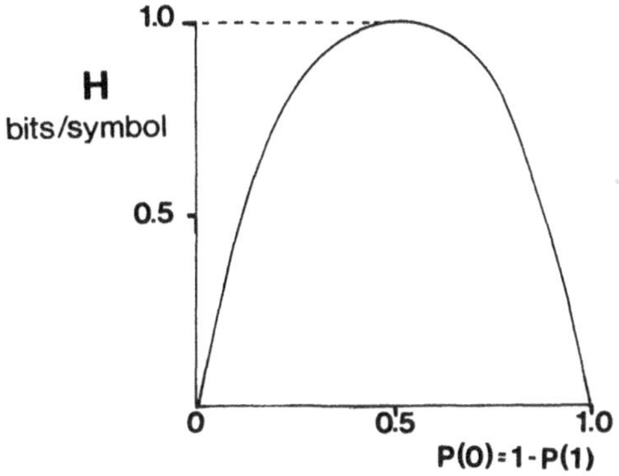

Figure 2.1. Variation of entropy with symbol probability for binary symbols.

of symbols). Figure 2.1 shows how H varies with $p(k)$ for the binary alphabet. The ***redundancy, R,*** of a source is formally defined as

$$R = (H_{\max} - H)/H_{\max}$$

2.2.2. Encoding to Minimize Redundancy

Natural sources of information (such as written text, speech, or the pixels which make up an image) are notorious for being hopelessly redundant (e.g.), $R = (4.7 - 1)/4.7 = 0.79$ for text); the symbols are far from being equiprobable, and are not mutually independent. The consequence is that the processing necessary to analyze them, or the number of binary digits necessary to store them, is much larger (say, 5 times as large for written text, to more than 20 for gray-scale images) than it need be. To improve efficiency, we can use a *source code*,[3] which is a mapping between the alphabet of symbols generated by the source, and a new alphabet termed the *code alphabet*. Often, the source and code alphabets are the same; it is still possible to find unique and reversible mappings which will reduce redundancy very effectively. The codes redistribute the information between symbols so as to make them more equally probable. The consequence is to reduce the average length of messages, i.e., given messages conveyed on average by J symbols; they are recoded to occupy only K, where K is less than J.

Source codes are of two kinds; reversible codes, in which all the symbolic information originally generated by the source is preserved, and irreversible codes, which dump some symbolic information but preserve the higher level information (semantic, pragmatic, etc.) which is useful.

The performance obtainable from an encoder is limited by Shannon's first law. This states that, given that the coder receives N symbols/second with entropy $H(s)$ and generates M symbols/second with entropy $H(1)$, then

$$N.H(s) \geqslant M.H(1)$$

That is, the encoder can at best only pass on the information it has received; it cannot create information. A possible definition of an intelligent system, incidentally, is one which violates Shannon's first law by generating more information than it receives.

An optimally efficient (and extremely simple) algorithm is available for reversible source coding; this is the Huffman code.[4] To design it, the probabilities of occurrence of the various symbols are used. Design starts by rearranging the original symbols into decreasing order of probability of occurrence. Then, the two least probable symbols are combined; their sum is inserted into a new table (termed the "first reduction" for the code), that is rearranged to restore the descending order of probabilities. Next, the two least probable terms are combined once again, and the table is rearranged (if necessary) to maintain descending order, giving the "second reduction" for the table. This procedure is repeated until only two entries remain in the reduced table; it is illustrated in Table 2.1.

A "tree" is then drawn in which each branching represents a

Table 2.1
Construction of a Huffman Code

				Reductions		
Symbol	*Probability*	*Symbol*	*Probability*	*1st*	*2nd*	*3rd*
A	0.1	E	0.35	0.35	0.35	0.65
B	0.2	C	0.3	0.35	0.35	
C	0.3	B	0.2	0.3		
D	0.05	A	0.1	0.15		
E	0.35	D	0.05			

combination of symbols. Each original symbol is represented by the termination of a branch. At each branching, one alternative path is assigned the symbol "0," the other the symbol "1." The binary codeword for each original symbol is represented by the sequence of code (binary) symbols proceeding from the "root" of the tree to the appropriate terminal node (Figure 2.2). The code produced is shown in Table 2.2.

In the case just described, the source (e.g., A,B,C,D,E) and code (0,1) alphabets are different, but even if the source had emitted binary codewords of roughly equal length and with highly unequal probabilities, a worthwhile improvement in efficiency would have been achieved. The encoding produced by this procedure is of minimum average length, where average length L_{av} is defined by

$$L_{av} = \sum_{i} l(i).p(\mathrm{i})$$

Here, $l(i)$ is the length (number of code symbols) of the ith codeword, and $p(i)$ is its probability of occurrence. Thus, the number of digits required to represent the message has been minimized. This is true for all code alphabets, not just the binary alphabet. The codewords are generally all of different lengths, which may in practice be a nuisance. A codeword is decoded by tracing a path through the code tree, with decisions at the various nodes which correspond to the successive symbols in the codeword, until a terminal node is reached.

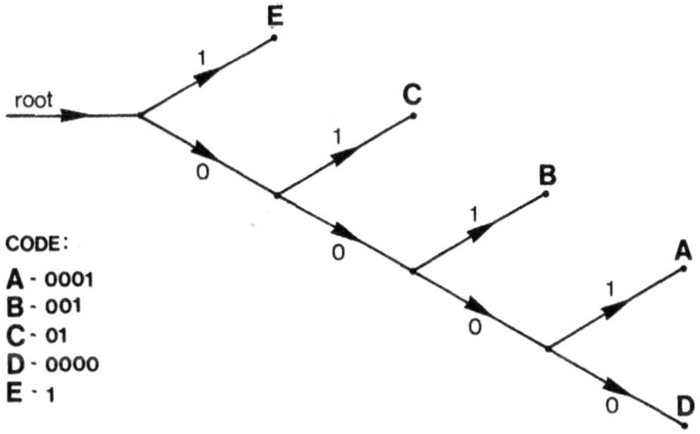

Figure 2.2. Tree for a Huffman code.

Table 2.2
Example of a Huffman Code

Symbol	*Code*
A	0001
B	001
C	01
D	0000
E	1

Huffman coding has been applied, for example, to facilitate the storage and communication of documents, since white areas are much more common than black areas. Other reversible codes, less efficient than Huffman, are more convenient, and we now illustrate how these may be applied to improve the efficiency with which binary images may be handled by considering run length codes and Freeman chain codes. Figure 2.3 shows a silhouette (binary image); suppose this is contained on a rectangular array of pixels 1000 × 1000 points. The total number of cells (binary symbols when the image is considered as a message) is 1,000,000; 1,000,000 binary digits are being used to store the image. But the image really contains much less than 1,000,000 bits of information, and can thus be represented using far fewer binary digits.

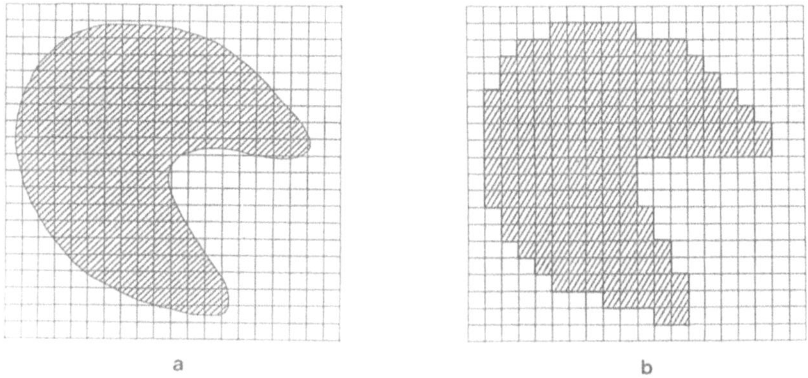

Figure 2.3. Binary image stored on a grid of cells: (a) before quantization; (b) after quantization.

To show this, we use the property that a silhouette image can be represented completely by specification of its boundary, and that such boundaries form closed curves. Each cell in the boundary has exactly two nearest neighbor cells which are also in the boundary. The Freeman chain code representation (illustrated in Figure 2.4) exploits the property that only eight relationships are possible between a particular cell in the boundary and the next boundary cell moving in a given direction. Three bits are required to specify which of these is present. Thus, a boundary may be represented by a sequence ("chain") of numbers from 1 to 8 (more usually, from 0 to 7), with one number per cell. Figure 2.5 shows the silhouette of Figure 2.3 chain encoded; starting from point A and moving clockwise, the chain is:

4,2,2,3,4,4,4,4,5,4,4,5,4,4,5,5,5,5,5,6,0,0,0,0,0,0,7,6,6,5,5,6,6,5,6,
6,0,1,0,0,1,0,0,1,1,1,2,1,2,2,2,2,2

Our silhouette on the 1000 × 1000-element grid will have perhaps 4000 cells in its boundary, each requiring 3 bits. It can thus be specified using only 12,000 bits, which can be contained in 12,000 binary digits. This is less by almost two orders of magnitude than the 10^6 bits required to specify the image in its original form. Further reduction may be obtained using the property that the boundaries of real objects are generally smooth. Thus, the Freeman chain will consist of runs of links of the same kind. These runs may in turn be encoded, for example specifying first the number (in brackets), then the kind

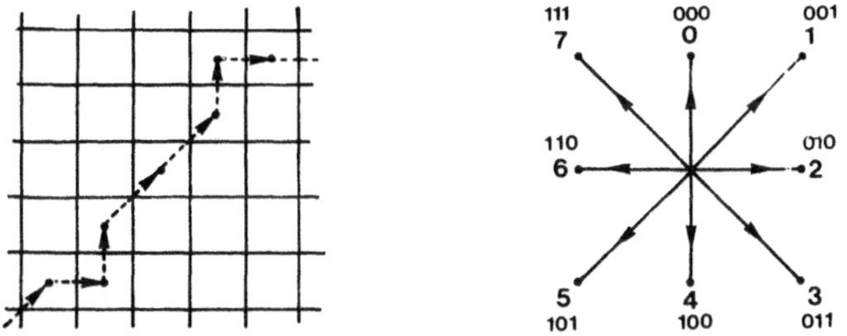

chain representing segment is: 1,2,0,1,1,0,2,2

eight directions-3bits

Figure 2.4. Freeman chain code principle.

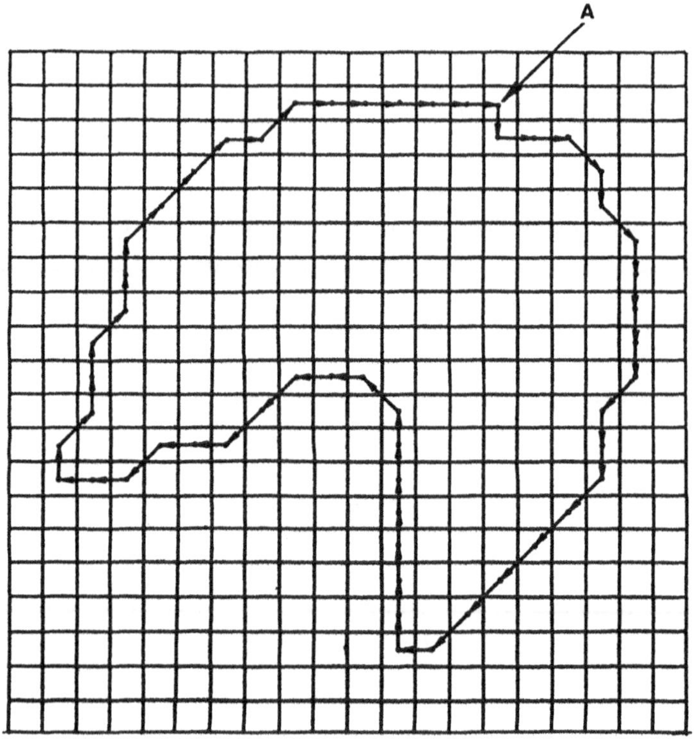

Figure 2.5. Binary image described by a Freeman code.

of links in each run, i.e., for Figure 2.5, with the "1" multiplier not stated explicitly:

4,(2)2,3,4,3,(4)4,5,(2)4,(5)5,6,(6)0,7,(2)6,(2)5,(2)6,0,1,(2)
0,1,(2)0,(3)1,2,1,5(2)

The Freeman chain code is relatively complex to compute given the silhouette, and it is also complicated to reconstruct the silhouette given the chain code. For a square quantization grid, the Freeman chain code representation of a shape is altered by translation (because the spatial quantization affects the edge shape) and rotation, and no simple algorithm exists for correction. The actual position of the silhouette on the grid is also lost, though this is normally of no consequence. It may be preserved by including a few extra digits to specify the position of a reference point in the chain.

The *contour* code obtained by taking the difference between successive links in the Freeman chain modulo 8 is much less sensitive to orientation, but is susceptible to error from noise; it is explained fully elsewhere.[5] The contour code for Figure 2.5 is

2,0,7,7,1,7,0,0,0,7,1,0,7,1,0,7,0,0,0,0,7,6,0,0,0,0,0,1,1,0,7,0,7,0,1,7,0,6,7,
1,0,7,1,0,7,0,0,7,1,7,0,0,0,0

Note that symbols 3,4, and 5 have not occurred and 6 is very rare, implying no sharp corners.

The *run length* code is an alternative which eliminates many disadvantages of the Freeman code, particularly in being simple to encode and decode. The run length encoding of a silhouette (Figure 2.6) involves the number of pixels between successive transitions from white to black, and vice-versa, measured along successive rows. The

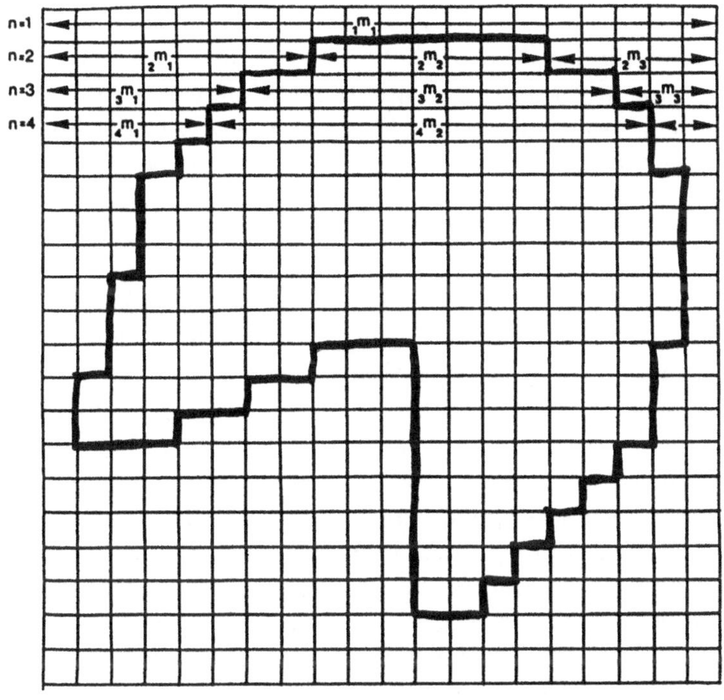

Figure 2.6. Binary image described by a run length code. The encoded description for the shape shown would be: 1*m*1, 2*m*1, 2*m*2, 2*m*3, 3*m*1, 3*m*2, 3*m*3, 4*m*1, 4*m*2, . . . etc.

data reduction thus obtained is estimated as follows, again considering a simple silhouette shape on a grid 1000 pixels square. The largest possible distance between successive transitions is now 1000 pixels, which is codable with 10 bits. Assuming an average of 3 transitions per row, 30 bits are required to encode each row, i.e., about 30,000 bits for the whole image. The compression achieved of about 30 times is only slightly less than for the Freeman chain code. If, however, the image dimension is increased 10-fold in each direction, the data in the Freeman code representation increases about 20 times, and the data in the original silhouette, 100-fold, while the data in the run length has increased only $\log_2 10 = 3.3$ times. This property, that the information in the pixel grid representation increases with the square of grid size while the Freeman code increases linearly with grid size and run length encoding only as the square root of grid size, ensures that the latter will become optimal for silhouette images as resolution increases. Still more compression may be obtained by exploiting boundary smoothness, e.g., by storing the differences between the run length encodings for successive lines.

Evidently, if a system working with simple binary silhouettes insists on using the crude cell grid representation, it will store and process much more information than is necessary. The redundant data must be rejected by preprocessing in a system as early as possible.

2.2.3. Coding to Prevent Error

Information theoretic ideas have also been applied[(6)] to detect and correct errors introduced during transmission of digital messages, and the approach used can be applied in order to correct errors arising when the properties of an image (or other signal) are measured incorrectly. Redundancy in data is exploited, so that information lost due to noise corruptions is recovered from other parts of the message. Natural signals such as speech are highly redundant, and hence very tolerant of errors, but redundancy is inserted more efficiently using so-called channel codes. The simplest such code is the Hamming code, which operates as follows. Suppose we are processing a word N binary digits long. Following communication on storage, some of the symbols may be corrupted; when the symbols are binary, this results in a binary "1" being flipped to a "0," or vice-versa. We assume first that corrupted symbols are very rare, so that the probability of more than one error in our word of length N is rare enough to be ignored. In this case, following processing, our word will contain an error in the first symbol, or in the second symbol, or in the third symbol (and so on), or in none of the symbols. Thus, there are $N + 1$ possible cases to be distinguished. This is possible using a minimum of $\log_2 (N + 1)$ checks, suitably disposed over the symbols. This scheme is most efficient when $\log_2 (N + 1)$ is an integer, i.e., for $N = 3,7,15,31$, etc.

Encoding is implemented by setting up "parity checks" over blocks of symbols; extra symbols are added to an original message to give even parity over each block, as indicated for the message 0101 in Table 2.3.

In Table 2.3, symbols in positions designated "m" are from the original message; those designated "p" are extra symbols inserted for parity checking. The redundancy of the data is thus increased by the encoding, since only $M - (\log_2 M)$ of the symbols in each encoded word are from the original message.

The message is decoded upon receipt (to reject the parity check symbols and correct any single error) by repeating the parity checks. The location of any single error is then given by the pattern of the outcomes of the parity checks, counting a "1" if the check fails, and "0" if it succeeds, i.e., if the block still has even parity. See, for example, the parity check for the received codeword 0110101 in Table 2.4.

In Table 2.4, the pattern of parity checks shows an error in digit 3. Thus, the original message with parity digits removed was 0101. The Hamming code will alternatively detect (without correcting) up to two errors present in a codeword; the errors simply cause at least

Table 2.3
Construction of a Hamming Code

p1	p2	m3	p4	m5	m6	m7	
		0		1	0	1	Original message
0		x		x		x	First parity digit
	1	x			x	x	Second parity digit
			0	x	x	x	Third parity digit
0	1	0	0	1	0	1	Encoded message

Table 2.4
Decoding of a Hamming Code

p1	*p2*	*m3*	*p4*	*m5*	*m6*	*m7*	
0	1	1	0	1	0	1	Received codeword
x		x		x		x	First parity check = 1
	x	x			x	x	Second parity check = 1
			x	x	x	x	Third parity check = 0

one of the parity checks to fail, but the pattern of failures can no longer be used to locate and hence correct the errors. A Hamming code is said to be of minimum distance 3, since all possible codewords in the code differ mutually in at least three positions:

0 1 0 0 1 0 1	message *A*
1 0 1 0 1 0 1	message *B*
x x x	positions in which words differ

For binary codewords separated by a minimum distance D, i.e., differing mutually in at least D positions, then $D - 1$ errors may be detected, or $(D - 1)/2$ corrected, using appropriate parity checks. This concept may be applied in pattern recognition, with the attributes used to specify and distinguish the patterns (termed "features") regarded as the symbols of the message. A pattern is identified by finding its match in a list of vectors of numbers. A perfect match is often not obtainable, and a "best fit" must be found. A common approach is to measure the Euclidean distance between the vector being recognized and all other vectors, and assign to the one which is closest, provided this minimum distance is less than a predetermined threshold. If the minimum distance exceeds this threshold, then no match exists. The chief difficulty with this approach is its computational complexity; for an N element vector with M entries in the match list, $N \times M$ squares and sums must be computed. An alternative and more economical approach is to compare each measure between stored and sample vectors, and count a "hit" (logical "1") if the measures are the same (to within a tolerance), and a "miss" (logical "0") if they differ. If the number of hits exceeds a threshold, a match is counted. By choosing the entries for the stored vector set to differ mutually in at least D entries, we can guarantee finding the correct match, provided $(D - 1)/2$ or fewer measures are corrupt. This approach is described further in Section 3.5.9.

2.2.4. *Other Applications of Information Theory*

Information theory may further be used to assess the power of image identification algorithms. For example, one approach that has been proposed for identifying silhouette images involves computing histograms of the number of cells in each row and column, as explained in Figure 2.7. An unknown image is identified by fitting its two histograms to a range of possible matches, and assigning to the one which is closest. But the process of taking histograms throws information away irreversibly, thus the approach is subject to confusion. Suppose the silhouette is contained on a grid M pixels square (i.e, $N = M$ as defined in Figure 2.7). The M squared cells contain M

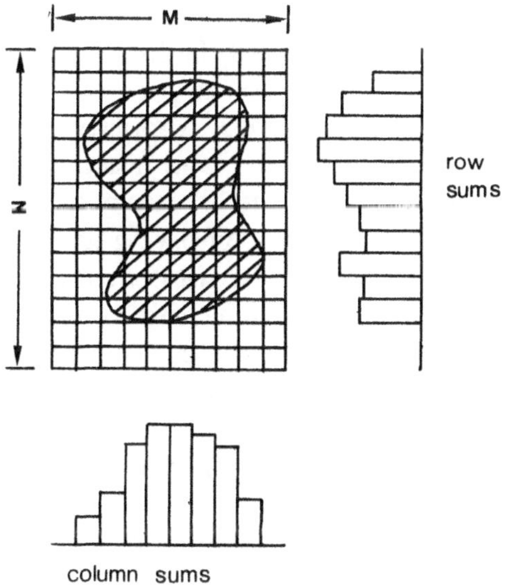

Figure 2.7. Image described by projection histograms.

squared bits of information. Since the number of possible cell counts in each row (or column) is $(M + 1)$, then the row histogram contains $M \log_2 (M + 1)$ bits, and so does the column histogram, giving a total of $2M \log_2 (M + 1)$ bits. Except for the trivial case $M \leqslant 5$, M is always greater than $2 \log_2 (M + 1)$, proving that taking histograms throws away information and explaining why the confusion occurs.

2.3. Efficient Computation

The efficiency with which a system processes information may also be improved by exploiting theorems of computer science,[7] such as complexity theory. This indicates the order of the number of cycles required to implement an operation. For example, it asserts that a list may be searched using $0 [\log_2 N]$ cycles (the notation $0[N]$ means "of the order of N"), where N is the number of items in the list. A crude bubble sort takes $0[N]$ cycles, and is hence unnecessarily long.

Consider the task of fitting a word to itself contained in a list in alphabetical order. In the bubble sort, we compare the word with each list entry in succession, until a fit is found. Assuming that the word is a random selection (all words in the list are equally probable) and that the list includes N entries, then an average of $N/2$ comparisons will be needed. If the words were not equiprobable (and words in real text are certainly far from equiprobable), then the average number of comparisons could of course be improved by ordering the list so that the entries were compared with the most likely first.

The alternative and faster way, even with equiprobable entries, is to first divide the list into two halves, and determine into which half the word falls by comparing its initial letters with the midpoint entry. Then the half of the list containing the word is itself divided into two halves, and the half of this in which the word lies is determined. The process is repeated ($\log_2 N$ times) until the fit is found. This process is even more convenient with numbers than with sequences of letters, particularly if assembler programming is to be used, since the "greater, less, or equal" comparison required is then very easy to implement. Differences in the probabilities of entries can be exploited to minimize the search time by configuring the search to be similar to the decoding of a Huffman code.

2.4. Signal Representation

2.4.1. Fundamental Properties

Information is conveyed by *signals;* a signal is inherently a physical observable, such as electrical voltage or light intensity. Signals may be one-dimensional (sound waves) or two-dimensional (images), functions of time or of space. All signals are single-valued functions, and some possess additional restrictions; for example, images represent the distribution of light energy over a surface, and hence cannot be negative.

Almost invariably, the processing used in automation applications is digital, i.e., signals are represented as sequences of numbers. This form is obtained from original analog signals by sampling (retaining the amplitude of the signal only at particular instants separated by a constant interval in time or space) followed by quantization, in which the amplitude of each sample is rounded off to the nearest whole number, and only a restricted range of numbers (say, 256) is available. The processes are performed using an analog/digital (A/D) converter. Sampling throws away no information provided the interval T between successive samples is less than $1/2f_{max}$, where f_{max} is the highest frequency at which significant energy is present in the original analog signal. This is the Nyquist criterion; it is strictly true only for signal sequences of infinite length. Quantization, however, results in an irreversible loss of information, manifest by the introduction of random noise. The original signal may be a function of either time or space; following sampling, the order of each sample is specified simply by a number, the *epoch.* If each sample is considered as a symbol, the nature of a signal as information provider is evident.

2.4.2. Signal Space

The properties of a sampled signal are best visualized using the *signal space* representation. The samples are considered to be mutually independent, and each represents one dimension of a hyperspace (i.e., a space having more than three dimensions). Each signal is represented by a different point in the space, specified by a vector of numbers which are the amplitudes of the various samples. Thus, the location of the vector is a unique representation of the waveshape of the signal (Figure 2.8). Increasing the amplitude of each sample by the same linear scaling produces a vector which points in the same direction as the unscaled vector, but is longer. Otherwise, different signals are represented by vectors which point in different directions. The similarity of two signals is measured by their closeness of their directions, i.e., by their normalized scalar product. We shall use this concept to explain orthogonal transformations, to analyze filters for contrast enhancement, and as a means for recognizing waveforms.

2.4.3. *Orthogonal Transforms*

Fourier's theorem states that any waveform which is single-valued, continuous, and finite within an interval may be represented uniquely within that interval as a weighted linear sum of sine and cosine signals whose frequencies are harmonically related. The weights represent the relative amplitudes of the component sinusoids. Uniqueness means that only one set of weights can represent a given signal. The value of the concept is most obvious when the signal is represented by its samples, i.e., by a vector in signal space. The vector of weights is an alternative representation for the signal; there are exactly as many weight coefficients as there were samples in the original signal. In fact, the sinusoids may be regarded as defining a new signal space; Fourier transformation moves the representation from the original (time or distance) space to the new Fourier space. As Figure 2.8 indicates, the signal vector is not changed by the transformation, which merely involves rotating the coordinate axes about the origin, whereupon the projections of the signal on the axes are changed. The transformation clearly preserves the lengths of all signal vectors and hence the *angles* between them, and is unique and reversible. It thus preserves relationships between the vector representations of the two signals, and is said to be *orthogonal*. The mathematics for computing Fourier representations is given in many textbooks.[(8),(9)] For a segment of discrete signal comprising a sequence of N samples $f(1)$, $f(2), \ldots f(1) \ldots f(N)$, the amplitudes $g(1), g(2), \ldots g(k) \ldots g(N)$ of the N discrete sinusoids which sum to represent the signal exactly are given by the Discrete Fourier Transform (DFT):

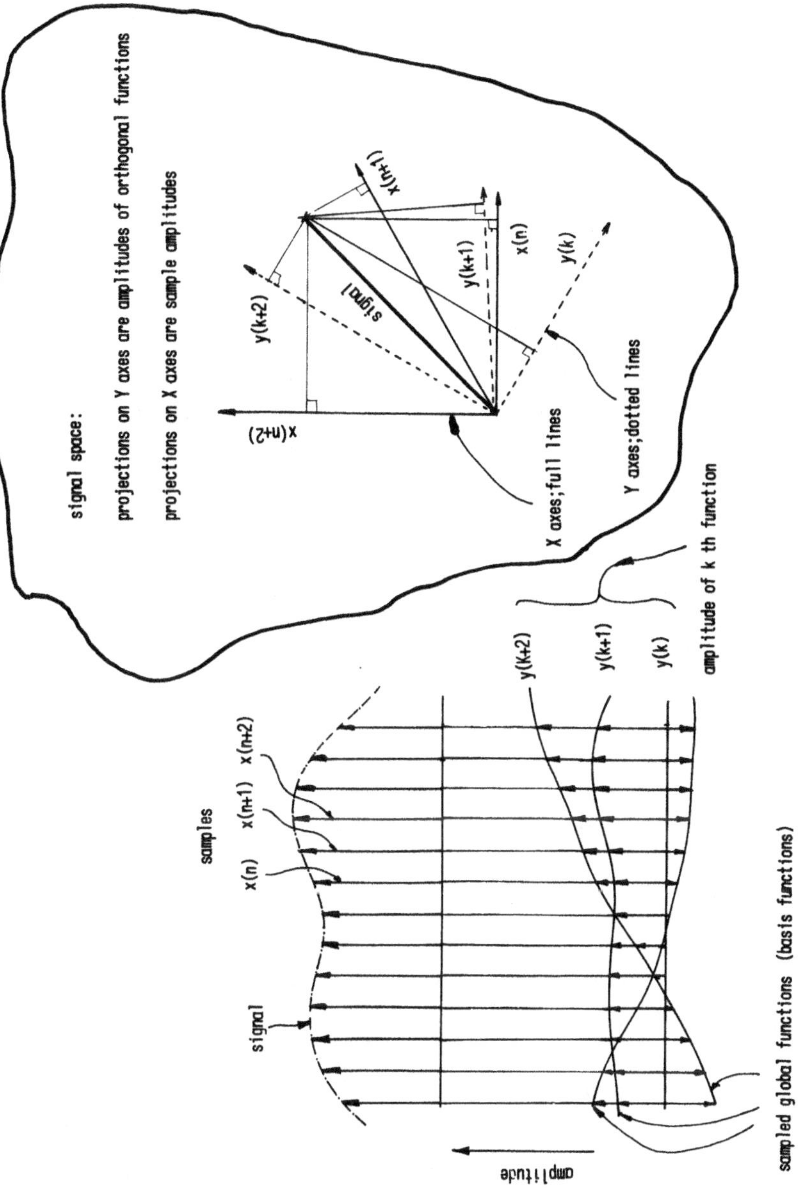

Figure 2.8. Signal and orthogonal transformation described in signal space.

$$g(k) = \sum_{\ell=0}^{N-1} f(\ell).\exp(-2\pi j.\ell.k/N)$$

The values of the signal samples may be determined from the amplitudes of the sinusoids using the inverse DFT:

$$f(1) = 1/N \sum_{k=0}^{N-1} g(k).\exp(2\pi j.\ell.k/N)$$

An attempt to reconstitute an original signal using only the lower M (of N) Fourier components causes (as expected) error in the reconstruction. For many natural signals such as images, most of the *energy* in the signal is concentrated in the lower spatial frequencies, while *information* is concentrated in middle spatial frequencies. Thus, though the mean square error may appear to be acceptably small, the degradation produced so far as interpretation is concerned may be appreciable.

Two- (and higher) dimensional signals also have a Fourier representation. The two-dimensional discrete Fourier transform $g(k,n)$ of the signal $f(\ell,m)$ (two indices are now needed to specify each sample) which has dimension L samples in the x-direction and M samples in the y-direction is given by

$$g(k,n) = \sum_{m=0}^{M-1} \sum_{\ell=0}^{L-1} f(\ell,m).\exp\left(-2\pi j \left[\frac{\ell k}{L} + \frac{mn}{M}\right]\right)$$

The multidimensional Fourier transform has the useful property of being separable, i.e., the transform of a two-dimensional function (such as an image) may be computed as a sequence of one-dimensional transforms along strips in the two perpendicular directions (Figure 2.9).

The Fourier transform requires the computation of complex trigonometrical functions, and although fast algorithms (notably the Fast Fourier Transform (FFT)) are available to speed things up, it is generally too slow to be used on-line (unless *optical* or *hardwired* processing is used). The speedup realized by the FFT is of the order of

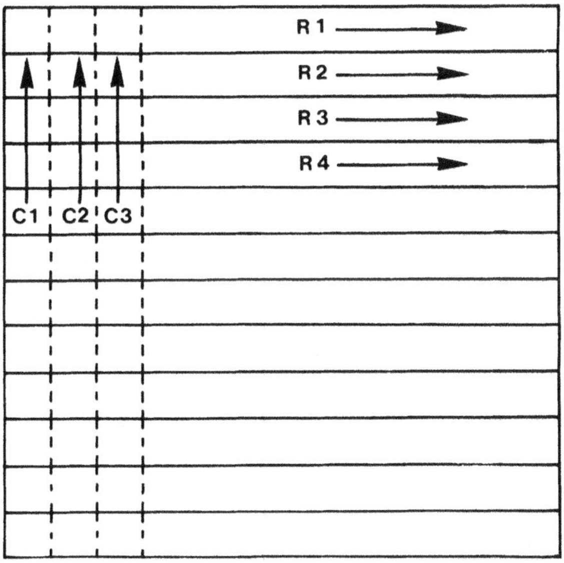

Figure 2.9. Two-dimensional transform computed by one-dimensional transforms. R1, R2, etc. represent transforms along rows; C1, C2, etc. represent transforms along columns. It is immaterial which are computed first.

$(\log_2 N)/N$ for one-dimensional signals, where N is the number of samples in the signal vector. The action of a hardwired Fourier computer (note that this is analog) is indicated in Figure 2.10.

The chief value of the Fourier representation of a signal in instrumentation applications lies in the tendency of natural systems to execute sinusoidal motions when excited. Another important property is that the squared amplitude of the transform is invariant to position within a window; thus, a constant representation may be obtained using Fourier descriptors for an object whose position within a field of view cannot be fixed.

Another potentially useful orthogonal transform is the Walsh-Hadamard (W/H) transform.[10] This represents the signal as a linear weighted sum of rectangular waves (i.e., waveforms which may be only +1 or −1) which like the sinusoids form a complete orthonormal set. These waveforms (termed "Walsh functions") form two groups which are respectively symmetric and antisymmetric about their center point,

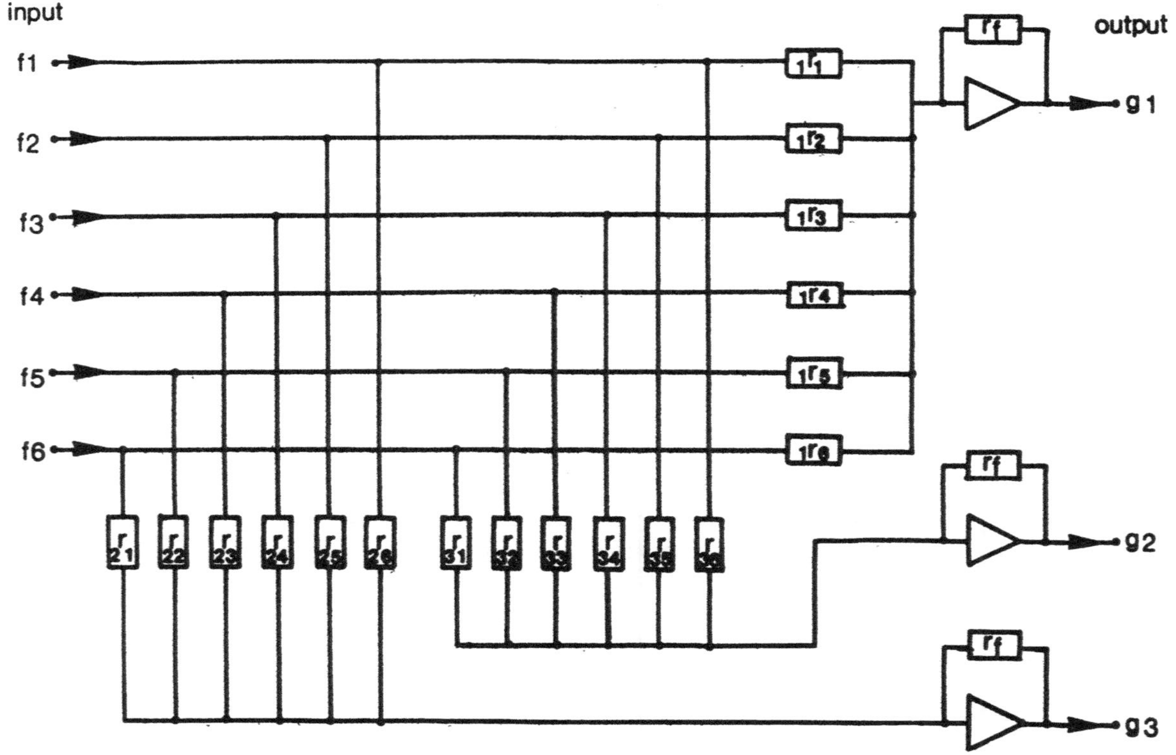

Figure 2.10. Hardwired discrete Fourier transform. Two operational amplifiers are needed for each frequency component "1" to be computed; one determines the real part, the other the imaginary part. The weight resistors *r* correspond to the multipliers sin 2π (1.*k*))/*N* and cos 2π (1.*k*)/*N*, where *k* specifies the input sample and 1 the frequency component. The feedback resistors r_f set the scale.

and are ordered according to the number of zero crossings (the "sequency," analogous to frequency in sinewaves) within the interval. Note (Figure 2.11) that Walsh functions are not merely infinitely amplified sinusoids, but include other more complex rectangular waveshapes. The weights $w_1, w_2, \ldots w_N$ (i.e., the relative amplitudes of the various components which must be combined to sum to the signal) are obtained from the Discrete Walsh Transform (DWT) which is analogous to the DFT for computing Fourier coefficients:

The Discrete Walsh-Hadamard transform $w(k)$ of a sequence $f(n)$ of N discrete samples is given by:

$$w(k) = \sum_{n=0}^{N-1} f(n)\, h(n,k)$$

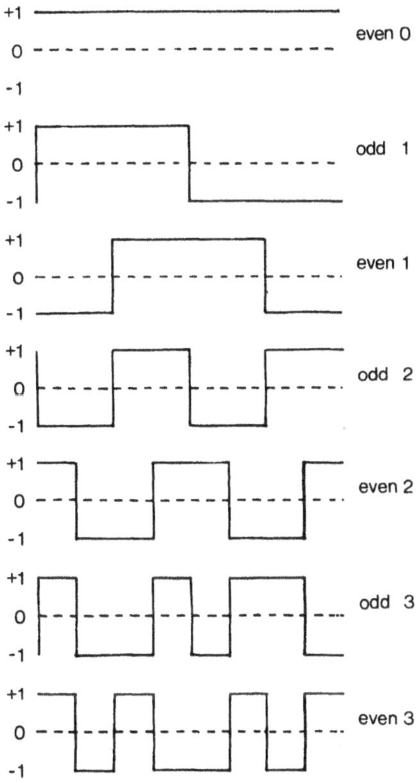

Figure 2.11. Walsh functions.

then

$$h(n,\mathrm{k}) \equiv \frac{1}{\sqrt{N}} (-1)\, \ell(n,\mathrm{k})$$

where $l(n,k)$ is a rather complicated function whose value is always $= \pm 1$, available, e.g., in ref 10.

The W/H transform may again be represented as projecting the signal vector onto a rotated set of coordinate axes. Its great merit is that it is very easy to compute, since the multiplications involved are all by ± 1. A fast algorithm exists for computing it, and it is separable and hence easy to compute for two-dimensional functions such as images. It has, however, two disadvantages: it lacks the position invariance of the Fourier transform, and the Walsh functions rarely represent natural behavior of actual systems. It has nevertheless been used in a commercial machine vision system for identifying shapes.

The Principal Moments (PM) transform[11] is a representation as a weighted sum of functions *peculiar to the signal being represented,* which are chosen to be orthogonal, and to give the fastest possible decrease in amplitude with increasing order. This latter property ensures that the approximation to the signal produced by including only the lowest M functions in the weighted sum is the best possible for any orthogonal representation, in the sense of having least mean square difference from the original signal. Two- (and higher) dimensional versions of the transform exist, but these are not separable (as for the Fourier and Hadamard transforms), and no fast algorithm exists for computing them.

When applied to an ensemble of signals whose properties are known only statistically, the PM transform becomes the Karhunen-Loeve transform. The orthogonal functions are then the eigenvectors of the covariance matrix for the ensemble, and the weights are the eigenvalues associated with each eigenvector. This finds practical application for assessing the independence of features in machine pattern recognition.

The Mellin transform[(12)] produces a description which is invariant to scaling (i.e., to dilation and shrinkage).

Finally, the Discrete Cosine Transform (DCT)[(13)] has the property of producing the best mean square approximation (when only the M

lowest coefficients are retained) for image signals. The orthogonal functions are Tschebycheff polynomials. The DCT is computed using the following equations:

$$g(k) = \frac{2\,c(k)}{N} \sum_{\ell=0}^{N-1} f(\ell) \cos \left\{ \frac{(2\,\ell+1)\,k\pi}{2N} \right\};$$

$$c(k) = \begin{cases} 1/\sqrt{2} & k=0 \\ 1 & k=1,2,\ldots,N-1 \end{cases}$$

Thus, the DCT produces the best compression of image information using a given number of components; it is not quite as good as the PM transform, but it is not specific to a single signal, and can be computed using a fast algorithm. It is becoming widely used for compressing images to achieve efficient transmission, but has yet to be applied in automation.

2.4.4. *Signal/System Interaction*

If a signal comprising a sequence of samples $x(n)$ is passed through a linear system whose impulse response is another sample sequence $h(m)$, the sequence $y(n)$ generated as output is given by the convolution summation[14,15]

$$y(n) = \frac{1}{k} \sum_{m=0}^{m=\infty} \mathrm{v}(m)\, h(n-m); \qquad k = \sum_{p=0}^{\infty} h(p)$$

For an ideal transducer system, the impulse response $h(m)$ would comprise only the single nonzero component $h(1)$, and the output sequence is simply a replica of the input sequence, possibly scaled. Most real transducers have deficiencies such as lag which produce an impulse response comprising several nonzero samples. The result is to blur the signal, concealing detail. It is possible to produce a linear filter which will exactly cancel "deconvolve" the blurring thus caused, but only if the signal is completely free of noise. To produce this filter, the impulse response $h(m)$ must be found experimentally, by applying an impulse or step signal to the input and measuring the output. Figure

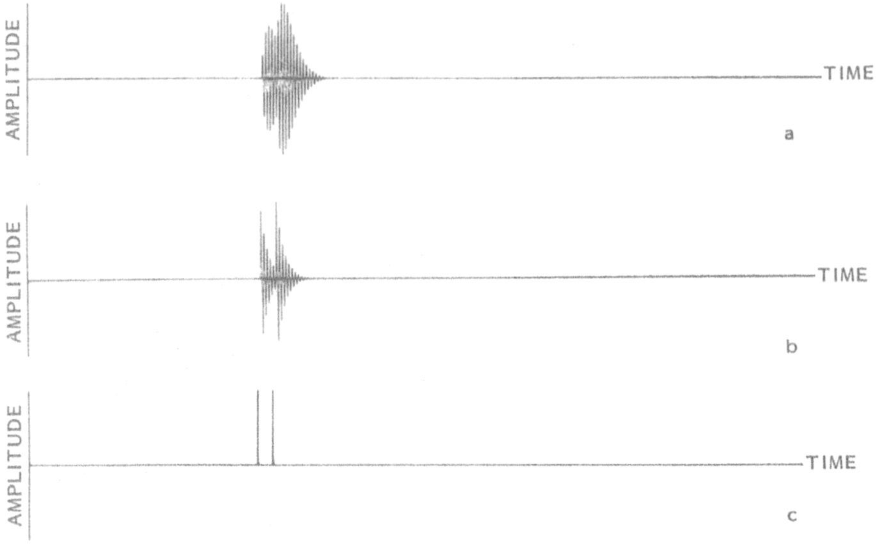

Figure 2.12. Simulation of deconvolution. Trace "a" represents echoes obtained from two point targets from a pulse echo ranging system, which cannot be resolved because the transducers which generate the pulses and receive them have nonideal impulse responses. Trace "b" shows the echoes improved following a filtering to correct the impulse response of the receiving transducer. Trace "c" shows the targets resolved as points following a second filtering which corrects the impulse response of the transmitting transducer. A perfect restoration is obtained because of the absence of noise.

2.12 shows a simulation in which a noise-free signal is first corrupted by convolution, then corrected by deconvolution. If any random noise is present, it is possible to achieve only an approximate restoration; the techniques are complex mathematically, and expensive computationally[16] and are unsuitable for implementation in real time.

2.4.5. *Time–Frequency Duality*

A signal may be described or processed equally in the time (or space) and frequency (or, spatial frequency) domains. A segment of sampled signal is represented by a sequence of N numbers $f(i)$ which are the heights of consecutive samples, or equivalently by another sequence of N numbers $g(k)$ which are the amplitudes of N harmonically related sampled sinusoids. The frequency domain samples $g(k)$ may be obtained from the sequence $f(i)$ using the DFT (Section 2.4.3), and the $f(i)$ recovered from the $g(k)$ using the inverse DFT. This duality has important practical consequences because some signal processing operations are much faster (and/or simpler) when performed in one domain than in the other. For example, the convolution relationship giving system output from input and impulse response becomes a simple multiplication of transforms in the frequency domain, i.e.,

$$Y(k) = V(k).\mathrm{H(k)}$$

where $V(k) \equiv \mathrm{DFT}\ \{v(i)\};\ Y(k) \equiv \mathrm{DFT}\ \{y(i)\}$

and $H(k) \equiv \mathrm{DFT}\ \{h(i)\}.$

Likewise, deconvolution reduces to a simple point-by-point division of transforms, which is very easy to implement providing simple precautions are observed to avoid the effects of ill-conditioning, such as not dividing by zero. The deconvolution relationship in the frequency domain is

$$V(k) = Y(k)/H(k)$$

Filtering (to remove or enhance energy at selected frequencies) can be implemented in the time domain using a transversal filter (Figure 2.13). This convolves input signal with impulse response using a tapped delay line, which is easily simulated computationally or constructed using a shift register in digital hardware. Successive samples of output signal are obtained by summing the weighted outputs from the various taps. A transversal filter is designed by selecting values for the tap weights, and standard procedures(17) are available for this. Any filter response which is obtainable in the frequency domain may be produced in this way, without danger of instability, though the finite wordlength of a digital processor may cause errors due to rounding.

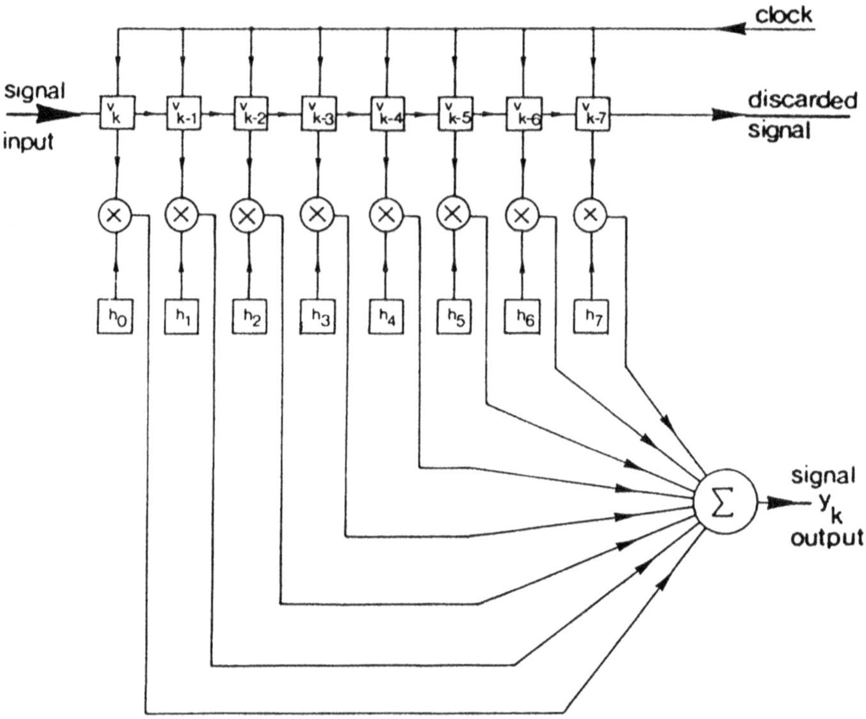

Figure 2.13. Transversal filter. The weights h_0, h_1, etc. are samples of the filter impulse response, and the input signal samples are v_1, v_2, etc. Following each clock pulse, the signal samples move one place to the right, and a new output sample y_k is generated.

2.5. Noise

2.5.1. *Nature of Noise*

All signals comprise two components, a *message* component comprising information which is useful and wanted, and a *noise* component which is unwanted and tends to mask the message component, making it more difficult to perceive. Noise may be either *deterministic* (in which future values of amplitude may be predicted), or *random,* in which this is not possible.

In automation systems, deterministic noise can be introduced as electrical interference (e.g., pickup of the mains frequency and harmonics, though radio signals in the range 100–1000 KHz are often a nuisance), or arise from systematic defects in sensing transducers such as fixed pattern noise and odd–even noise from CCD electronic cameras (explained in Section 4.3.1). Deterministic noise is easily removed by subtracting it out. Though this is the only solution for sensor noise, it is better to combat electrical pickup by preventing it from appearing at all, by good shielding and grounding, and by using optical isolation and fiber optic communication in environments which are electrically noisy.

2.5.2. *Statistical Descriptions of Noise*

Although it is impossible to predict the values of a random noise signal at future times, such signals can be described very usefully in statistical terms.[18] A stationary random signal is described completely by its autocorrelation function (from which its power spectral density, psd, may be obtained), and by its probability density function, pdf, which gives the probability that its amplitude lies within specified ranges. Noise for which all frequencies carry equal energy is termed *white noise;* its autocorrelation function is an impulse.

The form of pdf most commonly met is the Gaussian form with zero mean, in which the probability $p(v)$ of the noise taking an amplitude v is given by

$$p(v) = [1/(2\pi)^{1/2}\sigma].\exp(-v^2/2\sigma^2)$$

σ is the $v$ms amplitude where the probability $P(t)$ that the signal level equals or exceeds a threshold amplitude t is given by

$$P(t) = \int_t^{\infty} p(v).\mathrm{dv}$$

For the Gaussian pdf this integral has no exact solution; a series solution is available (useful when t is more than four standard deviations from the mean), and tabulated numerical values are available for the error function (*ERF*) for use otherwise. The relationship between $P(t)$ as defined and $ERF(t)$ is given by

$$P(t) = 1 - [ERF(t)/2\sigma)^{1/2})/2]$$

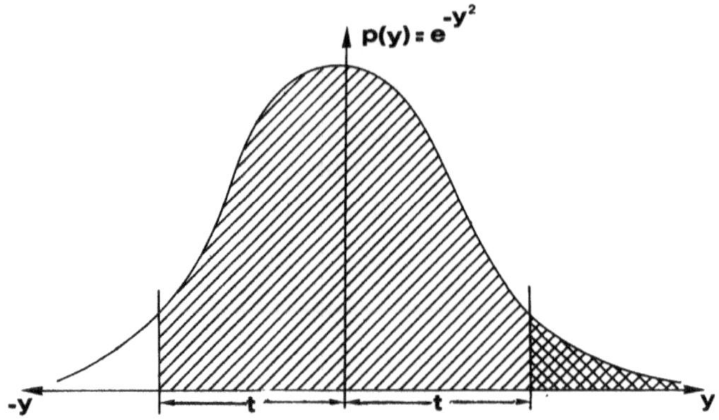

Figure 2.14. Relationship between error function and threshold crossing probability.

The relationship between $P(t)$ and $ERF(t)$ is illustrated in Figure 2.14. In this figure, $ERF(t)$ is indicated by the lightly hatched area in the center of the graph, and $P(t)$ by the more heavily hatched region on the right-hand side. This material is treated well by Schwartz.[19]

2.5.3. *Thermal Noise*

Random noise arising from the thermal motions of charge carriers is unavoidable in any electronic circuit not at absolute zero temperature. For a signal source whose output resistance is R ohms, the mean square amplitude of this Johnson noise is given by

$$v^2 = 4.\mathrm{k}.T.R.dF \qquad \text{volts-squared}$$

or, alternatively,

$$i^2 = 4.k.T.dF/R \qquad \text{amperes-squared}$$

In the above, k is Boltzmann's constant which is 1.6×10^{-23}, T is the temperature of the source in degrees absolute (about 300°K at normal temperatures), and dF is the bandwidth in which the noise is measured, expressed in Hz. The noise is Gaussian and white within its passband.

On substituting reasonable values for R (= 1000 ohms) and df

(= 1 MHz.), we obtain $v = 4.07 \times 10^{-6}$ volts and $i = 4.07 \times 10^{-9}$ amps; thus signals due to thermal noise are minute and are rarely a nuisance in automation systems. But there are other sources, such as quantization noise, which arises when a signal is rounded off to the nearest of a number of discrete values. This is white but obeys a uniform distribution. The message/noise ratio in dB due solely to noise introduced by uniform quantization is:

$$M/N = 4.8 + 20 \log_{10} N$$

Here, N is the number of levels.

There are many other sources of random noise. For example, noise is introduced into the signal acquired by an image sensor by surface roughness. The worst such source is speckle noise, arising when coherent (e.g., laser) illumination is used. Its magnitude is greatest when the roughness of the surface is greater than the wavelength of the illumination; it is then independent of the actual magnitude of the roughness. The ratio C between mean squared (noise) variation and squared mean level, given[20] by

$$C = \langle I^2 \rangle / \langle I \rangle^2$$

is unity.

When the roughness is not greater than the illumination wavelength, the variance of the speckle noise is proportional to the magnitude of the roughness, and may in fact be used to measure roughness.

The amplitude probability density function of speckle noise viewed with an aperture of finite size (as in a laser scanner) is close to a gamma distribution.[21]

2.6. Decision Making and Detection

2.6.1. Decisions

Making decisions is a common requirement in automation. It is often necessary to assign samples from two classes, denoted, for example, "good" and "bad," to one class or the other from observations or measurements made of the samples. Normally, we select measurements to be used for discrimination that differ as much as possible between the classes. But in practice, "noise" due to random variation in the measurements will introduce errors. Noise may arise due to unavoidable variation within a class, or to error in making the measurement due, for example, to rounding.

To explain these ideas, consider the task of discriminating screws with threads from those without threads. Assuming that the formation of a thread removes material, screws with threads will weigh less than those without. Thus, we could detect the presence of a thread from the weight of the screw. However, the weights of individual screws will vary for other reasons, such as variation of length within a tolerance. We must first establish that weight is a good measure before we can use it. To do this, we perform an experiment; we obtain batches of screws with threads (denoted class I), and of screws without threads (denoted class II). We weigh each member of each batch, and plot histograms giving the number of samples from each class which lie within a given weight interval. This generates an approximation to the "a priori" probability density function (pdf) for each class. Figure 2.15 shows a typical (hypothetical) example, and we note that the histograms overlap, i.e., for some values of weight, the screw might be either of class I or class II.

To inspect samples of unknown class membership, we select a threshold weight $w(t)$. Samples whose weight exceeds $w(t)$ are assigned to class II (no thread), and samples below, to class I (thread

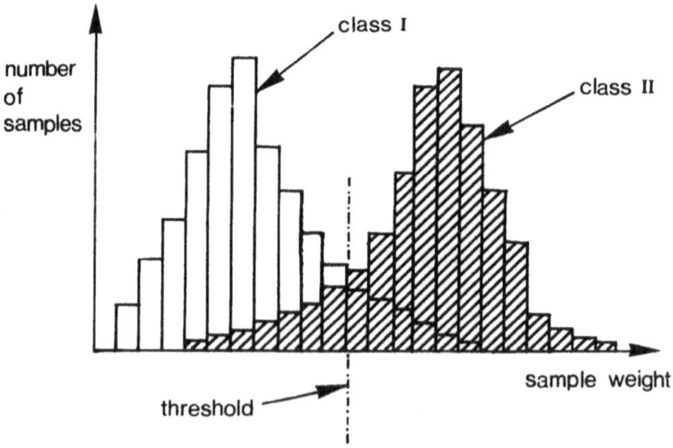

Figure 2.15. Histograms for two-class decision problem.

present). To assess the likely effectiveness of the method, we note that the pdf for class I (as approximated by its histogram) overlaps that for class II. The breadth of the pdf reflects the size of the noise. Some samples from class I will have weights exceeding $w(t)$, and will be assigned erroneously to class II; these are designated "false alarms." Also, some samples from class II will have weights below $w(t)$, and will thus be assigned to class I; these are called "missed detections." The probability $P(fa)$ of false alarm is given by the area under the a priori pdf for class I that lies above $w(t)$. The probability of missed detection $P(md)$ is given by the fraction of the area under the pdf for class II which is below $w(t)$. These quantities may be estimated directly from the histograms. The latter are merely measured approximations to the a priori pdfs. The decision process is analogous to the detection of a message accompanied by random noise in which the presence of the message causes a local increase in signal amplitude. A threshold can be used for detection, so that when the signal is below the threshold, noise alone is assumed present, whereas when the signal exceeds the threshold, a message is recognized as present also. The idea is explained in Figure 2.16. The analogy between message detection in noise (well explored by radar engineers) and the detection of defects or other changes in property among samples of material is very valuable. It is not accidental; the concepts used in radar detection were developed from those developed earlier for quality control in industrial manufacture.

In industrial inspection, missed detections are usually more serious than false alarms; they might, for example, cause automatic machinery using a defective product to jam, or allow a defective product to pass causing loss of goodwill in a customer. A small fraction of rejected good material is, however, generally acceptable. It is thus worthwhile trading $P(fa)$ against $P(md)$ by adjusting the position of the threshold $w(t)$ to minimize the cost of the errors, as follows:

In mathematical terminology

$$P(fa) = \int_{-\infty}^{w(t)} p(w/\mathrm{I})\, dw$$

and

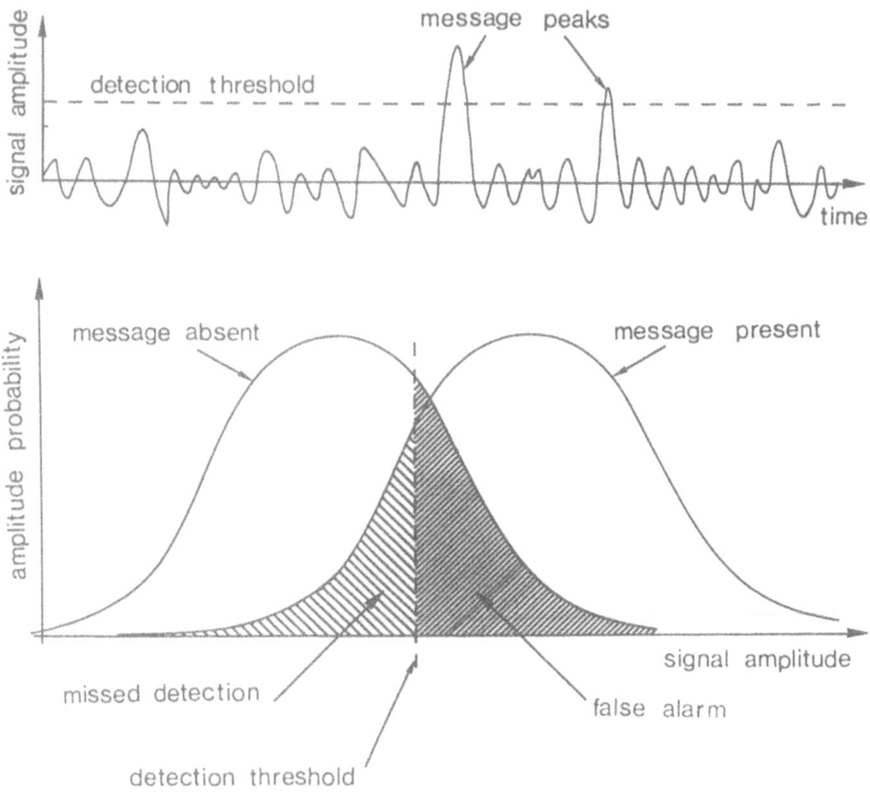

Figure 2.16. Detection of message masked by noise.

$$P(md) = \int_{w(t)}^{\infty} p(w/\mathrm{II})\, dw$$

The lower case p's are used to denote probability density functions, the upper case P's for total probability. Note that pdf's are mathematical functions, whereas total probabilities are numbers between 0 and 1. The notation $p(w/x)$ means "probability that sample has weight w, given that it is known to be (a priori) from class x."

In fact, we can adjust the threshold $w(f)$ to minimize the total loss L defined by

$$L = C(fa).(1-R[i]).P(fa) + C(md).R(i).P(md)$$

In the above equation, $C(fa)$ is the cost of a false alarm and $C(md)$ the cost of a missed detection. $R(i)$ is the probability that an inspected sample is from class i. This takes into account the fact that the two classes of sample are not equally likely to appear. Screws with threads are, we hope, much more likely than those without!

Often, several of the quantities in the above equation are not available, and "closed form" selection of a decision threshold to minimize total loss is not possible. The quantities most often absent are the costs $C(fa)$ and $C(md)$, and the a priori pdf for one of the classes, such as the fault class we are trying to detect. In this case, one of two alternative strategies is adopted. In the *Neymann-Pearson* strategy, $w(t)$ is adjusted to maintain an acceptable false alarm rate. If the probability of detection thus obtained is not adequate, the detection system must be improved. In the *inverse Neymann-Pearson* strategy, $w(t)$ is adjusted to obtain a satisfactory probability of defect detection, and other measures (such as visual reinspection of scrap material) are then used to ameliorate the consequences of an excessive false alarm rate. This latter course is the most usual.

The variation in system performance as $w(t)$ is varied may be illustrated using a *Receiver Operating Characteristic* (ROC), which shows how probability of detection varies with defect contrast, for different values of false alarm rate (and hence of threshold $w(t)$) (Figure 2.17a). However, in inspection problems the contrast of defects is generally not known explicitly, and the alternative form of ROC shown in Figure 2.17b is then more useful for comparing the performance of alternative detectors. The better the detector, the more closely will its characteristic approach the dotted line representing ideal performance.

2.6.2. *Complex Detectors*

If a simple detector does not provide adequate performance, there are several things that can be done. These are, however, more complicated and hence more costly; they include:

1. Use of an alternative feature measure, for example, silhouette outline in the case of screw inspection.
2. Use of a more powerful form of decision (e.g., sequential detection);[22,23] this generally requires further measurements over some of the samples, which may be inconvenient operationally.

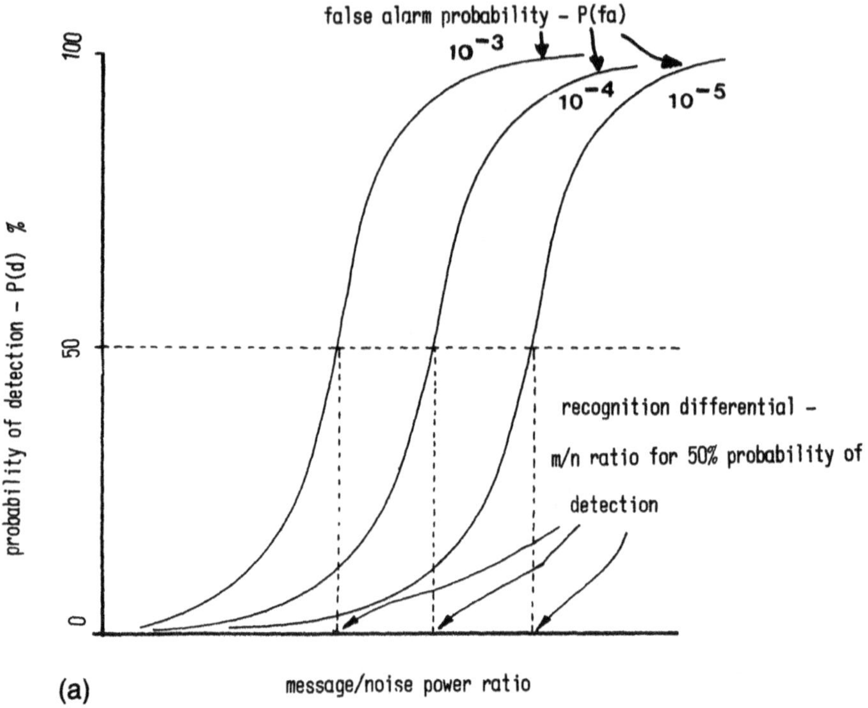

false alarm probability - P(fa)
10^-3
10^-4
10^-5
probability of detection - P(d) %
100
50
0
recognition differential -
m/n ratio for 50% probability of
detection
message/noise power ratio

(a)

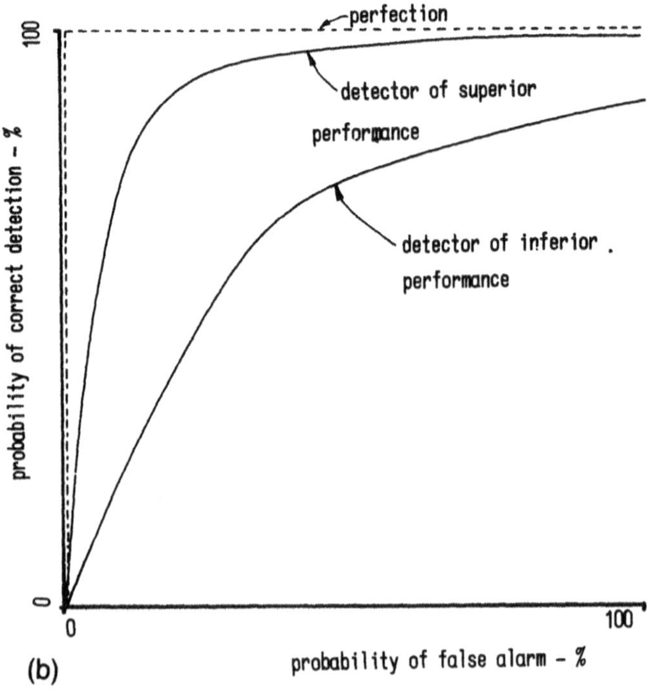

perfection
detector of superior
performance
detector of inferior
performance
probability of correct detection - %
100
0
0
100
probability of false alarm - %

(b)

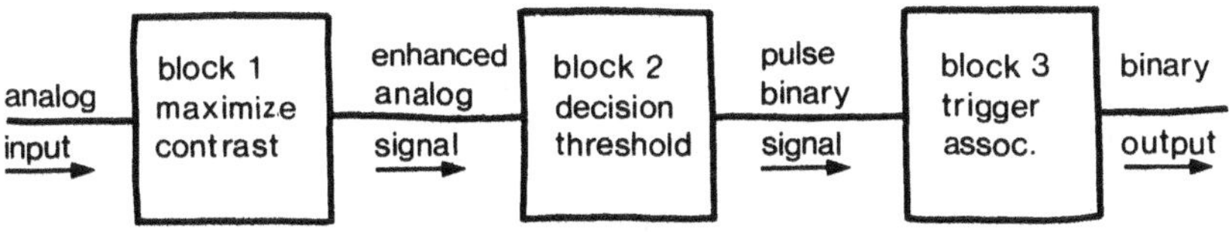

Figure 2.18. Composite detection scheme.

3. Reduction of the noise in the feature measurement (which controls the width of the pdf) before a decision is made, while retaining the same feature measure.
4. Use of several features simultaneously (e.g., both weight and silhouette outline for inspecting screws); if the features are carefully chosen, they will provide independent information. This approach leads to statistical pattern recognition, which is discussed in the next chapter.

The use of 2 and 3 together leads to the composite detection scheme whose canonic description is shown as a block diagram in Figure 2.18. The decision is taken essentially in block (2) and the performance of this decision depends on the separation between the values of the mean between the two classes, expressed in terms of the width of the pdf for each class. This may be regarded as the "contrast" for the decision. If performance is inadequate, we can try and improve the contrast. Figure 2.19 explains contrast as a statistical distance; the contrast between the two signals having means $m(1)$ and $m(2)$, and probability density functions pdf(1) and pdf(2) is

$$C(1,2) = D/([sd(1)] + [sd(2)])$$

Here, D is the separation between the means for the distributions, and $sd(1)$ and $sd(2)$ are the square roots of the respective variances.

2.6.3. *Contrast Enhancement*

For increasing the contrast of an analog signal, several approaches are available. The simplest is averaging; this exploits the property that

Figure 2.17. (a) Receiver operating characteristic (ROC) type "A;" (b) ROC type "B."

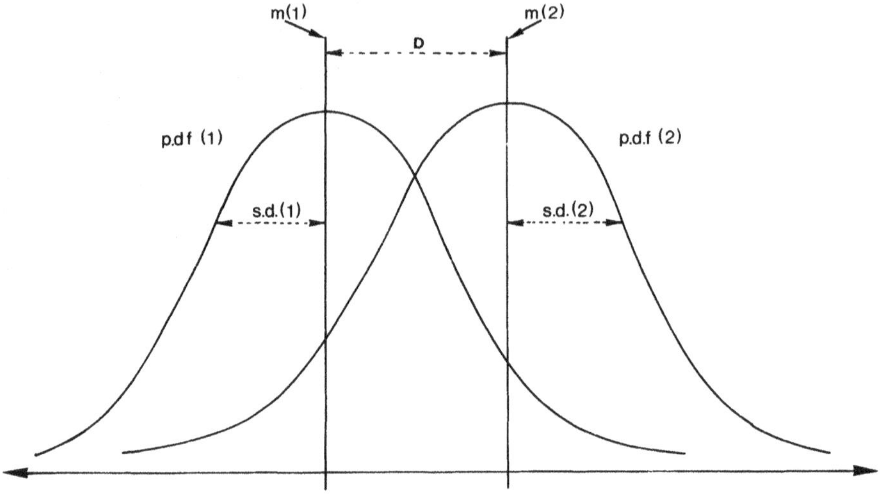

Figure 2.19. Statistical distance.

if two signals each containing constant (message) and uncorrelated random (noise) components are added, then the constant term in their sum is simply the sum of the constant terms in the two signals. The random noise components, however, combine so that their power is the sum of the powers in the components, i.e., as the square root of the sum of squares. If the message/noise amplitude ratio before combination was m/n, then after N signals are added it will become

$$N^{1/2} \,.\, m/n$$

If the noise components are partially correlated, the gain in contrast is smaller; when N signals each having a message/noise amplitude ratio of m/n and mutual correlation r are summed, the resultant signal has message/noise ratio

$$(N \,.\, m)/[N.\{1 + (N - 1)r\}]^{1/2}.n$$

This approach is useful (for example) for cleaning the output from line scanners making successive sweeps across moving sheet material, when the target defect appears in the same position in several successive scans. The accompanying noise due to surface roughness or texture is generally largely uncorrelated between successive scans.

It is also valuable for improving the contrast of two-dimensional images such as TV sequences, in which the frame rate is so high that the message component is effectively unchanged over many successive scans.

Random noise can also be removed using a linear filter. This exploits the property that the message component of a signal is generally concentrated in a particular part of the system passband, whereas the noise power is distributed more or less uniformly. The best improvement in message contrast when message and noise are uncorrelated (and the noise is white) is obtained using a *matched* filter, in which only those components of a signal known to contain significant message energy are allowed to pass (Figure 2.20). This unfortunately distorts the message signal, which may as a consequence become more difficult to identify. When message and noise are partially correlated, best contrast is provided by a Wiener filter.[24]

These ideas are best visualized using the signal space approach (Figure 2.21). The signal to be enhanced is a particular vector in the space. To enhance it, we multiply its vector by another vector which coincides in direction with the signal to be enhanced as closely as possible. Thus, the amplitude of the signal vector increases by more than all other possible signal vectors, e.g., noise. We see from Figure 2.21 that a signal whose vector is orthogonal to the match vector for the filter will be reduced to zero, and that a signal which is in the exact opposite direction will also be enhanced. Often, the form of the

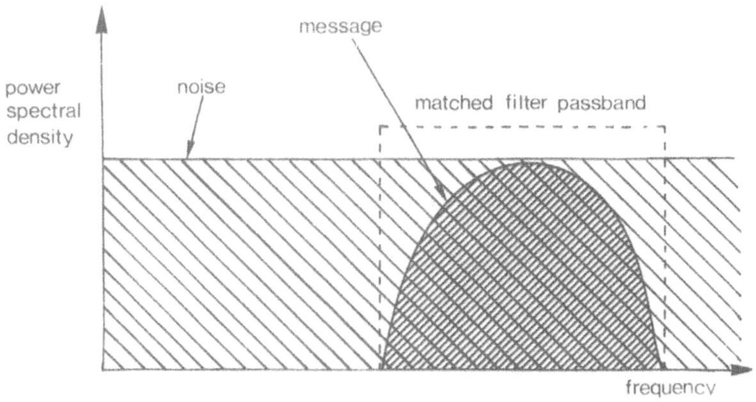

Figure 2.20. Matched filter described in frequency domain.

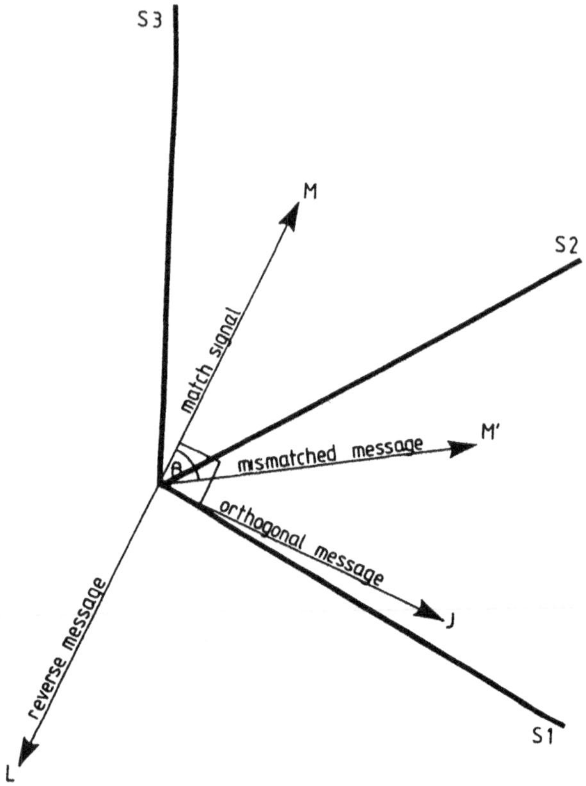

Figure 2.21. Matched filter described in signal space.

signal to be enhanced is not known explicitly; the solution here is to use a form for the filter giving maximum enhancement (i.e., maximum average match) over the ensemble of most likely forms for the signal. If the match between signals i and j is $m(i,j)$, and if the probability of the ith signal is $P(i)$, then we choose as filter waveform the jth signal which maximizes the function $M(i,j)$ given by

$$M(i,j) = \sum_{i} P(i).m(i,j)$$

The idea is illustrated (with $P(i)$ equal for all classes) in Figure 2.22. More sophisticated analysis would generally indicate a form for the match waveform somewhere between the signals actually encountered.

						average match
	1	0.78	0.77	0.86	0	0.68
	0.78	1	0.60	0.95	0	0.66
	0.77	0.60	1	0.63	0.12	0.63
	0.86	0.95	0.63	1	0	0.69
	0	0	0.12	0	1	0.22

Figure 2.22. Selection of a "best" matched filter for ensemble of waveforms.

The question then arises, is this single filter giving an overall best match better than a bank of filters each matched to a single waveshape, with the output triggers combined logically by ORing? It is necessary (and worthwhile) deciding such questions to produce efficient processing, but further discussion of this subject exceeds the scope of this book.

Finally, if the decision maker cannot generate an indication reliably without producing excessive false alarms, we can resort to the

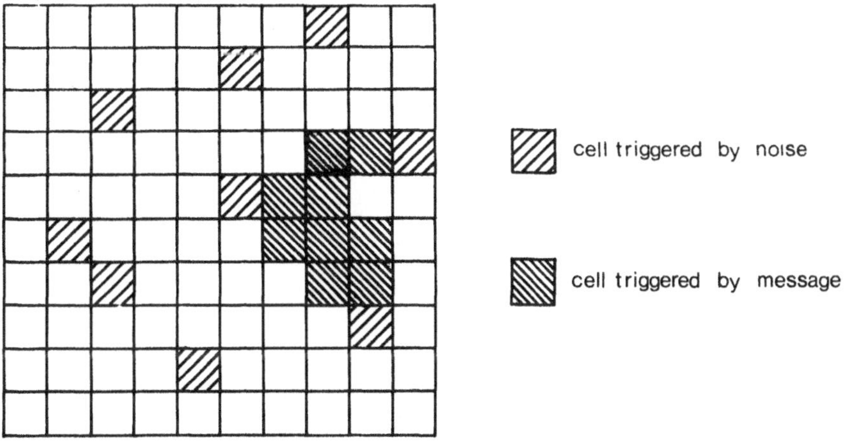

Figure 2.23. Principle for rejection of false alarm triggers using clustering.

trigger association indicated in block 3 of Figure 2.18.[5] Trigger association provides a method of eliminating excessive false alarm triggers which is particularly valuable in the automated inspection of moving strip material. Considerable practical difficulty arises in implementing an automated system for detecting defects on (for example) a rough surface which generates many noise triggers, such as steel strip, as follows. Suppose the surface is divided into blocks, each comprising an array 1000 pixels square. If the pixel separation were 0.1 millimeter, the block would be 10 cm square, containing 1 million pixels. Making the detection threshold low enough to be reasonably

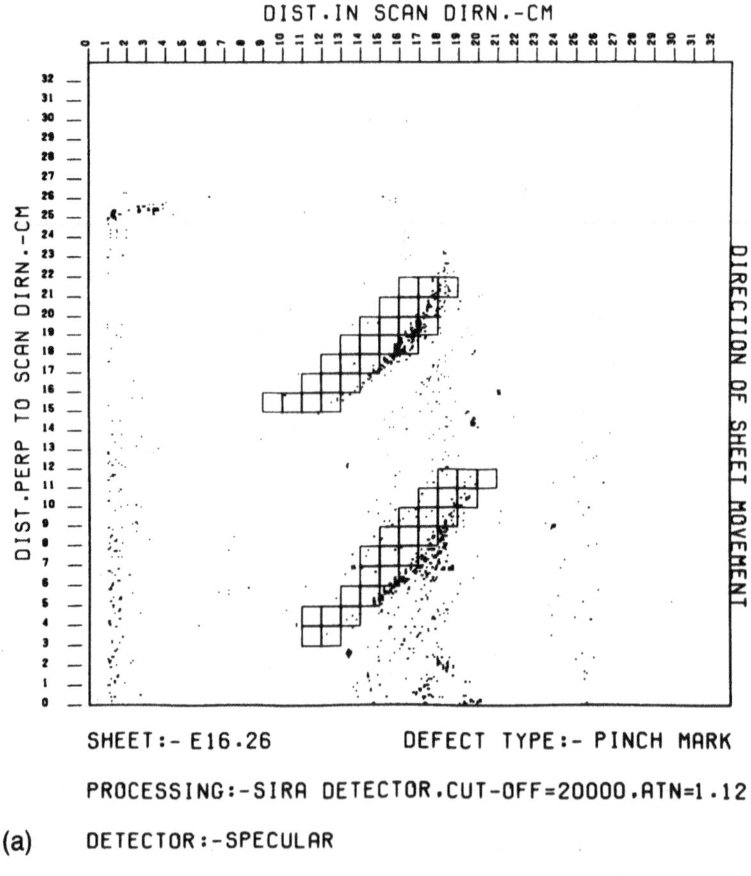

(a)

Figure 2.24. (a) An image containing many triggers generated by noise, in addition to triggers due to a defect. (b) The same image cleaned using the trigger association method.

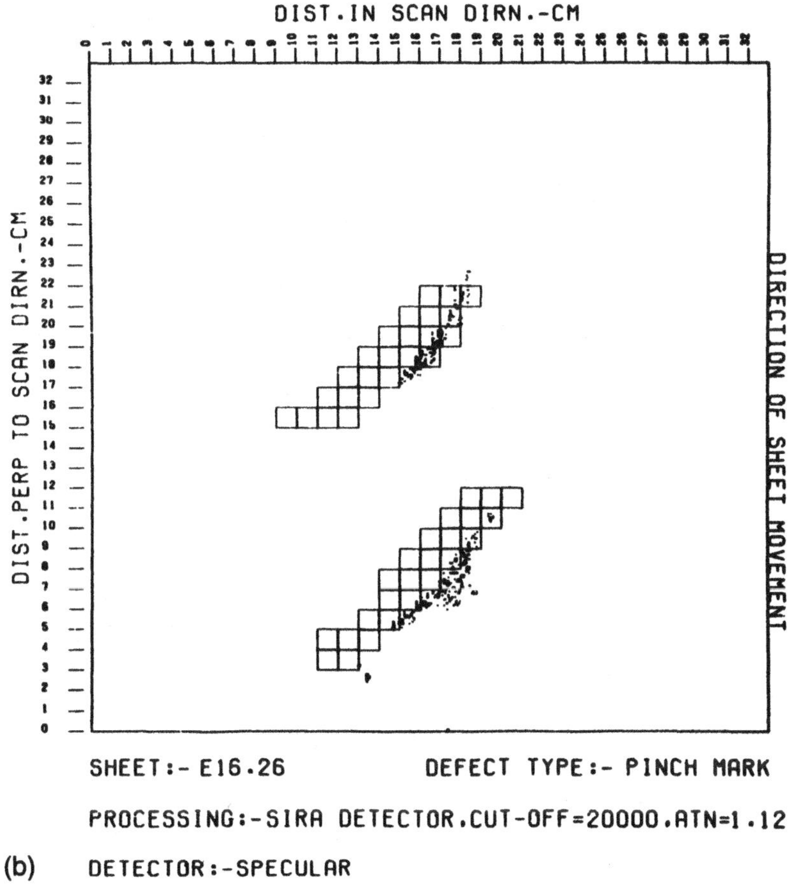

(b)

Figure 2.24. (*Continued*)

(say, 90%) certain of detecting each defect might make the probability of false alarm generation per trigger be, say, 0.001. Each block would then be virtually certain to contain several hundred false alarm triggers, and the whole of the surface would be designated incorrectly as being faulty.

The noise triggers would be distributed uniformly and at random over the surface, given that the noise was white. Triggers due to defects would on the other hand occur in compact clusters. This property could be used to eliminate selectively those triggers due to noise, as follows (see Figure 2.23). The pixels would be examined in groups of N, and the number within each group containing triggers counted.

The occurrence of some number K or more triggers within the group due to noise alone is considered an event so rare that it must be due to a defect. The threshold K is selected so that almost all defects generate this number of triggers or more. In fact, the probability $P(K)$ that K or more triggers occur due to noise alone is given by the cumulative binomial distribution:

$$P(K) = \sum_{L=K}^{L=N} N!/L!.(N-L)!.p(\mathrm{O})^L.[1-p(\mathrm{O})]^{(N-L)}$$

For typical values, e.g., $N = 100, K = 10, p(0) = 0.001$, then $P(K)$ is about 10^{-15}. Thus, instead of a defect being indicated due to noise for all clusters, it will now be so indicated for only 1 in 10^{-15}. Since almost all (this might reasonably be 99%) defects will contain 10 triggers or more, and hence be detected, this approach renders the inspection practicable. It could be implemented simply by dividing the surface into cells each 10 pixels square, though more complex methods (such as using lines along the preferred defect direction if there is one) may give better results. The improvement obtained by this method may be seen by comparing Figures 2.24a (detection threshold low enough to detect defect-too many false alarm triggers), and Figure 2.24b (trigger association applied to reduce false alarm rate to acceptable level).

References

1. L. Hyvarinen, *Information Theory for Systems Engineers*, Springer, New York (1968).
2. G. Raisbeck, *Information Theory—An Introduction for Engineers and Scientists*, MIT Press, Cambridge, MA (1963).
3. N. Abramson, *Information Theory and Coding,* McGraw-Hill, New York (1963).
4. R.C. Gallager, *Information and Reliable Communication,* Wiley, New York (1968).
5. M.J. Eccles, M.P.C. McQueen, and D. Rosen, Analysis of the digitized boundaries of planar objects, *Patt. Recog. 9,* 31–42 (1977).
6. R.W. Hamming, *Coding and Information Theory,* Prentice-Hall (1980).
7. E. Horowitz and S. Sahni, *Fundamentals of Computer Algorithms,* Computer Science Press, Potomac, Md. (1982).
8. R. Bracewell, *The Fourier Transform and its Applications,* McGraw-Hill, New York (1978).
9. N. Ahmed and K.R. Rao, *Orthogonal Transforms for Digital Signal Processing,* Springer, New York (1975).

10. H.F. Harmuth, *Transmission of Information by Orthogonal Functions,* Springer, New York (2nd ed., 1972).
11. J.J. Gerbrands, On the relationship between SVT, KLT and PCA, *Pattern Recognition 14* (1–4), 375–382 (1981).
12. Robert O. Mitchell, Global and partial shape discrimination for computer vision, *Proc. SPIE,* 38–46 (1984).
13. D.F. Elliott, *Fast Transforms—Algorithms, Analyses, Applications,* Academic Press, (1982).
14. P.M. DeRusso, R.J. Roy, and C.M. Close, *State Variables for Engineers,* Wiley, New York (1967).
15. S.J. Mason and H.J. Zimmerman, *Electronic Circuits, Signals and Systems,* Chapman and Hall, London (1960).
16. D.C. McSherry, Computer processing of diagnostic ultrasound data, *IEEE Trans. Sonics and Ultrasonics, SU21* (2), April (1974).
17. L.R. Rabiner and B. Gold, *Theory and Application of Digital Signal Processing,* Prentice-Hall, Englewood Cliffs, NJ (1975).
18. A. Papoulis, *Probability, Random Variables and Stochastic Processes,* McGraw-Hill, New York (1965).
19. M. Schwartz, *Information Transmission, Modulation and Noise,* McGraw-Hill, New York (1980).
20. J.W. Goodman, in *Laser Speckle and Related Phenomena* (C. Dainty, ed.), Springer, New York (1975).
21. G.T. Stansberg, On the first order probability density function of integrated laser speckle, *Optica Acta 28* (7), 917–932 (1982).
22. A. Wald, *Sequential Analysis,* Dover, New York (1973).
23. P.G. Hoel, *Introduction to Mathematical Statistics,* Wiley, New York (1971).
24. M. Schwartz and L. Shaw, *Signal Processing—Discrete Spectral Analysis, Detection and Estimation,* McGraw-Hill, New York (1975).
25. L. Norton-Wayne, On the removal of random noise from binary images—A comparison study, *Proc. 5th IJCPR (IEEE pub. 80CH1499-3),* 1180–1183 (1980).

3

Introduction to Machine Pattern Recognition

3.1. Overview

Machine pattern recognition is a very valuable technique for distinguishing objects and signals in intelligent automation. However, when considered in depth, pattern recognition becomes very complex mathematically. Thus, we provide here an introduction only, aimed chiefly at newcomers. References to textbooks and papers providing more detailed information appear at the end of this chapter.

The chapter is organized as follows. We begin by explaining the nature and terminology of pattern recognition, including the measures used to specify performance. Then the three approaches to PR in common use are introduced, and two (heuristic and feature space methodologies) are explained in further detail. Finally, advice is offered for selecting and assessing the applicability of pattern recognition methodology for specific problems.

3.2. Elementary Approach—Attribute Matching

Consider the task of distinguishing eight items which are typical articles which must be priced and added to a bill at a supermarket checkout. Items [1], [2], and [3] are tin cans; item [1], a can of sardines, is low and flat; [2] is a standard tall cylindrical can, containing perhaps baked beans, and [3] is a rectangular can with rounded corners of the shape used to contain corned beef. Item [4] is a bottle (e.g., of mineral water). Item [5] is a rectangular package (e.g., a soap powder carton), item [6], a flat package typical of butter, item [7], an inverted frustum of a cone, made of plastic and containing yogurt, and [8], a rectangular pack with soft corners, typifying a bag of sugar.

Table 3.1
Optimal Binary Tests for Distinguishing Eight Items

	Test		
Item	A	B	C
[1]	y	y	y
[2]	y	y	n
[3]	y	n	y
[4]	y	n	n
[5]	n	y	y
[6]	n	y	n
[7]	n	n	y
[8]	n	n	n

Since there are eight items, a minimum of three binary tests (i.e., tests whose outcome is either "yes" or "no") would suffice to distinguish the items, as shown in Table 3.1.

In Table 3.1, a 'y' entry indicates a 'yes' outcome for the test, and a 'n' entry, a 'no' outcome.

In practice, it is unlikely that we shall find three tests that provide the required orthogonal partition, i.e., that divide the list of items each into two equal halves in a different way. Tests we could actually make might be as follows:

A: Is the item metallic?
B: Is the weight greater than 500 grams?
C: Is the height greater than 10 cm?
D: Is the item transparent?
E: Is the item soft?

Applying these tests to the items yields the results shown in Table 3.2.

The process is inefficient; instead of three tests we have to make five. Instead of the 4:4 partition which is optimal, no feature does better than 3:5, and D gives only 7:1, which is very poor. Although tests C and E yield useful information, the result given by C will depend upon which face the item is resting on, e.g., item [6] will exceed 10 cm if stood on end but not otherwise. Test E might cause damage; unless very carefully designed, it is not nondestructive. It is just pos-

Table 3.2
Tests We Might Actually Perform

	Test					
Item	*A*	*B*	*C*	*D*	*E*	
[1]	y	n	n	n	n	
[2]	y	y	y	n	n	
[3]	y	y	n	n	n	y = yes
[4]	n	n	n	y	n	n = no
[5]	n	y	y	n	y	
[6]	n	n	n	n	n	
[7]	n	y	y	n	y	
[8]	n	y	n	n	y	
	3 : 5	3 : 5	3 : 5	7 : 1	3 : 5	Partition

sible to distinguish the items using these tests since the "row vector" of outcomes is different for each. But there is little room for error; in many cases, inversion of a single "y" to an "n" or vice-versa will change the vector to that characteristic of some other item.

To identify an unknown item, a *parallel search* may be used, in which the vector of "yes" and "no" decisions measured is compared with those for all items stored, until a fit is found. The unknown item is assigned to the class of the vector which fits. If N classes are stored, then (assuming the classes are equiprobable) we must make on average N/2 comparisons between vectors to find a fit. If the items are not equiprobable, this can be reduced by trying the most likely matches first.

The alternative *serial* matching process involves a sequence of decisions based on the outcomes of the tests. This may be represented by a tree structure (illustrated for the perfect partition in Figure 3.1 and for practicable tests in Figure 3.2), in which nodes represent decisions; branches emerging from a node, alternative outcomes for decisions; and terminals, final assignments. If there are k outcomes for a decision, k branches must emerge from the appropriate node; often k is the same for all nodes, and in our example k is constant and is 2. The identification process starts from the root, and proceeds in the direction of the arrows until a terminal is reached. Arrival at a terminal constitutes assignment. Note that in some cases the tree may

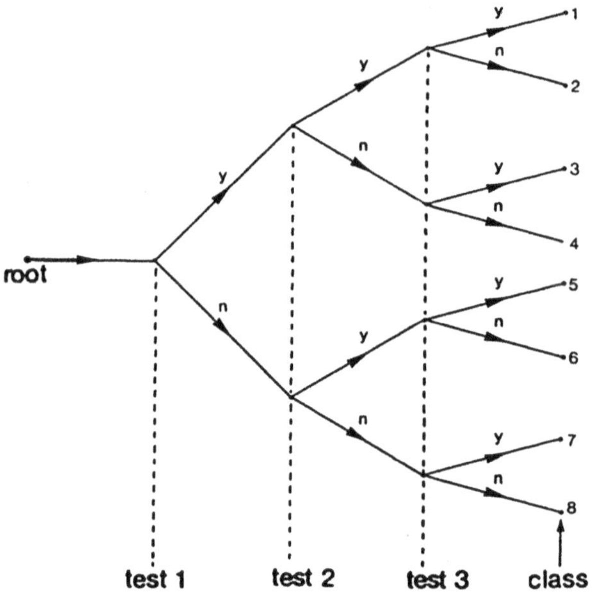

Figure 3.1. Identification tree—ideal features.

be pruned, i.e., the search may be stopped before all five tests have been made. Branches which may obviously be pruned away are indicated in Figure 3.2 by a broken line.

The merit of the serial approach is its speed. Given optimal design, only $\log_k N$ decisions are required to identify N classes, given that the number of outcomes is the same for each decision and is k. For a search with 32 classes and 2 outcomes per decision, only 5 decisions (comparisons) will be needed, on an average, to identify an item. If the 32 classes are not equiprobable, the average number of decisions may be reduced to less than 5, using an approach similar to the Huffman code (Section 2.2.2) to make the alternative outcomes at each decision as nearly as possible equally likely. This is much better than the 80 decisions required if a parallel matching process is used on the same attribute data. The disadvantage of the simple serial search described lies in its sensitivity to error; if even a single decision is incorrect, the correct outcome cannot be reached. It is possible to modify the process to provide error protection, at some cost in speed and complexity.[(1)]

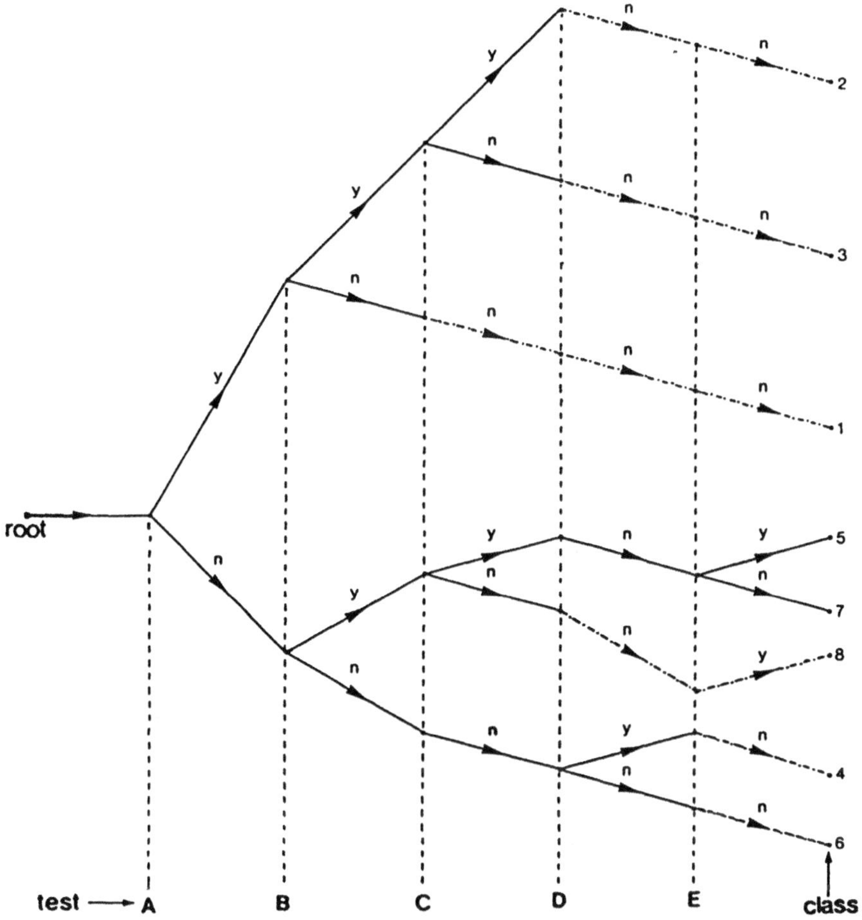

Figure 3.2. Identification tree more realistic features.

3.3. Principles and Terminology

3.3.1. Desiderata

In the very simple example just quoted, it is possible to select attributes as features, demonstrate their utility, and produce tables and trees for the identification process intuitively. In most practical problems, this is not possible. The number of classes of pattern is too large, and it is not obvious which attribute measures will be most useful as features. Thus, recourse must be made to computer algorithms for the design process. Two further complications are inherent even in identifying supermarket merchandise. The system must learn to recognize items using a "training by showing" procedure, which may be performed by an unskilled operator, and it must be possible to add new items to the list of articles to be recognized without having to rescan those already stored. Thus, powerful and general techniques are required for recognizing patterns by machine, in which the statistical nature of many pattern recognition tasks, and the unavoidability of error, can be taken into account.

In any pattern recognition problem, it is necessary to distinguish between two classes, though normally there are more than two. Thus, a process of *decision* is involved. The screw sorting problem discussed in Section 2.6.1. is a simple two-class pattern recognition problem; more complex problems possess essentially the same characteristics in that:

a. Some incorrect decisions are inevitable, and the classifier must be designed to minimize these, or, even better, their harmful consequences.
b. Prior ("a priori") information regarding the properties of the various classes is required to design the classifier. This usually involves making measurements on substantial numbers of samples from each class, with the class assignment of each sample known. The data used to design the classifier is termed the *training set.*
c. The classifier operates on a small number of attributes (normally measurements) which have been selected very carefully. These are termed *features.* The several measurements made on a sample may be presented as a group of numbers, termed a vector; e.g., if there are five feature measurements $m1$, $m2$, $m3$, $m4$, $m5$, then the feature vector for a sample will be ($m1$, $m2$, $m3$, $m4$, $m5$). This can be regarded as defining a point in a vector space, in which the measurements $m1$, $m2$, etc. are the projections on the orthogonal feature axes $f1, f2, \ldots$ etc. The vector space is termed a *feature space.*

Pattern problems can be categorized according to the properties of the feature vectors representing each class. If each class can be represented by a single vector, the problem is said to be *deterministic.*

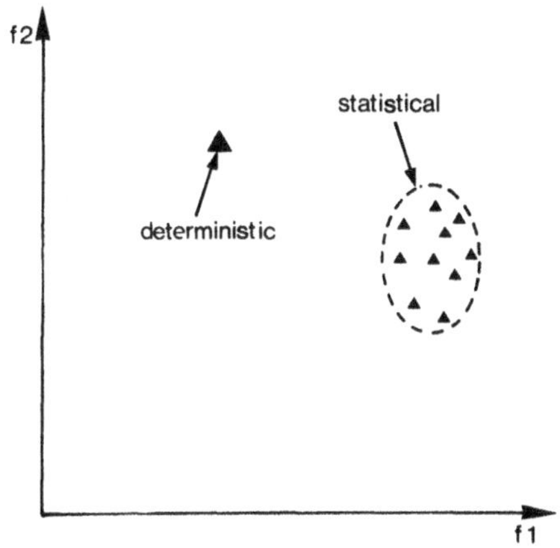

Figure 3.3. Deterministic and statistical problems in a two-dimensional feature space. The feature measures are *f*1 and *f*2.

This does not imply that classification will always be error free, since errors may occur in measuring features. However, in a deterministic pattern recognition problem, the properties of each pattern class may be learned from only one prototype sample. In many problems, however, "within class" variations in the properties of samples mean that each class may be represented only by a cluster of vectors, which occupy a region in the vector space of finite size. To determine the properties of this cluster, it is necessary to examine many training sample patterns. This concept is explained in Figure 3.3. If the feature clusters representing different pattern classes do not overlap, error-free classification should be possible.

In scientific research, the technique of *unsupervised learning* is occasionally used, in which data vectors are processed without any a priori class assignments. Computer algorithms are used to group the data into classes, on the bases of similarities and differences found within the data, such as a tendency for vectors to group into clusters. Vectors falling into the same cluster are then assumed to arise from the same class. This approach in which classes are designated *by the processing* is of little use in automation.

3.3.2. Feature Extraction and Selection

The first stage in any pattern recognition task is to extract a number of attribute measures to be used as features. The measures represent the patterns much more efficiently than would the patterns themselves; a useful compression of information is involved, which leads to efficient processing. This is, however, irreversible; the information thrown away when features are extracted is not recoverable. As an example, Figure 11.2 shows how a silhouette shape is represented by features which are the lengths of radii measured from centroid to boundary. Measures must be selected which are easy to compute and are unaffected by errors; thus area is a good feature for specifying silhouette shapes, whereas boundary length is less useful (Figure 11.3).

When (as is usual) it is not possible to select features providing low-error recognition using intuition only, computational tests must be used. One starts off with a set containing a large number of features which are considered *likely* to be useful, from which a subset will be selected. The original large feature set is termed the *candidate* feature set. There is no benefit in using all available features in a classifier, since though performance initially improves quickly as new features are added, it soon flattens off and eventually declines (Figure 3.4). The additional features add little new information and a lot of noise.

Systematic evaluation of all possible combinations of features is an ideal which is rarely achievable because the number of combina-

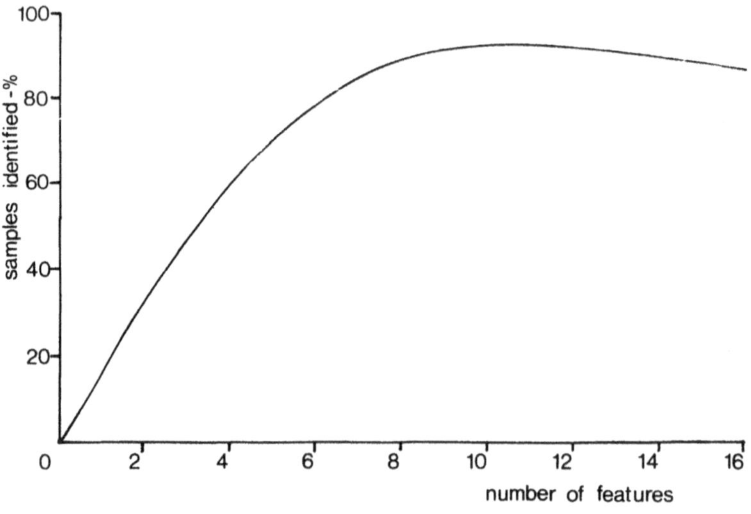

Figure 3.4. Variation in performance of a typical (hypothetical) classifier with number of features.

tions which must be examined is astronomically large. Given N candidate features, then $N!$ combinations are possible; even for $N = 10$, $N!$ is about 3.6 million, and N is often much greater. One possibility is to use a suboptimal search, in which the best individual feature is found first, then the best pair incorporating this first feature, then the best triple incorporating this pair, and so on. The quality of the features is of course determined by trial, using computational simulation of the classifier. Some idea of the extent to which feature sets found by suboptimal searches are inferior to those found by optimal searches can be gained from Figure 3.16.

Another approach is to use tests which assess the extent to which the features possess desirable properties. A first requirement for features is that they be mutually independent; if not, then the new information contributed by adding a feature is reduced. This is a necessary requirement, not a sufficient one; mere independence in the statistical sense does not guarantee that the extra information provided will be useful for the task at hand. The amount of independent information provided by a feature set may be determined by constructing the covariance matrix for the set, in which the I,Jth entry is the covariance between the Ith and Jth features. The covariance is the cross-correlation between the features, multiplied by the square root of the product of their autocorrelations. The eigenvectors of the covariance matrix specify new features which are linear combinations of the original features and are orthogonal, i.e., change in one feature does not affect the others. The magnitude of the eigenvalue corresponding to each eigenvector is proportional to the variance of the eigenvector, and hence to the information therein contained. Although this process generates the same number of features as were contained in the original set, only a few of the new features will contain sufficient information (known now to be completely independent) to justify further interest.

Another approach is to use the distance measures introduced in Section 2.6.1., in which, for example, the spacing between class means expressed in terms of class variances gives an indication of discrimating power. This is valid for statistical problems in which features form unimodal clusters; it is most valuable when the clusters are Gaussian. Several sophisticated measures are available; these are explained in Reference 2.

3.3.3. Performance Specification and Evaluation

The errors made by a classifier are of two types. One is to *substitute* a sample by assigning it to the wrong class; this is generally a fairly serious error. The other is to *reject* a sample, saying that the system cannot assign it to any class with acceptable reliability. Rejection is normally much less serious than substitution; the offending item may then often be withdrawn from a processing sequence without causing harm, and possibly resubmitted, for instance in a different orientation.

Classifier performance as measured experimentally is generally specified using an "a priori" confusion matrix, in which each row is devoted to a "true class" and each column is devoted to an "assigned class." The entry in the *I*th row and *J*th column is the number of samples from Class *I* assigned by the classifier to class *J*. If the classifier were perfect, all samples from a particular class would be assigned to that class, and all entries not on the main diagonal would be zero. The entries for each row sum to the total number of samples which have been examined from the appropriate class. Figure 3.5 shows a typical a priori confusion matrix, and we see that, for example, of the 35 samples (the sum of all the entries in row 1) from class 1 presented to the classifier, 25 were actually identified as being from class 1, 5 were substituted to class 3, and 3 to class 6. Two samples were rejected as being unassignable to any class with sufficient confidence. On totalling the entries on the main diagonal (which gives the total number of samples correctly classified), dividing this by the sum of all entries in the confusion matrix (which gives the total number of samples

true class \ assigned class	1	2	3	4	5	6	reject class
1	25	0	5	0	0	3	2
2	0	40	1	1	0	2	1
3	1	0	49	0	0	0	0
4	0	1	0	37	2	0	5
5	2	3	0	5	42	3	0
6	1	0	1	0	2	40	6

Figure 3.5. A priori confusion matrix.

		true class					
		1	**2**	**3**	**4**	**5**	**6**
assigned class	**1**	86	0	3	0	7	3
	2	0	91	0	2	7	0
	3	9	2	87	0	0	2
	4	0	3	0	97	0	0
	5	0	0	0	4	91	4
	6	6	4	0	0	6	83
reject class →		14	7	0	36	0	43

Figure 3.6. A posteriori confusion matrix.

examined), and multiplying the result by 100, we obtain the percentage of input samples correctly classified. This is a useful measure of the overall performance of the classifier; for the matrix in Figure 3.5 it is 83%.

The "a posteriori" confusion matrix illustrated in Figure 3.6 gives the percentage probability that a sample identified as being from class I (the row class) is from class J (the column class). Again, for a perfect classifier, all samples identified as being from a particular class will actually be from that class; the matrix will have nonzero entries only on the main diagonal. The a posteriori confusion matrix may be computed from the a priori matrix, by transposing it, summing the entries in each column, dividing each entry by the sum for the appropriate column, and multiplying the result by 100.

In evaluating a classifier, optimistic results are obtained if the sample data used to design the classifier are also used to evaluate it. Use of an independent set of samples of known class termed the *testing set* is highly desirable. To overcome the practical difficulty of gathering sufficient samples to ensure statistical validity, the leave-one-out technique of Lachenbruch and Mickey[(3)] may be used. Here, given a training set containing K samples, the classifier is designed K times using the K possible combinations of $(K - 1)$ samples, and its performance in identifying the Kth sample which has been omitted is evaluated. For a typical problem involving 50 samples from each of 10 classes, K will be 500 and the classifier must be designed 500 times. Though this may appear very time consuming, it is quite practicable using currently available computers.

3.4. Heuristic Techniques

Heuristic techniques comprise a collection of "ad hoc" techniques, often devised intuitively to solve specific problems, and generally unrelated to one another. The simplest heuristic technique is mask matching, which we illustrate by considering its application to the automated identification (reading) of alphabetic characters. The principle of the method is illustrated in Figure 3.7 in which sketches A–C show different characters displayed as binary matrices. Sketch D shows an unknown character which is one of types A–C, but is identical to none of them, due to corruption by noise. It is identified by assigning to the class which gives the best fit. We compare the unknown to each prototype element by element, and score +1 for each element which is similar, and −1 for each element which is different. The calculations are shown in Table 3.3; the fit to the letter "L" has the highest score, and the unknown is therefore identified as a letter L.

Performance may be improved by weighting the elements, thus assigning more importance to selected positions such as the "tail" which distinguishes a "Q" from an "O." But this increases sensitivity to noise.

Despite its simplicity, this approach has so many weaknesses that it is difficult to apply to practical problems. For example, a slight rotation between sample and prototype can destroy a match completely; distortion has the same effect. Different type fonts (typefaces such as pica, elite, and so on), and styles such as capitals and italics, are effectively different patterns, and a separate mask must be provided for each, increasing complication and the risk of confusion. Some relief is obtainable by using specially designed fonts for specific tasks

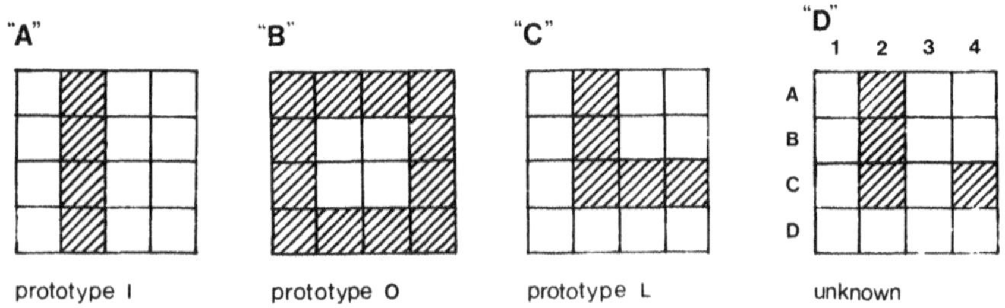

Figure 3.7. Mask matching example.

Table 3.3
Identifying by Mask Matching

	Similar	*Different*	*Total*
Comparison with I	14	−2	12
Comparison with O	4	−12	−9
Comparison with L	15	−1	14

where machine reading is particularly valuable (as in check sorting), and where the font used can be controlled.

A better approach to character identification, and many other problems, is to exploit the property with which characters are identified by humans using specific features in their structure which are highly invarient to distortions and changes of font. Each character is identified by specifying a number of primitive features, plus their relationship to one another. This is illustrated in Figure 3.8, in which the digit "5" is seen to comprise some singularities termed "nodes" (numbered 1–4), connected by lengths of curve. Nodes may be places where the curve has a corner, where the curve ends (a "terminal" node), or where two or more branches of curve join. The two "5"s are easily identified despite the distortions in the (b) version. Their representations on a grid of squares would on the other hand be quite different. Moreover, the "9" shown in (c) can be distinguished from both 5's merely by counting the number of nodes; it is not necessary even to classify them. The numbers "5" and "9" can be represented by the graphs shown in Figure 3.8 (d) and (e) respectively. These represent the digits uniquely as nodes connected by lengths of curve. The difficulties in this approach lie chiefly in the complexity of the processing required before features can be extracted; the characters must be digitized to a high resolution, and a computationally expensive skeletonization process must be applied before the extraction of nodes can begin.

Other useful heuristic pattern recognition approaches particularly applicable to images include logical decomposition,[(4)] row and column sums,[(5)] and Fourier (and Walsh/Hadamard) descriptors.[(5)] These are explained in the references listed.

Linguistic pattern recognition is a powerful general methodology,

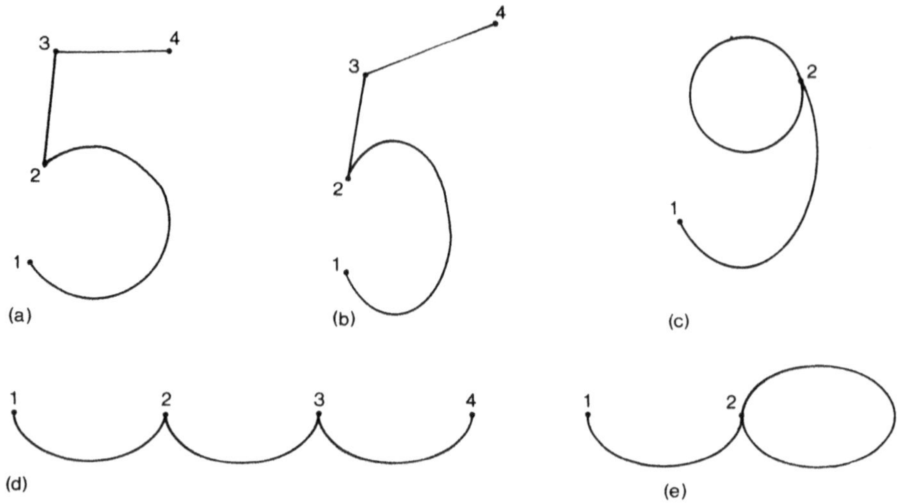

Figure 3.8. Identifying a character using nodal relationships.

which is mentioned among heuristics only because nowhere else seems as appropriate. It involves characterizing patterns as strings of primitive symbols, whose relationships within the string are governed by rules ("productions") similar to the rules of grammar in a language. The method described above for identifying the digit "5" is a simple form of linguistic methodology. Although the linguistic methodology can be applied to two-dimensional patterns and to noisy patterns, it has been little used in intelligent automation—but this does not mean it should be ignored! Its most successful applications have been, for example, in the analysis of slowly varying waveforms (one-dimensional signals) such as electroencephalograms, which are ideal for description as a symbol string. Linguistic pattern analysis is well documented.[6,7] Reference 4 describes an attempt to apply linguistic analysis to an automation problem (itself described in Section 11.5) better solved using other approaches.

3.5. Feature Space Analysis

In this popular and very general approach, the feature measures define the axes of an orthogonal space. Pattern classes are specified by points in the space for deterministic problems, in which all samples from a particular class are identical, and by regions in the space for statistical problems, in which the properties within a class may vary. Many varieties of feature space classifiers have evolved, and are explained below.

3.5.1. Maximum Likelihood

This form of feature space classifier is a simple extension of the statistical decision process introduced in Section 2.6.1; it is appropriate for statistical problems. The a priori probability density functions for the pattern classes are estimated by, for example, plotting histograms. Suppose there is only one feature, and an unknown pattern is to be identified, for which this feature takes the value $f(u)$. The a priori probability functions $p(f/1)$ for class 1, $p(f/2)$ for class 2, . . . $p(f/k)$ for class k, are compared at $f = f(u)$. The sample is assigned to the class j, for which $p(f/j)$ is greatest. The idea is illustrated in Figure 3.9; it is simple but highly effective. Since the pdf's $p(f/1), p(f/2), \ldots$ are fixed, it is possible to specify threshold values $f = t1$, and $f = t2$, etc. which separate the classes. To identify, we simply see where the unknown sample falls relative to these thresholds. As described in Section 2.6.1, the utility of the maximum likelihood detector is improved by adjusting the positions of the thresholds to minimize the harmful consequences of incorrect decisions.

If a single feature does not provide adequate discrimination between classes, then using several together may well prove effective. Figure 3.10 illustrates how this works. Figure 3.10(a) and (c) show two classes, each operating on single features, called $f2$ in (a) and $f1$ in (c). Neither classifier is very satisfactory since the a priori pdf's for the classes overlap considerably. When, however, the joint a priori pdf's are used and plotted on the same graph (Figure 3.10(b)) it can be seen that these show much less overlap, and a discrimination threshold is now possible, in the form of the line labelled "Threshold 3" in Figure 3.10. Note that the pdf's are shown as contours in (b). There is no *guarantee* that two features taken together will provide better discrimination than either taken independently, but this may well happen and is certainly worth trying. Figure 3.10(a) may be regarded as (b) viewed along the $f1$ axis, and (c) may be regarded as (b) viewed parallel to the $f2$ axis.

The chief difficulty in applying the maximum likelihood classifier

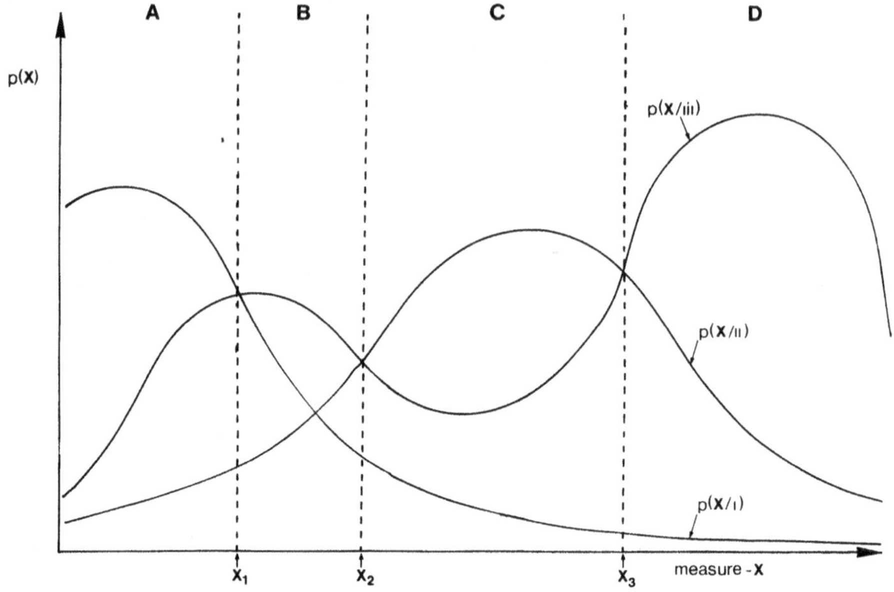

Figure 3.9. Maximum likelihood classifier using the single feature X. There are three classes—(*i*), (*ii*), and (*iii*). The a priori probability densities for the three classes are $p(X/i)$, $p(X/ii)$, and $p(X/iii)$, respectively. The value of X is measured for an unknown sample, which is assigned to the class having the maximum a priori probability for the value of X. For example, in region A for which X lies between O and X_1, class (*i*) has the greatest a priori probability, hence an unknown sample with X in region A would be assigned to this class.

to practical problems lies in determining the a priori pdf's, and thence the decision thesholds. These require examination of vast quantities of training set sample data, which is laborious and expensive. The magnitude of the task may be appreciated by considering the histograms used to obtain approximations to the a priori pdf's. The number of samples falling into each bar obeys (to a good approximation) a Poisson distribution; if the number counted in a particular bar is N, the variance is also N. If the histogram contains 100 bars, and the rms error in estimating the height of an average bar is not to exceed 10%, then the average bar will have to contain 100 samples, and 10,000 will be needed for each class. This is not unreasonable for many problems. If there are two features, then the number of bars for each class pdf is squared; 10,000 bars each containing 100 samples gives a total of one million per class. Since the number of training samples increases

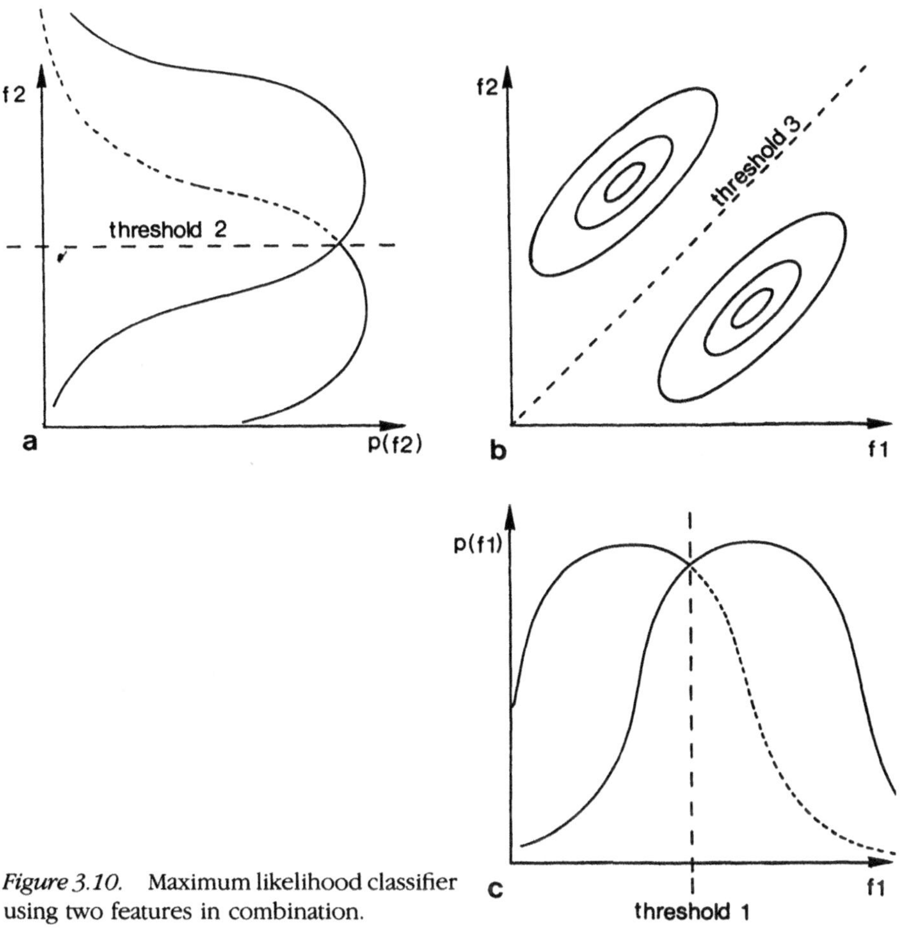

Figure 3.10. Maximum likelihood classifier using two features in combination.

exponentially with the number of feature measures, this direct method of estimating the a priori pdf's from histograms is of little practical use, unless the number of classes and the number of features are both small. It is ideal for problems involving two classes and a single feature, like the tack inspection problem described in Section 11.3.

3.5.2. *Parametric Approach*

This is a method for designing a maximum likelihood classifier which is much less laborious than histogram generation, but is applicable only in certain special cases. A form is assumed for the a priori probability density functions; the Gaussian form is by far the most common. This is specified by a relatively small number of coefficients, even when the number of dimensions is large. An N-dimensional Gaussian distribution is specified by N mean values and $N(N + 1)/2$ variances and covariances. These can be extracted from the same number of training set samples; if more are available, errors resulting from sampling can be reduced. For a problem with 10 features, only 65 samples are required per class; this is evidently far more reasonable than histogram plotting. The procedures are described in more detail elsewhere.[(8,9)]

The weakness of the method is that the pdf's may not be sufficiently close to a Gaussian or indeed any other common distribution for it to be applicable. The pfd's may be skewed and even multimodal. A solution applicable in some cases is to regard the pdf as the resultant of a small number of Gaussian distributions; best values for the parameters of these are estimated in the usual way.

3.5.3. *Kernel Discriminant Approach*

This is yet another way to estimate a priori pdf's economically. The idea is illustrated for one dimension in Figure 3.11. The pdf's are synthesized as a summation of symmetric elementary functions (called kernel functions) such as triangles, rectangles, or even Gaussians. One elementary function is provided for each training set sample, with its mean located at the sample position, for example at $v1$, $v2$, and $v3$ in Figure 3.11. All the functions have the same width, which is adjusted by trial to give best classification performance.

This approach has found considerable application; several program packages are available for designing kernel classifiers, and monographs[10] have been written providing analysis in depth. But it is really appropriate only for off-line pattern recognition problems, such as those occurring in medicine and taxonomy. For use on-line, the pdf's must be approximated by mathematical functions that can be stored in a computer, such as a multidimensional Taylor's series. This tends to cause horrendous errors to occur.[11]

3.5.4. *Linear Classifier*

This uses a feature space but abandons all attempts to estimate the a priori pdf's. Discriminant functions are simply inserted auto-

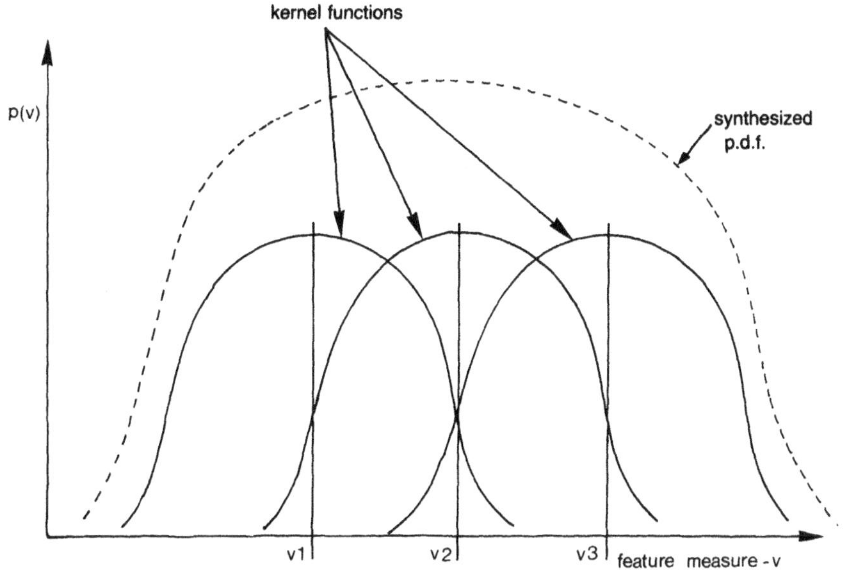

Figure 3.11 Synthesis of a pdf using kernel functions.

matically by algorithm until best classifier performance is obtained. The discriminant functions are analogous to the a priori pdf's in the Maximum Likelihood classifier, in that regions of the space are assigned to the various classes according to which discriminant function is greatest. The intersections between the dominant discriminant functions form the decision thresholds.

If there are N classes, N discriminant functions are generated, one for each class. Each function is of the form

$$W = w1 + w2 + w\,3 \ldots wm \ldots . wM$$

where the coefficients wm are weights determined during the training process; there are M features.

To identify an unknown pattern, the measured values of the features $f(1)$, $f(2)$, $f(3)$ are inserted into the discriminant function for each class k on which the machine has been trained, and the sum $S(\mathrm{k})$ computed, where

$$S(k) = w\,1.f(1) + w\,2.f(2) + w\,3.f(3) \ldots . + w\,M.f(M)$$

The pattern is assigned to the class for which $S(k)$ is greatest. For two classes, a sample is assigned to one class if $S > 0$ and to the other if $S < 0$.

To design a linear classifier, the values of the weights are determined using training samples of known class assignments. The most famous algorithm is the Perceptron algorithm of Rosenblatt. In this, initial values are guessed for the weights, which are modified iteratively as each new sample is considered, until values are found giving the desired discrimination. For example, consider a two-class problem in which the classes are CI and CII. An initial value $W(0)$ is assumed for the weight vector, which is then tried on each training set vector in turn. The kth training set vector is $F(k)$. If the kth trial results in correct classification for the vector considered, the weight vector is unchanged. If classification is incorrect, then the weight vector is modified as follows

> If $F(k)$ is from CI and $W(k).F(k) > 0$, then the vector $W(k+1)$ to be used at the next iteration is $W(k) + c.F(k)$.
> If $F(k)$ is from CII and $W(k).F(k) < 0$ then $W(k+1)$ is set to $W(k) - c.F(k)$.

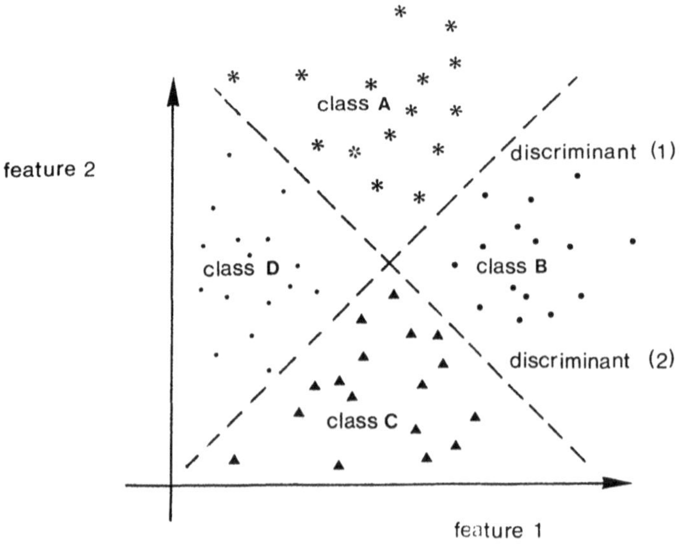

Figure 3.12. Classes separable using linear discriminants.

In the above, c is a positive constant, often unity. The procedure is repeated for all training set patterns until a value is found for W which classifies all training samples correctly. It may be proved mathematically that the algorithm will converge to a partition provided one exists (as in Figure 3.12), but if a linear partition is not possible, i.e., if it is not possible to partition the space into disjointed regions with linear boundaries (Figure 3.13) so that all training samples in each region are from a single class (as in Figure 3.12), then the algorithm will not converge but will oscillate infinitely. Even a single point which is badly placed will cause this to happen. The Ho-Kashyap algorithm[12] is a development of the Perceptron which overcomes the convergence problem, but is still iterative, and therefore wasteful of computer time.

The least mean squares (LMS) classifier of Wee[13] uses discriminant functions which minimize least mean square error (rather than providing a perfect partition). It considers each class to be represented by a point in a space whose axes are the weights, and the discriminant function is selected so as to cluster the training sample points as tightly as possible (i.e., with minimum mean square distance) about the appropriate class point. A sample of unknown class is assigned to the class whose point in weight space is closest. The design algorithm is

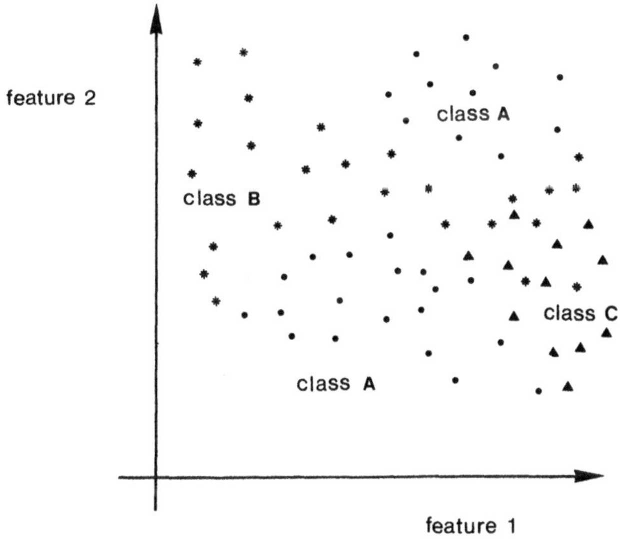

Figure 3.13. Clusters awkward for a linear classifier.

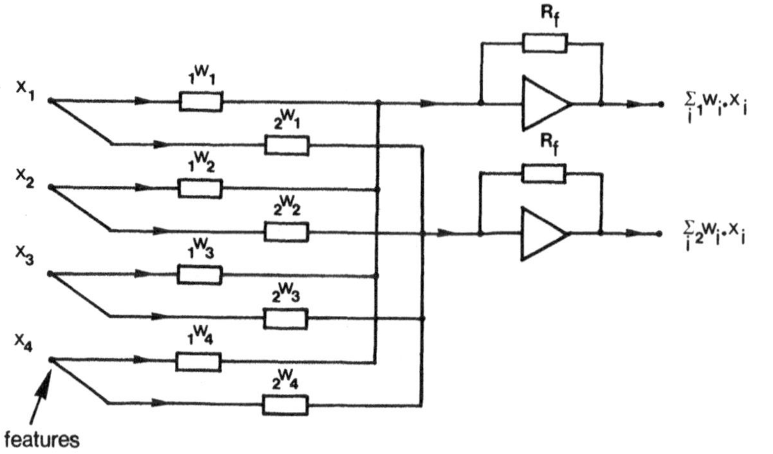

Figure 3.14. Analog hardware implementation of a linear classifier using four features to distinguish two classes. The inputs *X*1, etc. are feature measures. The resistors *iWj* correspond to the weight to be applied to the *j*th feature for the *i*th class. The feedback resistors *Rf* set the scale.

noniterative (it involves computation of a matrix pseudoinverse) and therefore expedient, but classifier performance is generally somewhat inferior to the basic linear classifier.

The kind of feature space classifier just described is potentially very valuable in on-line automation, because in operation it is necessary merely to compute a number of linear weighted sums and determine which is largest. To illustrate how easy (and how quick and inexpensive) this can be Figure 3.14 shows an analog processor which computes the linear weighted sums. The largest is found using comparators followed by simple digital logic. The speed is limited by the settling times of the operational amplifiers, and can be less than 1 microsecond. Design of the classifier (i.e., finding the weight vectors) is performed off-line using a fast general-purpose computer and standard program package. Though laborious, it is a "once for all" operation, completed before operation on-line commences.

3.5.5. *Sequential Analysis*

Two kinds of sequential analysis are useful in pattern recognition for automation. The first methodology is used when it is possible to perform multiple tests over the *same data*. For assignment to two classes, each test has three outcomes: class I, class II, and "try another test." Tests are performed until a definite assignment is reached. The approach was devised by Wald[(14)] who showed that the effort required to identify with given reliability is much reduced by allowing decisions to have three outcomes, and not performing some tests until they are needed. The mathematics involved is explained clearly by Hoel.[15]

The second sequential approach[(16)] is used when given data may be accessed only once, but when the entity to be classified (such as an elongated defect on a moving web of material) is continually providing new and largely independent data. Chittineni[(17)] provides an excellent example of its application. Each block of data (e.g., from successive lines of scan across the web perpendicular to the direction of movement) is analyzed using, for example, a linear classifier to provide probabilities of the data representing the different classes. This data is then combined for successive scans to provide a definitive classification; the consequence is again a substantial improvement in performance.

3.5.6. *Cluster Evaluation*

The chief difficulty with the methods already described is that the clusters for the various classes may overlap, may be disjointed, or may not be separable using a linear surface (Figure 3.13). The dimensionality of the feature space is generally too high (greater than three) to display graphically, thus the clusters cannot be observed visually. It is thus almost impossible to determine the reason for poor performance and hence take the appropriate corrective action. Some help is available using non-linear maps, which map clusters from the original hyperspace to a two-dimensional space so they can be plotted graphically for visual examination. Sammon[(19)] and Kittler[(18)] have produced useful algorithms. The distortion claimed is less than 5%, and both algorithms have found wide practical use and are understood to be available as standard packages.

3.5.7. *Nearest Neighbor Classifier*

This solves the problem of cluster integrity by abandoning any attempt to partition the feature space into regions characteristic of the various classes of pattern. Instead, a sample of unknown class is assigned according to the closeness of its feature vector to vectors from samples of known class assignment which have been stored during training. The latter are designated "prototype" vectors. One procedure is to surround the tip of the vector from a sample being identified by a hypersphere of specified radius, and to assign the sample to the class for which the greatest number of prototype vectors fall within the sphere (a "majority vote"). Another approach is to increase the radius of the sphere surrounding the tip of the unknown vector until K prototype vectors are included; again, assignment is by majority vote. A nearest neighbor classifier is trained by storing sufficient prototype vectors to define the class distributions adequately.

In practice, nearest neighbor classifiers often perform adequately, and are easy to understand and to program. They have hence proved to be the most popular type of feature space classifier for application in instrumentation and automation. The nearest neighbor classifier is applicable equally to deterministic and to statistical problems. Editing algorithms[(18)] enable the number of stored prototype vectors to be reduced by about 2.5 without significantly degrading performance. In some circumstances, it is possible (even for statistical problems) to represent each pattern class by a single vector; in this case, a sample vector is assigned simply to the class of the stored vector to which it is closest (Figure 3.15).

The chief weakness in nearest neighbor pattern recognition, so far as on-line application is concerned, lies in storing the vectors from the training data, and in computing the distances. The computation required is horrendous. The distance measure normally used is Euclidean distance, which is computed by extending Pythagoras' theorem to the hyperspace, i.e., by computing the square root of the sum of squares of distances for each individual dimension, and summing for all dimensions of the space. For example, if the coordinates of a vector are $x(1)$, $y(1)$, and $z(1)$, then its Euclidean distance from the point $x(2)$, $y(2)$, and $z(2)$ is

$$[(x(1) - x(2))^2 + (y(1) - y(2))^2 + (z(1) - z(2)^2\ )]^{1/2}$$

For problems having only one prototype vector per class, to identify an unknown pattern we have to perform this computation between its feature vector and those of all stored prototype patterns, and find

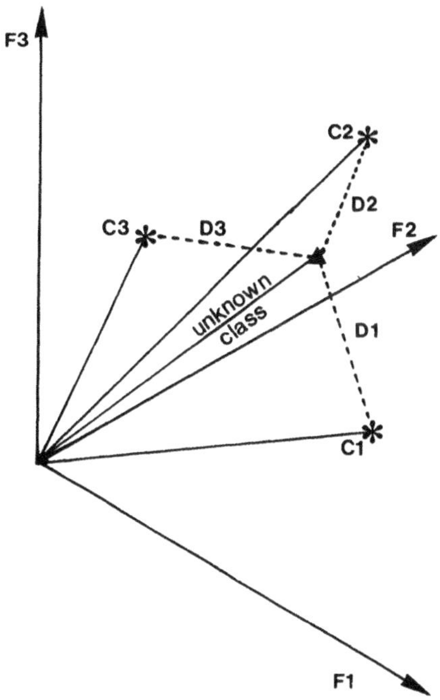

Figure 3.15. Nearest-neighbor classifier. The stored vectors representing classes 1, 2, 3 are *C*1, *C*2, and *C*3. The unknown class vector is distant *D*1 from vector *C*1, *D*2 from *C*2, and *D*3 from *C*3. Since *D*2 is smaller than *D*1 and *D*3, we would assign the unknown to class 2 in this case.

the minimum (some relief may be obtained by omitting the square root).

3.5.8. Coding Theoretic Classifiers

This approach exploits concepts* developed for correcting errors introduced into binary words during transmission or storage, explained in Section 2.2.3. It is suitable for deterministic problems only, in which each pattern class may be represented uniquely by a single point in the feature space. Its chief advantage over the nearest neighbor classifier lies in much reduced computation when finding a match between a sample vector and stored prototypes. With N classes and L features, $(N.L)/2$ comparisons between numbers are required on an average to find a match, compared with $N.L$ subtractions, squarings, and additions if the nearest neighbor approach is used. Other advan-

* These concepts for correcting errors in pattern recognition and its applications and related tasks are the subject of patent protection in the name of BUSM Co. Limited and its affiliate companies in a number of industrialized countries, including UK patent no. 2067326 and US patent no. 4360214.

tages include easy optimization, whereby parameter values are adjusted to minimize the possibility of substitution.

The classifier is designed as follows. We make J measurements on each pattern during training, where J is much larger than the number N of features which should suffice to distinguish the classes. We then divide each measure u by a tolerance t, to yield a quotient v as a potential feature. This occupies a much smaller range of values than the original measure u, and is consequently much easier to store. Further, measured and stored v will much more frequently coincide exactly, even in the presence of errors in measuring features, and the v may be integral for convenience in storage and comparison. From the J candidate features, we then select a subset containing N such that the vectors for the various classes differ mutually in at least D measures. Thus, redundancy has been inserted, which is exploited to eliminate misclassification due to errors in feature measurement. From the analogy with coding for error correction, D is termed the minimum distance for the feature set.

An unknown pattern is recognized by comparing its toleranced features $v(\mathrm{k})$ one by one with the corresponding measure $v'(k)$ for each stored class of pattern. It is assigned to the class (there will be no more than one, assuming that fewer than $[D - 1]$ comparisons are erroneous) from whose prototype it differs in $[D - 1]/2$ entries or fewer. If no such class exists, the sample is rejected. If there are more than $[D - 1]$ errors, then the sample pattern may differ in fewer than $[D - 1]$ entries from the wrong class, and a substitution will result. Provided the probability of mismatch per feature can be kept reasonably small, and provided the prototype samples are distributed sparsely in the feature space, then the probability of misclassification can be kept extremely small. It is found in practice that matching sample to prototype to within $[D - 1]$ entries is preferable since this yields a lower rejection rate without causing a noticable increase in substitutions.

The classifier is designed by building up a feature subset of minimum distance D. Feature vectors are added one by one, and after each addition, the feature vector for each class is compared with those for all other classes. If the vectors for any pair differ mutually in fewer than D measures, a new feature is added and the comparison is repeated. The process is continued until (if necessary) all J candidate features have been included.

Two schemes are available for building up the subset. The first simply adds features in the order in which they appear in the candidate set. Features which do not increase the distance are not discarded. This approach is simple to program, quick in execution, and guaranteed to find a partition having a specified *D*, provided one exists. The features are not necessarily independent so it is suboptimal in that most feature sets will include more features than are required to achieve the specified *D*. The consequence is to increase data storage requirements and lookup times for identification, and to increase the possibility of misclassification.

The second approach which is optimal involves trying all combinations of features in turn, i.e., individual features and then all pairs, triples, and so on until the smallest set having the required *D* is found. Combinations having fewer than *D* features are not searched. The features selected are in this case independent. But the number of combinations which have to be investigated will approach *J!*, where *J* is the number of features in the candidate set, and this may well be too large to be feasible.

This classifier has a great merit in being easily optimized, i.e., the minimum distance *D* and cell size *u* may be traded to minimize the possibility of rejection. The procedure is as follows. The probability of error per feature $p(f)$(assumed the same for all features) is measured using an experiment with a particular value of *D* and *u*. The probability of rejection $P(r)$ is the probability that *D* or more features are in error, and is given (assuming that errors in the various features are mutually independent—see Section 11.2) by the cumulative binomial distribution

$$P(r) = \sum_{\ell=D}^{N} [N!/\ell!.(N - \ell)!]p(f)^{\ell} (1 - p(f))^{(N - \ell)}$$

Using the relationship that p(f) is inversely proportional to u, the number of features N required for each distance D is determined computationally using the feature selection program. The value of $P(r)$ as a function of *N*, *D*, and $p(f)$ [hence *u*] is then computed using the equation above. The results can be plotted on a graph (Figure 3.16 is an example) and the parameter combination having minimum probability of rejection determined. Application of the method to a practical

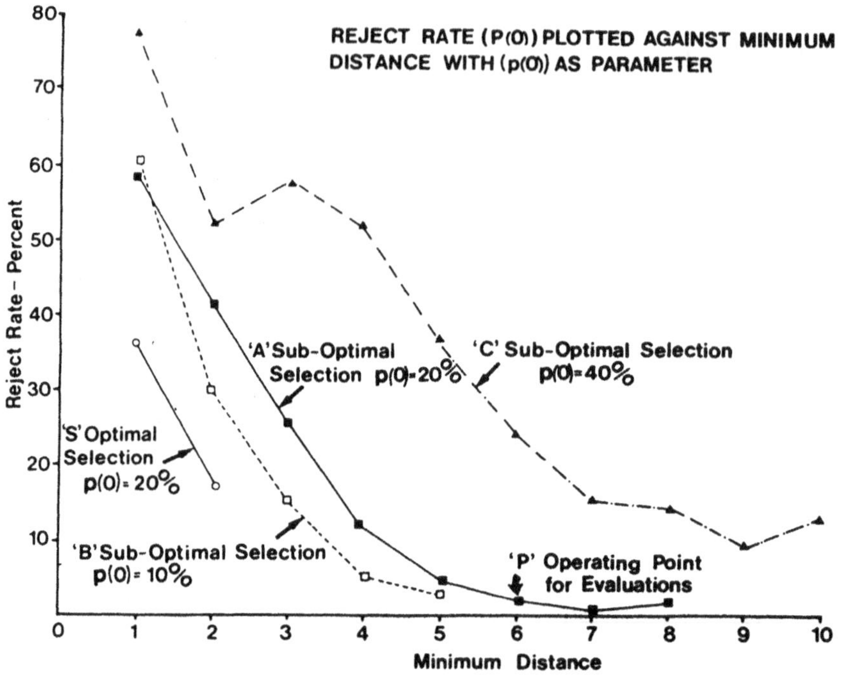

Figure 3.16. Optimization curves for coding theoretic classifier.

task in shape recognition is described in Section 11.2.; the method itself is described in further detail elsewhere.[20]

3.6. Selecting an Approach

The advice offered is pertinent to automation applications of pattern recognition only; application to medical, taxonomic, and other essentially off-line problems requires a somewhat different approach. Intending users should read Kanal's classic paper[21] on the hazards of applying machine pattern recognition (PR).

The initial task is to specify candidate feature measures; it is best to consider everything possible at the outset, to give the feature selection process the best start. There is unfortunately no substitute for intuition in proposing an initial feature set, and the measures to be evaluated are guessed using guidelines such as ease of computability, stability to noise, and so on. The most comprehensive approach is to start by using ALL the data from the samples; if the latter are one-dimensional waveforms, then the sample sequence satisfying the Nyquist criterion (Section 2.4.1) provides a complete representation and can be used as a candidate feature set. But more complex measures are generally better as features.

In choosing a methodology, the following considerations must be taken into account: the number of pattern classes, whether the problem is statistical or deterministic, if the classes to be recognized are fixed or may be varied during operation of the system, whether "training by showing" is required, the (real) time available for identification, the error rates tolerable, and the costs and time delay acceptable for designing the classifier and for implementing it for operation on-line. Most methods require considerable computer simulation involving large databases and sophisticated program packages, to produce the design for a classifier which is then surprisingly simple to implement for operation on-line. The desiderata for a good classifier include low cost, high speed, low rejection rates and minute substitution rates, and, above all, predictability; the performance promised by simulation should be obtained or even exceeded.

For any problem, it may be possible to find a heuristic approach which provides a neat and economical solution, but risks may be involved, for example in that peculiar samples not anticipated and considered during design may turn up in operation and confuse the system. The standard approaches have been thoroughly investigated and documented, and surprises are likely to be rare.

If the number of pattern classes is small, if the classes are fixed, and if the problem is deterministic, it is often possible to determine suitable features intuitively, and to use the attribute comparison approach described at the beginning of the chapter. If errors in measuring features are unlikely to be a problem, then the sequential (tree search) procedure should be used to give high speed; sequential classifiers can be made to identify patterns in microseconds without much difficulty! Otherwise, the parallel search technique should be used.

For deterministic problems in which features cannot be determined intuitively, or for which errors are a significant problem, or where the patterns to be recognized may change, then the nearest-neighbor or coding theoretic approaches should be used. The latter is recommended as being a more powerful yet more economical method, particularly when the number of classes is large. Its main drawback lies in its being difficult to understand.

If the problem is not deterministic and the number of features and classes is small, then the maximum likelihood classifier can be considered. Although design of the classifier is laborious in collecting and processing the large number of training samples necessary to generate the histograms which approximate the a priori pdf's, the design eventually obtained is very reliable and virtually certain to work with whatever error rates are predicted. If (after a small number of samples have been examined) it is clear that the pdf's are unimodal and, say, Gaussian, then the parametric approach may be used to reduce the labor involved. The on-line processing required is also very simple, amounting merely to thresholding a linearly-weighted sum of the feature measures. The extensive effort required initially for design is a "once for all" process, which may turn out to be negligible when spread over a large number of instruments.

When the number of classes, and/or of features, is not small, then a more complex classifier will have to be used. If the feature measures form well-behaved clusters (and if you can demonstrate that they do!) then the linear classifier, preferably the LMS version, should be considered. If this does not perform well enough, a nonlinear map should be used to investigate the clusters; if the clusters are reasonably non-overlapping but have nonlinear boundaries, then extensions of the linear classifier having more complex discriminant functions (such as the quadratic classifier) can be tried. Otherwise, there are a number of feature space classifiers which partition the space into rectangular cells and so on which might be considered as a last resort, but these are rather experimental. The nearest-neighbor classifier might however be used here, with editing to reduce the number of stored patterns, and possibly clusters represented by prototype vectors to simplify comparisons.

The modern trend in machine pattern recognition is to imitate human reasoning and use high-level (semantic) information. This approach is embryonic, and no systematic methodology has as yet emerged from research.

References

1. D.A. Bell, in: *Pattern Recognition—Ideas in Practice*, B. Batchelor (ed.), Chapter 5, Plenum Press, New York (1978).
2. C.B. Chittineni, Efficient feature subset selection with probabilistic distance criteria, *Inf. Sciences 22,* 19–35 (1980).
3. P.A. Lachenbruch and M.R. Mickey, Estimation of error rates in discriminant analysis, *Technometrics 10* (1), 1–10 (1968).
4. V. Popovici, Application of syntactic pattern recognition to defect classification, Ph.D. thesis, The City University, London (1976).
5. K. Nakada, Y. Nakano, and Y. Uchikura, Recognition of chinese characters, *Inst. of Physics Conf. Pub. 13,* 45–52 (1972).
6. D.C. Gonzalez and M.G. Thomason, *Syntactic Pattern Recognition—an Introduction,* Addison-Wesley, Reading, Mass. (1978).
7. K.S. Fu (ed.), *Syntactic Pattern Recognition—Applications,* Vol. 14 of series "Communications and Cybernetics," Springer, New York (1977).
8. R.O. Duda and P.E. Hart, *Pattern Recognition and Scene Analysis,* Wiley, New York (1973).
9. P. Devijver and J. Kittler, *Pattern Recognition: a Statistical Approach,* Prentice-Hall, New Jersey (1982).
10. D.J. Hand, *Kernal Discriminant Analysis,* Wiley, New York (1982).
11. W.J. Hill, Defect recognition in automated surface inspection, Ph.D. thesis, The City University, London (1977).
12. J.T. Tou and R.C. Gonzalez, *Pattern Recognition Principles* (Sect. 5.3.3), Addison-Wesley, Reading, Mass. (1974).
13. W.J. Wee, Generalised inverse approach to adaptive multiclass pattern classification, *IEEE Trans. Comps. C-17* (17), 1157–1164 (Dec., 1968).
14. A. Wald, *Sequential Analysis,* Dover, New York (1973).
15. P.G. Hoel, *Introduction to Mathematical Statistics,* 4th ed., Wiley, New York, Chap. 13 (1971).
16. J. Raviv, Decision making in Markov chains applied to the problem of pattern recognition, *IEEE Trans. Information Theory,* Vol. IT-17, No.4, 536–551 (Oct. 1967).
17. C.B. Chittineni, Signal classification for automatic industrial inspection, *IEE Proc.,* Vol. 129, Pt.E, No.3, 101–106 (May 1982).
18. P. Devijver and J. Kittler, *Pattern recognition—A statistical approach,* Chap. 3, Prentice-Hall, New Jersey (1982).
19. J.W. Sammon, A Non-Linear mapping for data structure analysis, *IEEE Trans., Comp.* Vol. C-18, 401–409 (1969).
20. L. Norton-Wayne, A coding approach to pattern recognition in J. Kittler, K.S. Fu, and L.F. Pau (eds.), *Pattern Recognition Theory and Applications,* Reidel, Dordrecht (1982).
21. L. Kanal, Patterns in pattern recognition, 1968–74, *IEEE IT-20* (6), Nov. (1974).

4

Scanning Systems

The means by which information about a scene is transferred to a processing unit involves, in general, three factors. These are the illumination of the scene, the conversion of the optical pattern into electrical form, and the transfer of the electrical information to the processor. The conversion of the pattern is performed by sampling the brightness or reflectance of individual points in the scene and usually will involve a scanning action either within the optoelectronic converter or before it. In some cases the illumination of the scene is incorporated into the scanning system. In many systems an optical unit, e.g., lenses, will be required to form an image of the scene within the optoelectronic converter. Optical systems may also be required in the illumination of the scene.

Many factors will affect the choice of acquisition system besides the ever present problem of cost. These include the number of picture elements, pixels, into which the scene is to be divided; the presence of ambient lighting; the nature of the objects in the scene; the time available to acquire and process the picture; and the environment. The latter raises problems of risk of damage, dirt on the optical system, electrical or other interference, and temperature.

The aim of the following sections on scanning is not to provide an exhaustive examination of the topic, since this can be found elsewhere in literature specific to the subject, but to give those working in the field of Vision Processing an understanding of the hardware that is available and of the way in which it can be used. For this reason details of specific proprietary devices are not given.

Scanning systems can be divided into two groups, those that scan a light spot across the scene and detect the strength of the returned light, and those in which an image of the scene is formed within the optoelectronic converter and scanned electronically. The second is by far the larger and more useful group.

4.1. Properties

4.1.1. Scanning Patterns

Scanning patterns comprise the following:

The *linear* pattern is suitable for position sensing, but is also used for scanning scenes where the scene itself is moving, for example on a conveyor belt.

The *raster* pattern is the most common pattern and is a format suited to picture processing in that it can be stored and processed in computer memories in a flexible and logical way. It is the pattern used in television systems but generally in the form of two interlaced scans. Usually it is simple to convert the system to superimposed scans. By more complex changes to the deflection circuits of television cameras it is possible to vary the line and frame periods. In some systems it may be advantageous to avoid the time lost in the line flyback. This can be done by using a boustrophedal scan in which the line scans are alternately in opposite directions.

The *circular* pattern can be used to determine orientation with repect to the center of the scan. Special solid state arrays exist in this format.

The *spiral* pattern covers an area. This pattern is useful in determining the orientation of an object if the scan can be centered on a specific point of the object.

In the *random* pattern, usually the scan is directly under the control of a computer and consequently has a slow pixel rate. On the other hand it scans only the interesting points and therefore is more efficient in that it acquires only the required information. The scan may follow any desired path or jump from one part of the field of view to another. A method used to trace the boundary between a bright and a dark area is to perform a small circular scan. The signal will alternate between high and low. The phase of the transitions indicates the direction of the edge and the direction in which the scan should move to follow the edge.

4.1.2. Resolution

Basically, resolution refers to the effective size of the scanning element at the scene. Usually this element does not have sharp boundaries and may vary in size across the scanned area. If the rate of decay of its effect away from its center is known, e.g., is Gaussian, it may be possible to improve the resolution of the system by electronic or computer aperture correction. The resolution can be expressed in terms of the MTF, modulation transfer function (see Section 5.1.1.4), which is the variation of the peak-to-peak output against spatial frequency, obtained when bar patterns of those spatial frequencies are scanned. For a television system the resolution quoted is usually that for 50% modulation.

Resolution can also be expressed as the number of elements that can be resolved within the limits of the scan, and in this form is a more useful parameter in the selection of a scanning system. In most systems the resolution is a function of the spectral wavelength. It may also be affected by the scanning rate, possibly by the response times of the detector and the video circuits, and by the light level, due to its effect upon the response time and the effect of noise and other processes not produced by the illumination.

In systems in which the scene is sampled spatially, the normal situation for picture processing by computer, the highest spatial frequency in the scene that can be correctly detected is limited by the spacing of the samples and the phasing of the sampling with respect to the scanned pattern. Usually the Nyquist criterion is quoted, namely that at least two analog samples must be taken in each of the cycles of the highest frequency in the scene. This is useful if the samples coincide with the antinodes of the signal but nothing is obtained if they coincide with the nodes. At slightly higher sampling rates the high frequency component is detected only over a span of many cycles. In picture processing it is usually necessary to detect the presence of a component with only a few cycles and consequently the sampling frequency should be three or four times that of the maximum spatial frequency.

When spatial frequencies exist in the scene which are higher than that corresponding to the Nyquist criterion, a phenomenon called aliasing occurs. The output indicates the presence of a frequency which is as much below the limit frequency as the actual frequency is above.

4.1.3. *Scanning Rate*

Two factors affect the scanning rate, the time required to move from one scanning element to the next and the minimum value of either the dwell time on each element or the time between successive scans of each element or both. The minimum dwell time may be related to the correct operation of the device, e.g., the correct recharging of a photosite, or to the signal-to-noise (S/N) ratio of the resulting signal. The minimum time between successive scans applies to those devices in which integration of the light occurs in a photosite. This integration time is related to the sensitivity and the S/N ratio.

4.1.4. *Sensitivity and Spectral Response*

In a scanning system the light source will have an output, and the sensor will have a sensitivity, which vary with the spectral wavelength. For the total system these must be combined for each wavelength. It will also be necessary to include the variation with wavelength of the reflective properties of the objects being scanned and the transmission of the optical system. The overall sensitivity of the scanning system may be different for different elements within the total picture. For example, the sensitivity of a two-dimensional sensor, e.g., a vidicon, will not be the same at all points. For variations within the sensor the term "shading" is used.

The magnitude of the video output from a scanning system is a function of the input light level. This relationship, which may not be linear, is called the gamma and is defined by

$$\text{output} = \text{input}^{\text{gamma}}$$

$$\text{whence gamma} = \log(\text{output})/\log(\text{input})$$

that is the gradient of the curve relating the logarithms of the output and the input. Electronic circuits can be used in the system to convert the gamma to that required for the picture processing. Usually the desired gamma is unity but in many television cameras the gamma is adjusted to be about 0.7.

4.1.5. *Signal-to-Noise Ratio*

The noise superimposed on the output of a scanning system arises from several sources. Noise is generated within the electronic circuits and there is a random function in the collection of photons. Both of these effects are worse when the light level, and hence the signal level, is low. Noise may also be introduced when the sensitivity of the system, or dc offsets, vary with the position of the scanning element. For this reason it is known as spatial or fixed pattern noise. By its nature it can be measured and cancelled from the signal by processing. Given an optimum output circuit design, the effect of true random noise can be reduced only by increasing the integration time. This may be done by reducing the scanning speed or by the point-by-point summation of several scans. The circuit should have a bandwidth sufficient for the response time required. A bandwidth greater than this will increase the noise component.

The noise level affects the dynamic range of the system, that is the ratio between the largest signal that can be generated and the smallest signal that can be detected. The noise level has a bearing on the size of this smallest signal. Related to this is the number of discrete steps that can be resolved in the total output range. This may be equal to the value of the dynamic range but in some systems the noise will increase with signal level, e.g., in the image dissector, or decrease, e.g., in the image orthicon, and the values will not be equal.

Frequently the video levels are converted to equivalent digital values for computer processing. The number of levels of digitization that can give a meaningful result is related to the dynamic range and the variation of the noise with the output level. The act of quantizing the signal adds an extra noise component, quantization noise, to the final result due to the rounding up or down that occurs in the least significant digit. Although a linear conversion is usually used, other functions are possible to produce gray-scale (gamma) correction or to provide greater intensity resolution at some output levels at the expense of other levels.

4.1.6. Other Properties

The susceptibility of the system to the environment and "rough" handling should be considered. The usual environmental problems are temperature, dust, and local electrical or magnetic fields. The use of suitable enclosures may provide a solution in each case and air curtains are a useful way of keeping dust from sensitive parts of the light path. Devices using magnetic deflection of electron beams are naturally prone to disturbance by external magnetic fields. Rough handling includes vibration and impact, and also the subjecting of the detecting devices to excessive light. The latter may produce a temporary overload, as in a solid state sensor, or produce an irreversible change, as in a vidicon.

4.2. Vacuum Tube Cameras

The large group of devices known as television camera tubes can be divided into two basic types.[1–3] One type has a light-sensitive screen and a means of scanning an electron beam across this screen. The other has a stage which transfers the electron image from a light-sensitive screen onto an intermediate electrode which is scanned by an electron beam. They differ in the type of screen, in the intermediate electrode (if present), and in the means of extracting the electronic output, and, consequently, in their properties. The image intensifier tube, although not a scanning device, has been included in this section because it is used with television tubes and is an integral part of the second type of the group.

In referring to types of tubes, trade names may have to be used, but in general, the comments apply to equivalent tubes by other manufacturers. The information given below can be only a guide to the use and performance of the tubes, due to the very large number of variants that exist. Manufacturers data should be consulted for specific tubes.

4.2.1. Properties

4.2.1.1. Sensitive Screen. In the tubes containing an electron imaging stage (Figure 4.8), the light sensitive layer on the window of the tube is photoemissive. The majority of photons, which have sufficient energy to exceed the critical value for that material, will cause an electron to be released from the layer. This will be accelerated away by the positive field applied to the system. The thickness of the layer is a compromise for the optical band to be used. For those wavelengths which are rapidly absorbed, normally the shorter wavelengths, the electrons have to pass through the layer and will be recaptured if the layer is thick. For wavelengths having low absorption, the light may pass through a thin layer without producing electrons.

In the tubes without an imaging section (Figure 4.4), the light-sensitive layers are of three types: an amorphous photoconductor, an amorphous photoconductor with a junction layer, often called a blocking contact, and a junction photoconductor. For each, a high conductivity transparent film is deposited on the window of the envelope followed by the photosensitive layer or layers. The effect of the scanning electron beam is to bring the potential of the side of the layer nearer to the gun, to the gun potential. A positive potential is applied to the other side through the electrode. The effect of light on the photoconductor is to raise electrons into the conduction band of the material. For the amorphous photoconductor, without a junction, the state of these electrons is similar to those in any conductor and the resistance falls as the light level increases.[(4)] The conductance is not a linear function of the light level and the gamma is about 0.7. The current through the layer depends upon the voltage between its surfaces and as a result the gain of the device can be controlled. In the junction photoconductor, and the amorphous photoconductor with a junction, the layers form a diode junction which is reverse biased by the applied voltage. This removes all majority carriers and the only current that can flow is that due to the minority carriers released by the light. The applied voltage normally used is sufficient to remove all such carriers so that the current is independent of the voltage and the gain is fixed. At low voltages there will be some loss of gain due to the recombination of the carriers before they pass through the junction. There can be a slight loss of charge at the highest outputs but the overall gamma is close to unity. Adjustment can be made, by the use of nonlinear amplifiers, to produce a gamma for the camera which is different from that of the camera tube.

In all these surfaces conduction can occur without the action of light and will be added to the output signal. This effect is temperature dependent and this dark current doubles for each 8 °C rise in temperature. Another effect of temperature is that above a specified level the photosensitive layer can be irreversibly damaged. For example, the face plate of a plumbicon should not be above 50 °C for an extended period and it may be necessary to provide cooling air for this reason.

The photoconductive layers are operated in the integrating mode. As the beam scans, it charges the surface nearer the gun to the gun potential. Conduction through the layer, due to the light or thermal effects, raises the potential of each element of that surface by an amount dependent upon the light received by that element up to the time of the next pass of the beam. The charge taken from the beam at that instant is the video signal for that point of the picture. Extraction of the signal is discussed in the sections on specific tubes. The thickness of the layer affects the resolution of the tube. The current does not flow exactly perpendicularly in the layer, and if the layer is thick the affected area will be larger and therefore the resolution will be less.

Camera tubes can contain small free particles which may fall on the sensitive surfaces for certain orientations of the tubes, i.e., with the axis near vertical. For the vidicon type of tube the danger occurs when the screen is downward but in tubes with an imaging section it can happen for both the screen up or down states. The particles may result in dark or light spots in the scanned picture.

4.2.1.2. Scanning and Focusing. The gun generating the electron beam is based on the standard triode pattern but variations exist for special purposes, e.g., the anticomet gun used in some plumbicons. Focusing and deflection of the beam may use electrostatic or magnetic methods, in any combination, the magnetic systems giving the better performance in both cases. Electrostatic deflection is usually avoided due to the lower resolution that is obtained; typically 90% at the center and 75% in the corners of that obtained with magnetic deflection. The accuracy of the scanning pattern is limited by the quality of the scanning circuit and the deflection plates or coils. Positional errors of a few percent of a diameter are not unusual although 0.1% can be achieved. These errors may not be stable due to variations in voltage, temperature, etc., or to varying external magnetic fields.

In most of the tubes the beam passes through a mesh electrode just before reaching the storage surface (Figures 4.4 and 4.8). This screens the flight of the electrons from potential variations that exist on the storage surface. This mesh may be connected to the metal tube which follows the gun and provides an electrostatic field free region in which the beam travels. In some tubes the mesh is at a slightly more positive potential than the tube and, forming a lens between them, causes the beam to arrive perpendicularly onto the sensitive surface. Although the velocities of the electrons will be very similar, if they land at an angle the perpendicular components of the velocities will have a greater spread. Uniform perpendicular components and direction can result in better resolution, higher output, and less geometric distortion. When excessive light has fallen upon part of the layer, the large potential change can produce fields which draw the electrons and in the video output the spot appears larger than it was, i.e., blooming occurs.

For high resolution along the line scan it is better to have a small scanning spot.[5] Unfortunately, such a spot will leave unscanned areas between the scanned lines. The buildup of potential in these areas can cause beam pulling onto the areas and away from darker parts of the picture, thereby apparently brightening these parts of the scene. Since the vertical resolution is affected by the line standard being used, the best solution would be to use an elongated spot, but the normal method is to use a large spot or to increase the beam current, both giving less resolution along the line.

4.2.1.3. Sensitivity. The luminous sensitivity is quoted as the average signal current per unit of light flux entering the end of the tube and falling uniformly on the scanned area. It is usually given in μA/lumen. The flux is equal to the illuminance, in lux, multiplied by the scanned area, in square meters. Because the lumen is involved, the measurement implies a specific distribution of the energy across the spectral band. Usually the test source will be stated. The radiant sensitivity, given in mA/watt, is similarly defined but each value relates to a single wavelength.[3] Consequently, it is often given in graphic form (Figure 4.1). The currents given are averages and as the scan is not continuous, due to flyback time, the average during the scan will be higher by about 30%.

The sensitivity will not be the same for all points in the scanned area, an effect called shading. Also there can be small area defects, spots of low, or occasionally high, sensitivity. Similar spatial variations can apply to the dark current and these become worse as the temperature rises. Some tube types are sold at several quality levels based on the degree of these faults. The assessment is, naturally, on the basis of the quality of picture that they will produce on a monitor but this may not be a satisfactory criterion in the selection of a tube for a vision application. Which faults are the most serious will depend upon the application and the picture processing to be used. Correction can be made in the preprocessing of the video to correct for the lesser faults and to ignore the major faults.

Many of the light-sensitive materials can be permanently affected

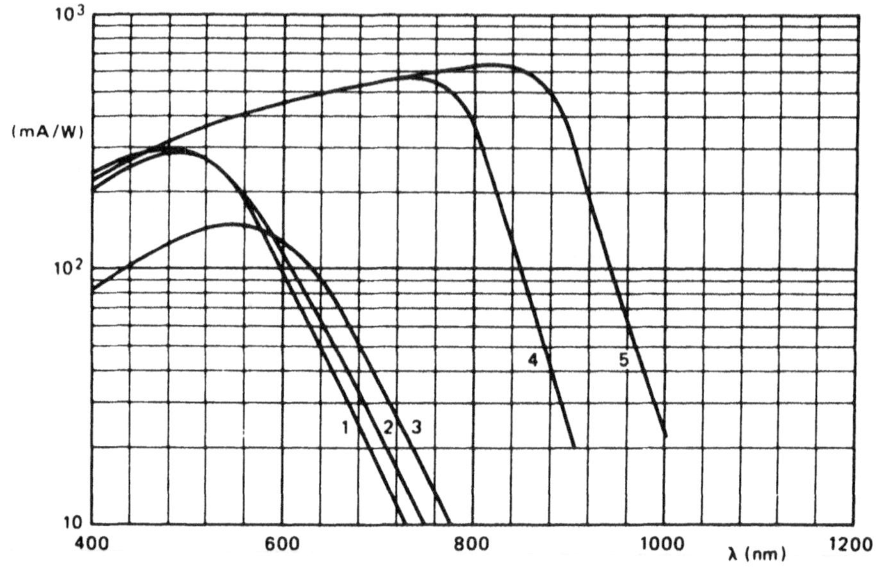

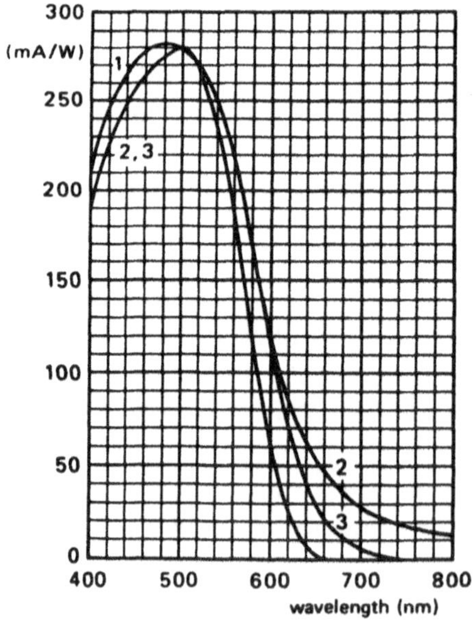

Figure 4.1. The spectral responses of the vidicon and related tubes. In the upper figure, curve 1 is for a Plumbicon, 2 and 3 are for vidicons, and 4 and 5 for Newvicons. The lower figure is for Plumbicons; 1 is standard, 2 extended red, and 3 is that of 2 with an infrared filter. (Courtesy of Mullard Ltd.)

by excessive light due to changes that take place in the material. This is called "burn in." Tubes are more sensitive to burn in when there are voltages applied, due to the additional affect of a high beam current, but higher light levels can cause damage when the tube is off. In a vision system it is essential to guard against bright lights or reflections being seen by the camera.

The effective sensitivity can be increased by turning off the scanning current for a few frames, thereby increasing the integration time. This will, for the shorter periods, improve the signal-to-noise ratio, but for very long integration periods the dark currents in the storage surfaces will introduce spatial noise.

4.2.1.4. Lag. Most tubes do not respond immediately to a change of illumination. This lag effect is in addition to the obvious consequences of the change occurring within the integration period, at a phasing which is different for each element due to the timing of the scanning beam. There are effects for both directions of change of illumination, known as buildup lag and decay lag. They are dependent upon the photosensitive material, its thickness, and the actual levels between which the change occurs, and they may be affected by temperature. Normally the effect is greater at the lower light levels particularly with respect to decay lag. For the decay lag, this is because, following a drop in illumination to a low level, the decay of the current expressed in absolute terms is almost independent of the original illumination level. An effectively shorter decay lag can be obtained by illuminating the sensor with a uniform bias light (Figure 4.7).[6,7] The light source can be in the lens to tube space, but in some tubes an optical coupling is built into the base of the tube so that light from a lamp in the connecting socket shines onto the sensitive surface from around the gun. The lag is dependent upon the capacity per unit area of the screen, and thin layers, giving the higher capacities, can have greater lag times. The lag is usually quoted as the signal level, a percentage of the final or original current, at specific times after the intensity change, e.g., 60 and 200 ms.

The lag effect can affect the use of the tubes in vision systems in two ways. For static scenes it may be necessary to perform several scans after a picture change before the previous picture has been purged sufficiently for it not to affect the result. For moving scenes it limits the rate at which the camera can pan or at which objects can move in the field of view. In each case it causes the greatest problems where the greatest change of light level has occurred.

4.2.1.5. Special Guns. The lag of a tube is reduced when the spread of velocities in the beam is minimized. This is accomplished by the diode gun whose grid is operated at a positive potential with respect to the cathode.[8] This avoids the curvature of the electron paths in the triode gun, which has a negative grid, and the consequent interaction of the electrons which spreads their axial velocities. This may benefit the resolution in the compromise between lag and resolution in the choice of layer thickness. The diode gun has a greater current reserve, which is useful when scanning extreme highlights. A circuit is used to raise the beam current when the video current exceeds a preset value. With the triode gun the current is usually limited to obtain a good resolution, and it may be insufficient to fully recharge, in a single scan, an area which has received a high illumination.

The anti-comet-tail (ACT) gun provides another method for removing the effects of excessive illumination.[6,7,9] During scan flyback a defocused spot is used with the cathode potential raised above that corresponding to the normal maximum white level on the photosensitive layer. Consequently, any areas that have gone beyond the white level are charged towards that level and the overload is removed before the area is scanned in the normal way.

4.2.1.6. Video Waveforms. International standards have been set by the International Radio Consultative Committee (CCIR), for the format of video waveforms used for television purposes.[10] These standards are applied in the majority of cameras used in picture processing systems. The standards are very similar and differ in most cases only in the timings of sections of the signals and in a few structural characteristics. There are significant differences in the techniques used to incorporate color information into a single signal, but this combined signal is rarely used in vision systems. Instead, color signals are carried on several conductors from the camera to the processor, usually corresponding to the red, green, and blue channels. Typical of the waveforms are those of the 625-line, 50-Hz frame rate systems and of the 525-line, 60-Hz systems. The raster scan pattern is used.

Besides carrying the video information from each line of the scan, the waveform contains timing pulses to mark the end of each line scan and other pulses to mark the end of a field scan. Following the start of these pulses, the scanning beam in the camera and monitor return to the start of the next line or frame. A typical line output is shown in Figure 4.2. The total excursion of the signal is from the base of the line sync pulses to peak white. Typically the pulse occupies 30% of this. At either side of the pulse, the signal is at the blanking level, the level of zero beam current. These are called the front and back porches. The former, before the pulse, allows the current to reach zero before the beam is switched back across the target of the camera or the screen of the display. The back porch allows the scan speed to stabilize before the electron beams are switched on. In some systems, the black level of the video is a few percent above the blanking level.

The field synchronization signals in most systems are complex for two reasons. The information is contained in the area below the blanking level, and to be disguishable from the line sync pulses, the field pulses have a longer duration. The line sync timing must not be lost during the field pulse, and line sync pulses are incorporated into the field sync pulse. Many systems interlace the lines of two fields to produce a frame scan. This is achieved by starting the field sync at the end of a line for one field and in the middle of a line for the other. For reasons similar to those for the porches, several lines before and after the field sync pulses do not carry picture information.

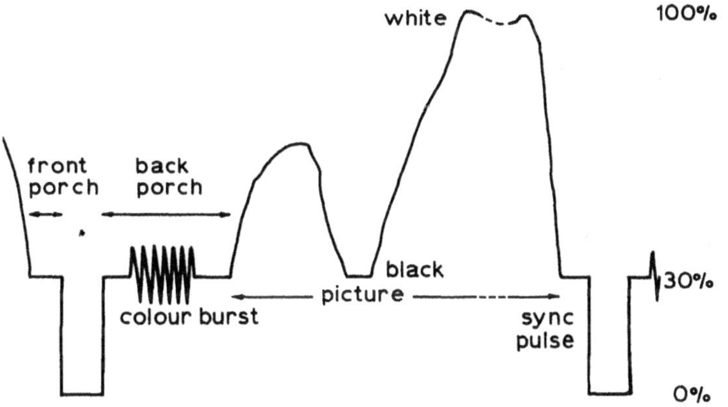

Figure 4.2. A line scan from the output of a television system. The duration of the front porch is 1.65 μs, sync pulse 4.7 μs, total picture blanking 12.0 μs, and line period 64 μs.

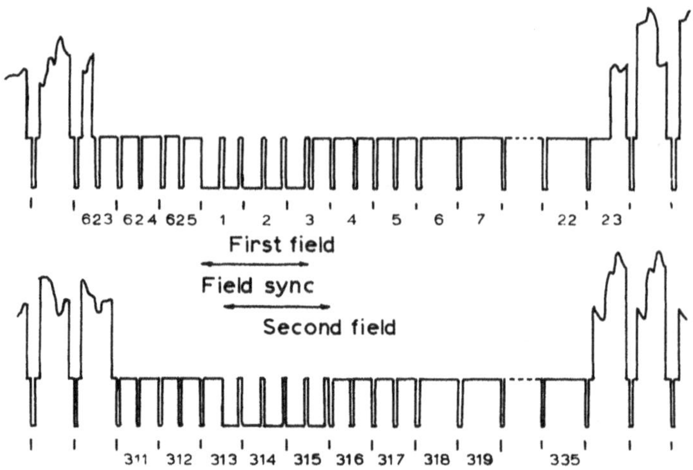

Figure 4.3. The two field synchronization periods from the output of a television system. Each field sync pulse contains five 4.7 μs positive pulses, and is preceded and followed by five 2.35 μs negative pulses.

These features can be seen in the typical field sync waveforms of Figure 4.3. If the normal line sync pattern of one pulse per line were maintained through the field pulse in each case, the field pulses would be different. Usually the field pulses are isolated by passing the signal through a low-pass circuit and the outputs for the two different pulses would not be exactly a half-line apart. For this reason, extra pulses, the equalizing pulses, are added midway between the line pulses. Since they are so far from the correct timing, they do not affect line synchronization. It will be seen that through the field sync period there is always a falling edge at the correct position for the line synchronization. In most systems the field pulse width is equal to the duration of 2.5 lines, preceded and followed by equal periods of equalizing pulses, but the 525- and a few of the 625-line systems have durations of 3 lines. There are variations in the number of blanked lines following the field pulse.

4.2.1.7. Dimensions. The standard sizes for tubes without imaging stages are 13, 18, 25, and 30 mm diameter, having scanned areas of 4.7 × 6.3, 6.6 × 8.8, 9.6 × 12.8, and 12.8 × 17.1 mm respectively. Tubes with image stages are 74 and 114 mm in diameter, with scanned areas of 27.5 × 36.6 mm and 24.4 × 32.6 mm, respectively.

4.2.2. Vidicon

Vidicons are the most common of the television scanning tubes. The sensitive surface is an amorphous photoconductor and consists of two or more layers of mainly antimony trisulfide with variations in the antimony/sulfur ratio.[6,7,11] This ratio determines several parameters of the tube. The target electrode is held at typically 20 to 60 volts positive with respect to the cathode. Conduction in the layer causes the gun side of the layer to move positively in a pattern corresponding to the incident light. The electrons passing through the mesh enter a retarding field of about 400 volts (Figure 4.4). At each point of the surface, electrons are collected from the scanning beam until that point is at the potential of the cathode. Remaining electrons return through the mesh (grid 4 in Figure 4.4) and are collected by the final anode of the gun. The collection of the electrons at each point on the surface is capacitively coupled to the target electrode and generates a current which is the output signal. The beam scans each point at regular intervals and the resulting current for that point is the integral of the charge movements, due to the light and the dark current, during that interval.

Because the photolayer is an amorphous photoconductor, the gain of the tube can be increased by increasing the voltage on the

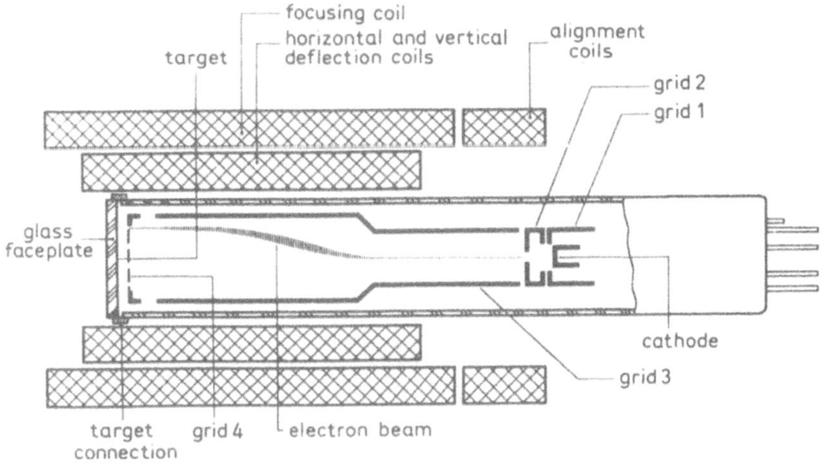

Figure 4.4. The electrode and coil arrangement of the vidicon and related tubes. (Courtesy of Mullard Ltd.)

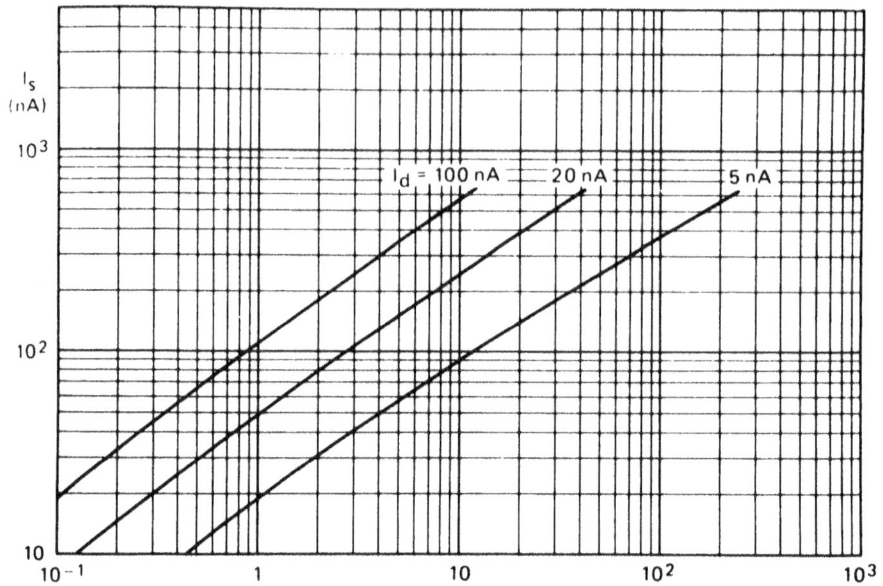

Figure 4.5. Typical light transfer characteristics of 25 mm vidicons for three dark currents. The horizontal axis gives the illuminance of the photoconductive layer in lux, for a light source of 2856°K (Courtesy of Mullard Ltd.).

target electrode. The camera may use this property to automatically compensate for variations in the average or peak brightness of the scene. As Figure 4.5 shows, the dark current increases more rapidly than the sensitivity. Since the layer is thin, it allows a high resolution to be obtained. Lag is significant, for example, the decay lag for total removal of the light is typically 10% after 200 ms, i.e., 5 frame scans. Also, the layer is susceptible to burn in.

4.2.3. Plumbicon, Leddicon, Vistacon

Plumbicon, Leddicon, and Vistacon tubes are basically the same as the vidicon (Figure 4.4). The photosensitive layer consists mainly of lead oxide in the form of a continuous PIN diode structure.[6,7,9,12] In the PIN structure, a thin layer of intrinsic, i.e., undoped, semiconductor exists between the p and n layers. A potential of about 45 volts on the signal electrode, positive with respect to the cathode, reverse biases the diodes. The material is a junction photoconductor and therefore, at the normal operating voltage, has a fixed sensitivity. The lead oxide is not sensitive to wavelengths longer than 650 nm, but the response can be extended beyond 800 nm by the addition of sulphur (Figure 4.1). For good color rendition in a color camera, it may be necessary to use a sharp cut optical filter to suppress the effect of the infrared. The gamma is nearly unity, typically 0.95. The layer has a very low dark current and a high sensitivity. Typical MTF curves (5.1.1.4) are given in Figure 4.6. for several plumbicons, measured along the line scan direction.

Since the plumbicon is used in high quality applications, e.g., studio work, several variants are available with improved performance. To reduce lag, a light bias can be applied to the face of the tube but some tubes are fitted with a light guide through which light may be fed from a small lamp in the base to illuminate the gun side of the layer, (4.2.1.4) (Figure 4.7). Also the tube is available with diode and ACT guns (4.2.1.5).

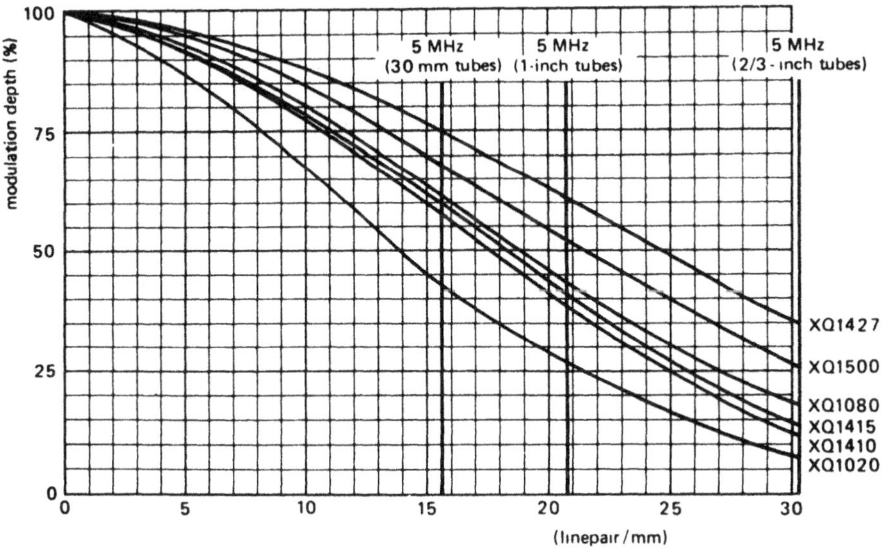

Figure 4.6. Square-wave modulation transfer curves of some Plumbicon tubes. The tubes, in order from the top, are an 18 mm, two 25 mm, and three 30 mm tubes. (Courtesy of Mullard Ltd.)

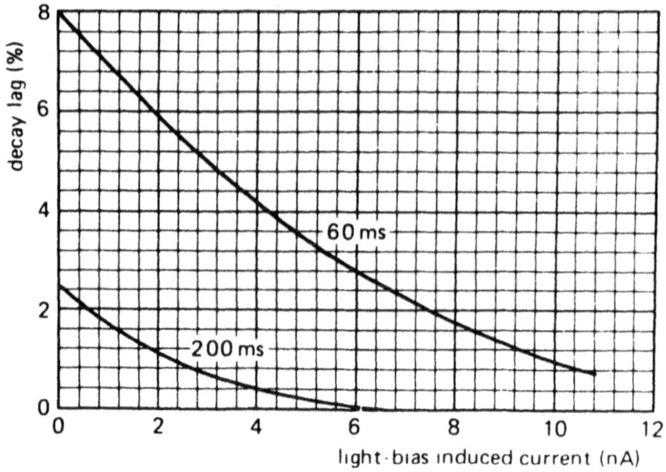

Figure 4.7. The influence of a light bias upon the decay lag of a Plumbicon. (Courtesy of Mullard Ltd.)

4.2.4. *Other Vidicon-Based Tubes*

The Newvicon has layers of zinc selenide and of a mixture of the tellurides of zinc and cadmium.[6,7,11] These form a junction photoconductive layer which is operated in reverse bias. The gain is fixed and the gamma is unity. The sensitivity, which is high, extends to 900 nm, which is further into the infrared than most of the standard tubes (Figure 4.1). The thinness of the layer gives a high resolution but a long lag, the latter being between that of a plumbicon and a vidicon. It is resistant to burn in but there is a significant dark current, although less than that of the vidicon.

The Saticon has a layer of amorphous selenium doped with arsenic and some tellurium and incorporates a junction.[11,13] Consequently the gain is constant and the gamma about unity. The tube has good resolution and very low lag.

4.2.5. *Silicon Vidicon*

The photosensitive surface is a thin layer of n-type silicon, formed on the transparent target electrode, into which boron has been diffused to form islands of *p*-type silicon.[14] These islands are typically on a 15-μm pitch in two dimensions and the layer thickness is about 15 μm. The photodiodes so formed operate in the junction photoconductive mode with the diodes reverse-biased by the positive voltage on the target, about 15 volts, and the cathode potential maintained on the p-type surface by the electron beam. The surface is resistant to damage by high light levels and there is no burn in. The sensitivity is similar to that of the newvicon. There is a significant dark current but due to the short carrier lifetimes, the lag is short. The resolution of the tube is determined by the size of the diode elements and resolutions up to 1000 lines are available. Due to the discrete nature of the surface, with its well defined pattern, the tube exhibits the aliasing effect (4.1.2).

4.2.6. *Image Intensifier*

Electrons released from a photoemissive surface, onto which the scene is imaged optically, are accelerated across a space in which a focusing field exists.[6,7] Electrons from a point on the surface are brought together at a corresponding point on a phosphor screen to produce a bright optical image of the original picture. The phosphor is covered by a thin layer of aluminum, transparent to electrons, which prevents light radiating to the photoemissive surface. When the wavelength of the output light is significantly different from that of the input, e.g., for infrared input and visible output, the tube is called an image converter.

Electrostatic focusing is used in most tubes although magnetic focusing also can be used. A disadvantage of electrostatic focusing is that it is very difficult to image between plane surfaces while maintaining focusing and linear scaling. It may be necessary to use optical correction lenses to overcome the curvature. Typically, gains of about a hundred can be achieved at 15 kv. Greater gains can be obtained by cascading several tubes. There is the problem of imaging from the phosphor of one tube onto the photocathode of the next. To avoid the use of complex lens systems, and the increase in length that their use would produce, it is usual to provide each stage with fiber-optic coupling. The photocathode and phosphor screen are formed onto high resolution fiber-optic plates whose surfaces have the optimum shape for the focusing field. The surfaces external to the tube are flat. Several units can be placed together with direct coupling between the plates, or the tubes can be made with single plates between the screen and the photocathode of consecutive tubes. A gain of 50,000 is possible with a triple tube. Flat image planes can be obtained when magnetic focusing is used and this simplifies the cascading of intensifiers. A multistage tube can be built into a single envelope, using magnetic focusing, by supporting the intermediate phosphor/photoemitter combinations on opposite sides of a thin mica sheet.

Instead of imaging the electron pattern onto the phosphor it can be imaged onto a microchannel plate. This consists of a glass plate with many fine tubes through its thickness, typically on a pitch of 40 μm, which are coated internally with secondary emitting material. A high voltage is applied between the faces of the microchannel plate, and when an electron enters at one end, electron multiplication occurs. The phosphor screen is placed close to the output of the microchannel plate so that little loss of resolution occurs. Gains of 50,000 can be achieved in a single stage.

A simpler but lower-resolution intensifier can be made by placing close together two transparent conductive plates, one carrying the photocathode and the other the screen. Typically, a voltage of 10 kV is used for a 1-mm separation. This is known as a proximity focused device although there is no focusing, simply direct transfer from input to output. In another tube, using proximity focusing, the two plates are placed on either side of, and close to, a microchannel plate. The microchannel plate blocks the feedback of light from the screen to the photocathode. Gains of 15,000 can be obtained using this method.

Image intensifiers, or converters, can be placed in front of any of the camera tubes. Provision must be made to image the output from the intensifier onto the input of the camera tube. This is simplifed if both have fiber-optic plates. The response time of the combination will be affected by that of the phosphor.

4.2.7. Image Orthicon

The orthicon differs from the vidicon type of tube in the means by which the output is taken.[15] the scanning beam neutralizes the positive charge on the gun side of the target as in the vidicon. Of the electrons of the beam which are not used for neutralization, some are specularly reflected back along the path by which they came and others are scattered (Figure 4.8). The majority of the electrons returning from the target pass into the deflecting field area. Consequently they are deflected near to the gun where they strike the back of the final anode

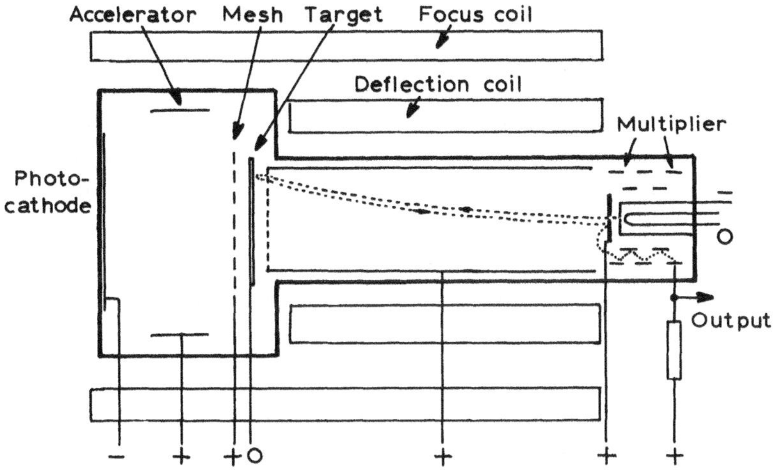

Figure 4.8. The electrode and coil arrangement of the image orthicon.

of the gun. This has an electron-emissive surface and secondary electron emission occurs. The electrons from this surface are attracted into an electron multiplier, often having five stages and a gain of 500. The output current is then proportional to the beam current less the signal current.

The normal application of the orthicon is in the image orthicon. In this, the scanning section is preceded by an integral electron imaging section, using magnetic focusing, similar to that of the image intensifier but with a target electrode replacing the phosphor. The surface of the target is photoemissive and the high-speed electrons cause secondary electron emission, leaving a positive charge on the surface equivalent to the original light intensity integrated through the period between scans. The secondary electrons are collected by the adjacent target mesh, held a few volts above the cathode potential. The target is a very thin sheet of resistive glass and a current flows through the glass to equalize the potentials on the two sides. The scanning beam neutralizes the total charge at each point of the glass sheet, as described above.

The use of the imaging section and the electron multiplier output give the tube a high gain and a high sensitivity. As the output is proportional to the difference between the beam current and the signal, the electron noise can be high for low light levels. It is advisable to set the beam current as low as is possible consistent with ensuring that the highlights are adequately neutralized. The characteristic of the tube is linear, gamma equal to unity, until the target potential is close to that of the target mesh, when saturation starts to occur. The secondary electrons from the target are then unable to reach the mesh, due to the reduced field, and return to the target, where they neutralize charge in surrounding areas. This has the effect of darkening the borders of highlights and sharpening the image. This can be useful in studio use but in pattern processing applications its use depends upon the task. The image orthicon is obsolescent but the basic structure survives in its derivatives (4.2.8, 4.2.9).

4.2.8. Image Isocon

In the orthicon, the electrons which are scattered by the target are proportional in number to those absorbed by the target. The specularly reflected electrons, having little radial velocity, are refocused onto a small area near the gun while the scattered electrons, having a range of radial velocity components, are spread over a large area. This difference is used in the isocon.[3,15,16] Only the scattered electrons fall on a secondary emitting surface to provide the input to the multiplier. As a result, the output is proportional to the integrated light intensity for each element of the picture. This avoids the introduction of noise from a high beam current when the video signal is small. The resolution is generally better than that of the image orthicon. In other respects the image isocon is similar to the image orthicon.

4.2.9. Targets with Gain

The sensitivity of tubes with image sections can be increased by producing electron gain in the target. In the normal tubes the electrons reach the target with energies of a few hundred electron volts and, by secondary emission, produce only a small increase in the number of electrons. By increasing the acceleration voltage to several thousands of volts, electron gains of several hundreds can be produced in suitable targets.

In the secondary emission conductivity (SEC) tube, the target is a porous layer of potassium chloride, about 20 μm thick, supported on an oxide layer on the imaging side.[15] A thin conducting layer exists between the two layers and is the output electrode. This electrode is held at a positive potential and, with the scanning gun restoring the surface of the chloride to zero volts, there are a few tens of volts across the layer. The electrons from the imaging section cause secondary emission within the pores of the chloride and, due to the high field in the chloride, these electrons produce more electrons.

In electron bombardment conductivity (EBC) tubes, the target construction is similar to that of the SEC tubes but the active layer is an amorphous photoconductor, a 1-μm layer of arsenic sulphide being a common choice.[15] The high energy electrons from the imaging section cause the release of many conduction electrons. It is also possible to use an array of silicon diodes, as used in the silicon vidicon, as the target of the tube. In this silicon intensified target vidicon (SIT) gains of over 1000 have been achieved.[3,14]

4.2.10. *Image Dissector*

In the electron imaging stage of the dissector, the electron image is formed on a metal plate which has a small hole at its center (Figure 4.9). Magnetic focusing is used. Electrons passing through the hole enter a high gain electron multiplier, typically of 12 stages, giving a gain of 1,000,000. Deflecting fields are applied to the imaging section so that the picture may be moved on the plate and the required part of the picture fall on the hole. It is usual to use magnetic fields for focusing and deflection due to the better performance that they give. The effect of moving the electron picture over the hole is equivalent to scanning an aperture across the picture.

There is no integration involved in the process; the output is directly related to the selected point in the scene. This has the dis-

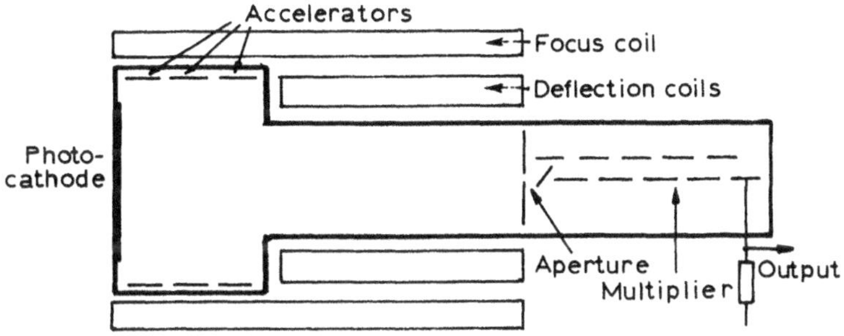

Figure 4.9. The electrode and coil arrangement of the image dissector.

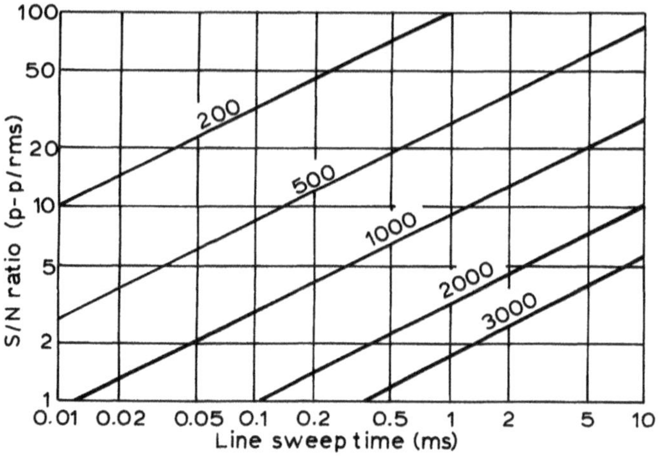

Figure 4.10. The signal-to-noise ratio of an image dissector for a photocathode current of 4 μA, with apertures giving limiting resolutions from 200 to 3000 TV lines.

advantage of reducing the sensitivity of the system, and as the maximum light intensity which the photocathode will withstand is limited, about 50 lux, high signal-to-noise ratios cannot be achieved with high resolution and high scanning rates (see Figure 4.10).

There are several advantages when the tube is used in picture processing systems. There is no lag and the output will follow rapid changes in illumination at the selected point. By using a very small aperture, e.g., 10 μm, resolutions of several thousand pixels per diameter can be achieved. Because there is no integration process it is not necessary to scan each element of the picture in sequence. The beam can sample any point as required, for as long as is required, and move in any pattern over the picture. For example, the outline of an object can be traced directly. This can save time and may compensate for the lower scanning rates imposed by the lack of an integration process.

The hole size is fixed for a given tube but holes having special shapes or multiple holes could be used.

4.3. Solid-State Cameras

Arrays of sensors, in a single line or multiple lines, can be integrated onto a silicon slice, together with means for connecting each sensor in turn, through appropriate circuits, to a single or limited number of outputs.[1,3,17–19] The light-sensitive element can be a photodiode, a phototransistor, or a depletion layer formed beneath an electrode,[4] although photoresistors have been used.[19] Typically, element spacings of 7 to 100 μm are used and linear arrays of 4096 sensors and area arrays of 400 × 600 sensors are produced. The arrays are usually packaged in standard IC packages with glass or clear quartz windows. The scene may be imaged onto the array by a lens system or delivered through a coherent fiber-optic bundle placed close to the array without an imaging stage. The main advantages of the integrated arrays, when compared with camera tubes, are the small size, absence of high voltages, resistance to damage from bright lights, resistance to physical shock, low sensitivity to the effects of magnetic fields, and fixed and known position of the sensors.

The diodes are pn junctions and when reverse-biased have a depletion region at the junction where conduction is by minority carriers. Incident light generates electron-hole pairs in this region which are separated by the fields of the depletion region. Besides being photosensitive, the diodes act as small capacitors, 0.1 to 0.5 pF. The diodes are, in most sensors, used in the integrating mode and at the start of each integrating period are charged, in the reverse bias polarity, to a preset potential. The effect of photoconduction is to reduce this stored charge. The differences between the various diode arrays are related to the means by which the change in stored charge is measured and the diode is recharged.

In the depletion devices, the application of a voltage to the over lying electrode creates a depletion layer below the electrode, i.e., a region with no majority carriers. Electron-hole pairs created by the light within this region, separate, and one component collects in the depletion layer. The charge so collected is proportional to the light received between each charge removal operation.

4.3.1. Properties

4.3.1.1. Sensitivity and Spectral Response. The basic quantum efficiency of silicon is high but, due to the presence of conductors on the surface of the chip and the separation of the sensors, the effective area that can be used is limited. Also there can be some attenuation if conductors are used over the sensor. Amorphous silicon (polysilicon) and some metal oxides, e.g., tin oxide, are conductors with low optical absorption. There will be small variations across the sensor in the pattern of the detectors and in the silicon and this will affect the sensitivity of individual detectors. Often variations of the pattern are cyclic, due to the way that the patterns are produced, and this may appear in the output.

The response extends from about 400 to 950 nm with a maximum at about 800 nm (Figure 4.11). Variation of the quantum energy with wavelength, an inverse proportionality, produces a fundamental reduction in sensitivity at shorter wavelengths when it is expressed in terms of energy. The penetration of light into silicon is a function of wavelength, with the longer wavelengths penetrating much deeper (Figure 4.12).[20] The effect of the light is to produce electron-hole pairs which are separated by the electrostatic fields within the sensor. At the extremes of the optical band it is possible for these separations to occur in low field areas, near the surface for the blue and deep in the silicon for the infrared. In these areas the chance of recombination in the contraflow streams of electrons and holes is greater and the sensitivity for these wavelengths will be reduced. The response can be affected by optical interference effects in transparent surface layers, e.g., of protective silicon oxide or of polysilicon electrodes. This appears as a ripple on the spectral response curve. The spectral response

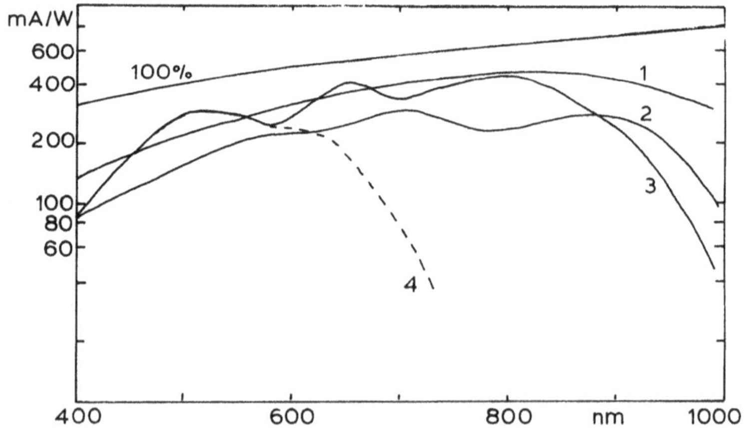

Figure 4.11. The spectral responses of semiconductor sensors, with the curve for a 100% photon-to-electron conversion. Curve 1 is for MOS, 2 for CID, 3 for CCD, and 4 for CCD with an infrared filter. Curve 3 shows the ripple due to optical interference in the thin surface layer. These curves are typical, but considerable variations can occur between different types and arrays of the same type.

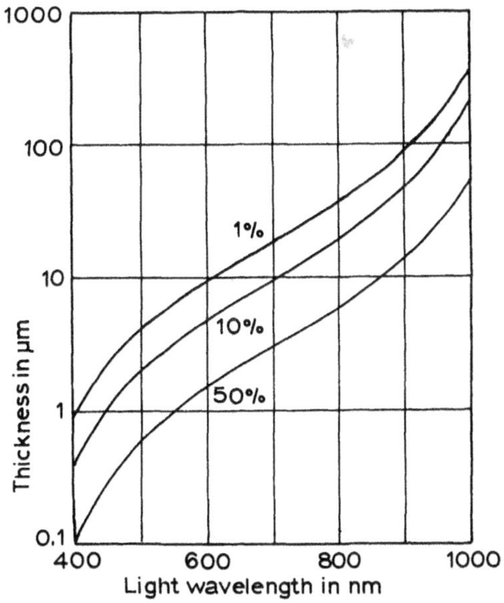

Figure 4.12. The absorption of light in silicon. The curves show the percentage of the light remaining after passing through the indicated thickness.

may be improved by thinning the wafer to less than a few tens of microns and imaging onto the rear of the wafer. This improves the blue response, since there are now no surface layers to absorb it, and reduces the infrared response due to the thinner silicon.[14] A useful conversion factor for silicon is that, in effect, 1 watt from a source at a color temperature of 2856°K is equivalent to 20 lumens, but this is only an approximation; for a more accurate figure the spectral response of the specific device must be considered.

The effect on sensitivity of defects in the silicon can depend upon wavelength, with the longer wavelengths suffering the greatest changes. This is one reason why infrared stop filters may be placed in the optical system of a camera. The sensitivity of arrays can be increased by preceding them with an image intensifier. For example, a microchannel plate can be placed before a fiber-optic bundle which transfers the intensified image directly to the array.[4,21]

4.3.1.2. Resolution. Basically, the resolution is determined by the spacing of the sensor elements.[19,22,23] At the shorter wavelengths, charge released in a sensor is collected in that sensor. At the longer

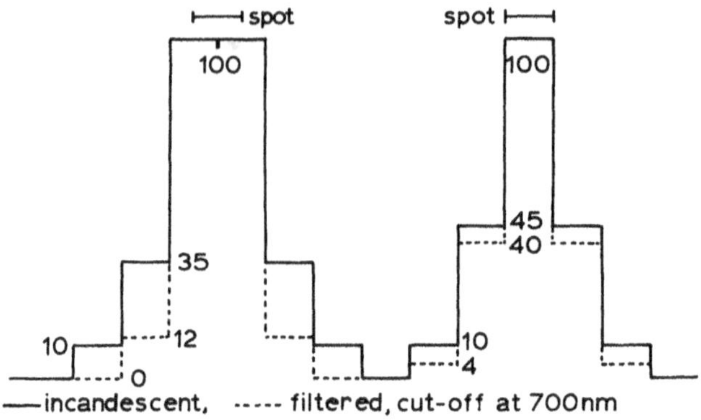

Figure 4.13. The video waveform obtained due to charge spreading in an array having a sensor pitch of 13 μm. The light spot, 13 μm wide, is positioned as shown.

wavelengths the separation can occur deep in the silicon and transverse motion can occur while the charge is moving towards the collection areas nearer to the surface. As a result it can be collected in an adjacent sensor site or even further away. The resolution in the infrared can be significantly less than that in the visible range and this is another reason why infra-red stop filters are often incorporated in the optical system. The effect is shown in Figure 4.13 for an array, a CCD, with a sensor pitch of 13 μm illuminated by a line of light 13 μm wide. The light is centered on a sensor or on a gap between sensors. The incandescent light source has a temperature of about 2850°K and the results for the adjacent sensors, given with respect to the outputs of the illuminted sensors, are for no filtering and an infrared filter cutting above 700 nm. The loss of resolution is increased if the light enters the array at an angle, for example, when a large lens aperture is used. Again, this is most significant at the longer wavelengths.

With linear arrays it is common to obtain a second scan direction by the relative motion of the scene and the array. As a result, the amount by which each element of the scene contributes to the integral in a scan depends upon the scan rate and the element's relative position to the scanner when the line scan starts. This is illustrated in Figure 4.14 for an array moving by a cell length in each integration period and one moving by less than a cell length. Some parts of the scene will contribute to two or more consecutive integrals. This modifies the effective resolution in the slow scan direction.

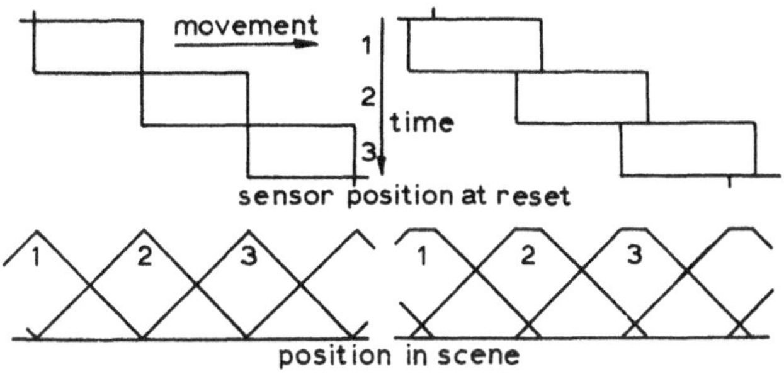

Figure 4.14. The contribution of each element in the scene to the integral in each cell of the array, for a moving array. The movements shown are of one cell width during the integration period, and of less than one cell width.

4.3.1.3. Linearity. The signal generated, as a function of the change in the stored charge, is proportional to the received light, gamma equal to unity, until the maximum storage capacity of each detector is approached. Near this limit the fields within that site become small and the fields from adjacent sites attract the charge. This spillover into adjacent sites is called blooming (see Section 4.3.1.7).

In those arrays where the output is directly related to the stored charge, the response is linear until saturation is approached. There are a few arrays which derive their output from the voltage across the diode before it is recharged. In these the reading and recharging are sequential processes. The capacity of a diode is voltage dependent, falling with increased bias due to the thicker depletion layer, and therefore when the voltage mode is used, the gamma falls continuously as the integration proceeds.

4.3.1.4. Dark Current. Thermal effects produce a leakage current within the detectors which doubles for every temperature increase of about 8 °C. The current is greater where the defect and impurity levels in the silicon are higher. Consequently it is more important in the manufacture of optical devices than of other integrated circuits that high quality material be used. The dark current is a greater proportion of the output when long integration times are used with low light levels. The dark current spatial pattern is constant for fixed operating conditions and is added to the picture output. It is called "fixed pattern noise." The dark currents can be reduced by cooling the sensor with cold fluids or with a Peltier cooler (a junction of two metals which becomes cold when a current is passed through in the correct direction). If this is done, care should be taken that the differential expansion between the substrate and the silicon chip does not cause fracture of the chip.

4.3.1.5. Noise. The main source of interference in solid-state arrays is breakthrough from the clock, i.e., control, signals into the video channel. If the clocks execute complete cycles for the output of each element of the array, then usually the interference is the same for each element. It appears as a high frequency transient on the signal and can be removed by filtering or by the use of the appropriate sampling time at the output. In some arrays, clocks are used which operate at a lower frequency than the output rate. For example, an output can be produced on both the rising and falling edges of a clock. In this case it is possible for the breakthrough to be different for the two edges and this appears as the addition of a small square wave to the video output. It is often called castellation.

Clock breakthrough is less significant in CCD arrays, but electronic noise from the output circuits can become important with the small signals in low light level applications. Some data sheets quote a figure for the dynamic range, typically 5000 to 1, and this is the ratio of the saturation signal to the rms electronic noise level. Note that it does not include diode-to-diode variations in dark level and sensitivity, which are much greater. For most applications, noise from the output circuit and the discrete nature of the photons is negligible.

4.3.1.6. Lag. Lag can occur in devices being used at high speeds. Time is required to reset the sensor to its original condition and if resetting is not completed, some of the picture will be retained and appear as part of the next scan. It is also possible for the output circuits to be incompletely reset with the result that some of the output from one sensor will be carried into the next.

4.3.1.7. Blooming. The generation of electron-hole pairs by the action of the light will continue even when the collecting location is saturated. What happens to this extra charge depends upon the structure of the device. At worst, it will be collected by adjacent locations and the highlight in the scene will spread in the image, an effect called blooming. The CCD tends to be most susceptible. It is usual to provide means for diverting the excess charge and to protect the adjacent cells from this overspill.[23,24] One method is to produce an implanted channel between the cells which collects the overspill. Protection against illumination of 10,000 times saturation level has been achieved.

4.3.1.8. Array Defects. Since the arrays are integrated circuits, localized faults can exist within the circuit or its surface layers. Two such faults have been mentioned with respect to sensitivity (4.3.1.1) and dark current (4.3.1.4). The fault may affect only one or a few sensors but in the case of CCD arrays (4.3.4), where the signal is passed along an analog shift register, the fault may affect all of the signal that passes through the faulty stage. This will appear in the output as a dark or bright line, starting from the fault position and running to the edge of the picture. Dust, etc. on the array window can appear as a fault but is distinguishable because its position changes as the direction of the incident light is changed.

It is difficult to compensate for array faults and they can be a serious problem in picture processing systems. One technique used is to set the analog value for a pixel lost due to a fault at the mean of the measured values of adjacent pixels. The criteria to be used in selecting arrays for picture processing systems differ from those applicable to entertainment or surveillance applications and this should be considered when specifying arrays.

4.3.2. MOS Linear Array

A linear array is shown in Figure 4.15. The sensor array is a line of diodes, usually with a width equal to the pitch of the diodes, but special arrays are available with the diodes much wider, e.g., 100 pitches. These wide arrays collect more light and are useful when the light pattern varies in one dimension, as in optical spectra. The capacity associated with each diode is that of the junction plus extra circuit capacity. Each diode is connected to an output line by a MOS transistor. These transisitors are driven by the stages of an MOS shift register. A single pulse, the start or scan pulse, is moved through the shift register by a single- or two-phase clock drive. At each stage of the register it sets the control transistor to low impedance so that the diode is connected to the output line. In some devices the output line is connected to a terminal of the package.

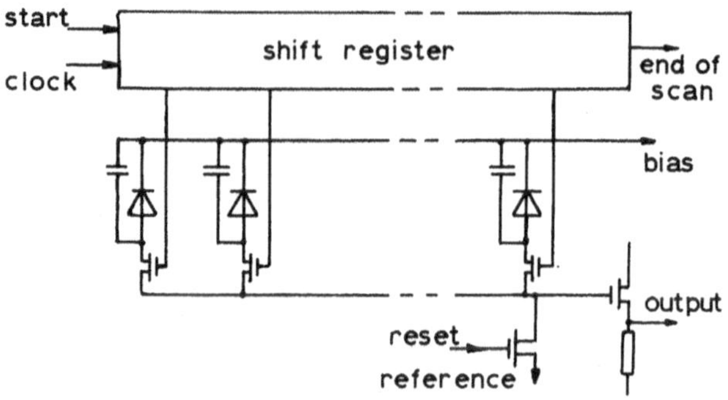

Figure 4.15. The basic structure of an MOS linear array scanner.

There are several methods for processing the output from the array. In the circuit shown, the reset transistor connects the output line to a low impedance bias source. Each diode is, in turn, charged to the bias voltage. After the integrating period for each diode, the stored charge will have been reduced. Charge sharing occurs between the diode and the output line which starts at the bias potential. The line has a capacity of the order of 100 times that of the diode and the change in line voltage is small. A source follower provides a buffered output of this voltage. Usually time is allowed for transients to subside and for charged sharing, etc. to be completed and then the output is sampled by subsequent circuits. Next the output line is connected to the reference voltage to bring the line and the diode to the full bias voltage. This is repeated for each diode in turn as the start pulse moves through the shift register.

In some arrays the circuit shown in Figure 4.15 is included on the chip. An alternative is to connect the output line to a charge measuring amplifier. In one form this is a virtual earth amplifier with resistive feedback and referred to the bias voltage. The output line is driven to the bias voltage and the amplifier output is a pulse whose area is proportional to the signal. Usually all the pulses from the amplifier have the same shape and duration, and therefore the amplitude is also a measure of the signal.

It is obvious that two pulses must not exist in the register at the same time unless some special effect is required. This sets a minimum integration time with respect to the clock rate but the integration time can be longer than that required for the pulse to pass through the register. The maximum integration time is set by the amount of dark signal that can be accepted. The minimum clock frequency is set by the shift register which is ac coupled. For frequencies less that a given value, usually about 10 kHz, there is a risk that the start pulse will be lost. The maximum clock frequency is determined by the time which must be allowed for the movement of charge within the sensor and the settling of the external circuits. Frequencies of the order of 10 MHz are possible. The maximum frequency given in some data seems to be that for the operation of the register because it is too high to allow for the complete transfer of the charge.

To obtain the maximum speed from the device it is necessary that the clock lines, which feed every stage of the shift register and have capacities of the order of 100 pF/1000 stages, are correctly driven. Typical voltage swings are of the order of 10 volts and power drivers are necessary. Some of the slower arrays, e.g., 2-MHz clock, have drivers built onto the chip and their clock inputs may be TTL compatible and have low input capacities.

In the higher speed devices, e.g., 40 million samples per second, several output lines are provided. Each stage of a single shift register causes the simultaneous connection of, for example, each of the first four diodes to the four output lines. The next stage of the register connects the next group of four, and so on. This allows a lower clock frequency to be used, with obvious advantages with respect to driving, and provides more time for the output on each line to stabilize.

One method used to allow the reduction of the effects of switching transients is to provide a duplicate array driven by the same shift register. The diodes of this array are covered by a metal layer so that no light is received. The principle is that the transients from both output lines will be similar and cancellation will occur if they are fed differentially into an amplifier.

It is not necessary that the linear array be in the form of a straight line. Arrays can be obtained having 64 or 720 diodes in a circle.

4.3.3. MOS Area Array

The basic principle of operation of the linear array can be applied to two dimensions.[3] Several configurations of the reading circuits are possible. In the first there are two MOS transistors connected in series in the output of each diode. Two shift registers are provided, one for the line scan and the other for the frame, each controlling one of the transistors of the diode output. Thus each stage of a shift register drives either a row or a column of diode circuits. A common output connection is provided for all of the diode circuits. By driving the shift registers at the appropriate rate each diode, in turn, is connected to the output. The main disadvantages of this system are the large output line with a high capacity, the high number of gating transistors, and the presence of line and column addressing lines across the array.

An alternative uses parallel linear arrays on the same chip. It avoids most of the above problems. The transistor gates of corresponding diodes of the arrays are controlled by one of the lines from a frame scan shift register (Figure 4.16). The output lines of each linear array are connected through an MOS transistor to a common output line, with these transistors being controlled by the outputs from a line scan shift register. The first shift register causes the connection of a diode in each array to its output line. The second register connects each line, in turn, to the common output line. The driving frequencies of the shift registers cannot be interchanged because, while the whole of one linear array was being scanned, diodes in the other arrays would be being connected to their output lines. The repetitive charge sharing would destroy the picture information.

In a variant of this system, the lines of the linear arrays are connected simultaneously to corresponding stages of a bucket brigade

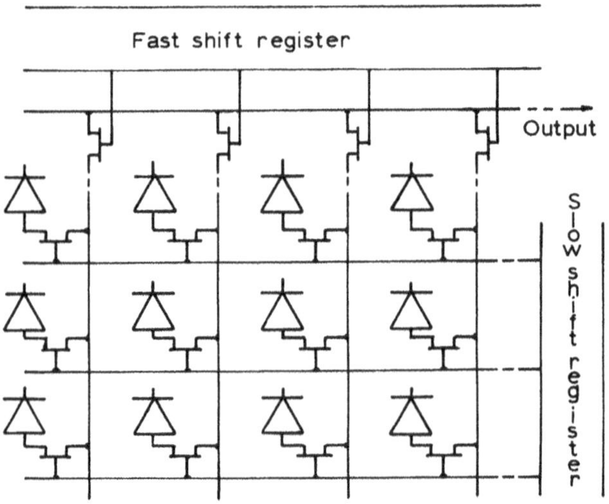

Figure 4.16. A two-dimensional diode array driven by MOS shift registers.

shift register. This is an analog shift register, using MOS transistors, in which charge packets can be transported to a single terminal. As before, the low-speed scan is applied to corresponding stages of the linear arrays.

4.3.4. CCD Operation

The charge coupled device (CCD) provides a means for tranferring charge packets along a shift register.[4,19,21,23,25,26] Typical structures are shown in Figure 4.17. In the three-phase device, *n*-type silicon is formed on a *p*-type substrate. This is covered by a silicon oxide layer onto which electrodes, usually of transparent polysilicon, are deposited. The electrodes are connected as shown and driven by a three-phase clock. A natural depletion layer exists near the junction but this is extended under one set of electrodes by connecting them to a positive potential. The potential distribution in the depletion layer is shown and is much deeper (lower energy although more positive potential) under the positive electrodes than the others. Consequently any free electrons will collect, and be trapped as separate packets, in the deeper "wells" under the positive electrodes. The charge packets

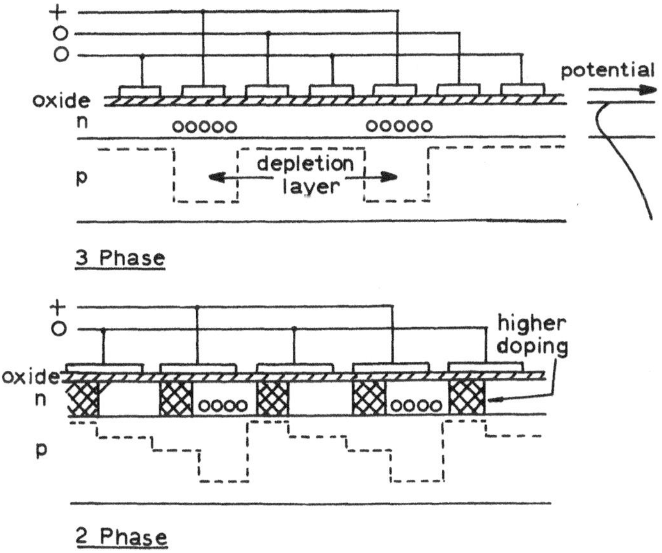

Figure 4.17. The basic structures of three- and two-phase CCD analog shift registers.

can be moved along the shift register by the appropriate variation of the electrode voltages. The voltage of an adjacent set of electrodes is raised and the packets can spread in that direction, under the effect of mutual replusion within each packet, although the packets remain separated. The electrodes that were high are brought to zero voltage with respect to the substrate. There is now a positive force moving the packet forward to the new set of electrodes. This is repeated with the third set of electrodes and so on, so that the packets are moved along the array.

The speed with which the packets can be moved depends upon the driving forces that are applied to the electrons and to their mobility. It is possible to construct a CCD without the *n*-type layer shown in Figure 4.17. In such a device the potential pattern would continue to rise right up to the oxide layer, and the well would be against the oxide. The mobility of the electrons near this surfce is reduced due to trapping of the electrons in the surface states, i.e., the imperfect surface structure. Also, the component of the field, in the required direction of motion, is small close to the electrodes. These effects increase the time for complete transfer of the electrons and in the limited time usually available the complete packet will not be transferred. A CCD with the *n*-type layer is called a bulk CCD (BCCD) and without the layer, a surface CCD (SCCD).

The value of the fraction of each packet that is transferred is called the transfer efficiency and is typically 0.99999 to 0.999999 at a frequency of about a quarter of the maximum. Incomplete transfer leads to blurring of the information along the register. Also, due to the random element that it contains, it increases the noise component of the signal. The efficiency of transfer is improved if the gap between the electrodes is made small. One way to achieve this is to overlap the electrodes. Some of the electrodes are deposited, using polysilicon, and their surfaces oxidized. The next electrodes are deposited so that they overlap the first with the thin oxide layer providing the insulation, and so on. The separation of the electrodes is then about 0.1 μm.

It is possible to produce a CCD with a two-phase clock system. By various means, the construction associated with each electrode is made asymmetric (Figure 4.17). As a result, the packet of electrons comes to rest under one side of the electrode. During manipulation of the clock signals the electrons can move only in that direction, whereas with the three-phase system it would be possible to reverse the direction by changing the relative timing of the signals. One method of producing the asymmetry is to increase the doping of the *n*-layer under part of each electrode. In this region the inversion layer will be smaller and the well will be shallower. The electrons will collect where the well is deeper. As the potentials of the electrodes are interchanged, the electrons will move forwards into the next stage but are unable to move backwards under the original electrode.

In an alternative system, the oxide layer is etched so that each electrode is deposited onto a thick and a thin area of oxide. The well will be shallower under the thicker oxide and a well pattern similar to that of Figure 4.17 will result. The operation of the CCD is the same.

In the three- and two-phase systems there is a limit to the size of the packets, and beyond this reverse flow can occur. In the three-phase system, the limit is reached when the electrons under a positive electrode have raised the potential of the well close to that of a well under a zero potential electrode. The retaining potential is then inadequate to overcome the thermal motion of the electrons and they will pass under that electrode to a preceding deep well. In the two-phase system, the limit is set by the relative positions of the two well levels under each electrode. The storage capacity is less than for the three-phase system when using the same potentials and doping level.

The two- and three-phase systems can be operated in the "one and a half" and "two and a half" phase modes. One of the electrode systems is held at a potential equal to the midpoint of the normal voltage swing. The others are driven normally. This technique has the disadvantage of reducing the charge handling capability of the register.

Light detection and charge integration can occur within the CCD itself or in separate photosites. When light enters the depletion region in the shift register, electron-hole pairs can be formed and the electrons collected in the wells below positive electrodes. It is moved from the register in the normal way. A disadvantage of this system is that the readout time must be short with respect to the integration time. As the packets are moved through the register they collect signal from light from parts of the scene to which they do not apply, i.e., blurring occurs, and this is worst for those packets passing through most of the register and for high shift time/integration time ratios.

Alternatively, separate photosites can be used to receive the light, with a transfer gate between each photosite and the appropriate stage of the CCD shift register. In this case it is necessary to cover the transfer gates and the shift register with an opaque layer, usually aluminum, to exclude the light. The photosites are covered by a polysilicon electrode held at a potential between the extremes of the CCD clock voltages. This creates a well in each photosite. During integration the transfer gate is at substrate, zero potential, and the photosites are isolated. At the end of the integration the transfer electrode is raised to a potential equal to that used for the CCD register. The phasing of the potentials of the register is such that the stages adjacent to the transfer channels also have positive electrodes. The electrons in the shallower well of the photosite are driven into the register. The transfer gate is returned to zero, integration starts again in the empty photosites, and the charge packets can be moved along the shift register.

Normally the register must be free of charge before the transfer of the charge from the photosites. This is achieved by ensuring that no charge can enter the register at the end away from which the charge is moved. There is a technique, called "fat zero," by which the effects of charge trapping in the SCCD can be reduced. A small quantity of charge is placed into each stage of the register, so that the zero level is raised, and any charge generated by the light is added to this. The principle is that a constant quantity of charge will be left behind at each transfer due to charge trapping. The resulting output will contain a fixed bias which can be deducted. The bias charge is generated by a biased diode diffused into the chip adjacent to the first clock electrode of the register. The well under this electrode is then filled to the bias level. An alternative method is to use a bias light, an LED mounted in the camera, to provide an extra uniform charge over the array.

The usual output circuit consists of a diode diffused into the chip, adjacent to the last electrode of the register. While the electrode is low the diode is biased to a present voltage. During the high potential on the last electrode the charge is transferred into the diode. The voltage across the diode is buffered by a source follower to provide the output.

4.3.5. Linear CCD Arrays

A typical linear array is shown in Figure 4.18, using separate photosites and a two-phase clock. Two shift registers are used, each receiving the output from alternate photosites. For a given detector

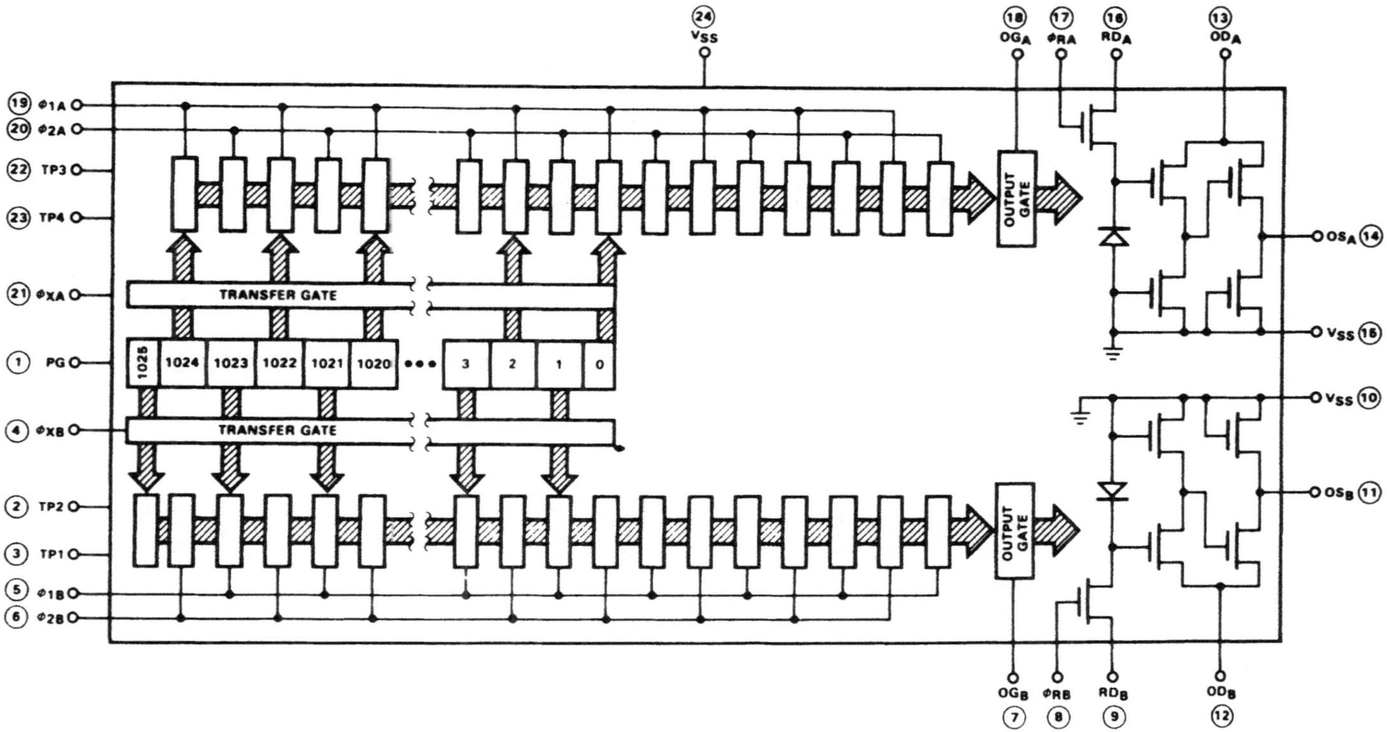

Figure 4.18. A CCD linear array with two separate output systems. (Courtesy of Fairchild Camera and Instrument (UK) Ltd.)

pitch this allows the spacing of the stages in the registers to be greater, the clock frequency to be less, and the number of stages through which the packets travel to be less. In this circuit the outputs are buffered by two-stage source followers. In some devices the outputs are combined on the chip by feeding the two registers, correctly phased, into a short register operating at twice the speed, or by having a twin output gate feeding into a single output diode and source follower.

It may be useful to have information about the dark signal level and maximum light signal level in the output from the array. This is produced by extending the shift registers and connecting some of the extra stages through transfer gates to voltage sources which fill those stages to a known fraction of the maximum permissible level. Other stages are connected to photosites beyond the used part of the array and protected from the light by an opaque layer. The former supplies the light reference level and the latter the dark level. These references suffer the same offsets, e.g., from the output circuit, as the video signal and any variations of these offsets due, for example, to temperature changes.

Arrays with 4096 sensors in a single line are available. An array with 5732 diodes has been made but this consists of two adjacent 5-μm wide arrays of 2866 diodes having a pitch of 10 μm. The arrays are offset by half of a pitch to improve the effective resolution. An MTF of 50% is quoted for 1800 cycles along the array length.(27) The resolution of these long arrays is approaching 100 line pairs per mm, and it is necessary that a very high quality lens is used to form the image. Using only small off-axis angles will help.

4.3.6. Area CCD Arrays

Several systems of operation are possible in area CCD arrays, and the light may be detected within the shift registers or in separate photosites.(17) Most of the systems consist of parallel shift registers feeding into a single shift register across their ends, but there are exceptions. The row access method of the MOS area arrays has been used, with the column conductors feeding to a CCD shift register.(28)

In the simplest system, the light is detected within a set of parallel shift registers. Each, in turn, is clocked into the output register. Although each main register enters the output register at a different point, this does not produce a problem if the clocks are correctly timed. Provided that the number of main registers is not small, the time during which each register is shifting is a small part of the integration time and blurring is negligible. The disadvantage of this system, besides the light attenuation in the register electrodes, is that separate clock lines must be provided for each main register. This uses space and reduces further the sensitivity of the device.

This disadvantage is avoided by the frame transfer system.[29] The array has two almost equal areas, one of which is covered by metallization to exclude the light. Each shift register runs across both areas but has separate clock signals for the open and covered sections, Figure 4.19. In each of the areas, the parts of the shift registers have a common clock system with the electrodes running across the array. Channel stops at right angles to the electrodes define the separate shift registers. At the end of an integration period, both clock systems are driven together so that the charge pattern is moved under the metallization. This takes a small part of the integration time and the blurring is negligible. The sensor area is now ready to start a new integration period. The registers in the covered area are stepped together, one stage at a time, to load the output register which is then driven to output one line of the picture. This is repeated until all of the stored picture has been output. In this system the registers run in the frame direction of the picture whereas in the previous system they were in the line direction.

Another system, known as interline transfer which also avoids separate clock systems for the registers, uses photosites to collect the

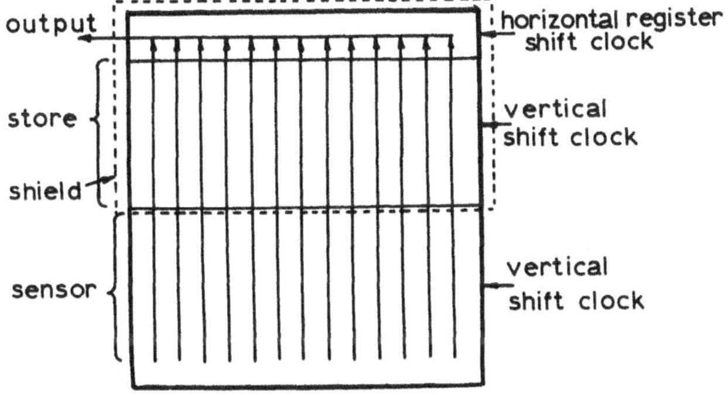

Figure 4.19. The frame transfer type of CCD area array sensor.

charge (Figure 4.20). The shift registers are protected from the light by a metal layer. A single transfer gate is used to move the charge pattern from the photosites into the shift registers. The registers are now driven in a manner similar to that of the frame transfer device. An array with 768 × 490 pixels has been made with pitches of 11.5 and 13.5 μm. The limiting resolution is stated to be about 500 TV lines.[30]

The array shown in Figure 4.20 illustrates some variants of CCD operation. The pitch of the electrodes in the registers, operating as two-phase devices, is equal to that of the photosites. Consequently the output from only a half of the photosites can be taken at one time. After this has been output the content of the other set of photosites is taken. In this way an interlaced scan pattern is obtained and also the necessary pitch of the register is doubled. The transfer from the photosites, normally under a positive electrode, is produced by bringing this electrode, the photogate, to substrate potential. Only those sites next to positive electrodes will transfer their charge. For the second, interlaced, field the clock potentials are interchanged during the transfer. In an alternative mode of operation, all of the register clocks are held high during the transfer and all of the charge pattern is transferred to adjacent register positions. After the photogate has returned to high, the potential of one set of electrodes goes low, ready for the shifting process, and the charge packets stored under them are moved on to join those in the adjacent stages. This process gives, in the frame direction, a resolution a half of that obtained with the interlaced process.

In systems integrating the charge within the registers, interlaced outputs are obtained by the appropriate phasing of the electrodes during the integration. In a two-phase system, either the electrodes can be set, alternately, in the two possible high/low arrangements or are all positive during integration and one set of electrodes or the other is lowered first prior to shifting. In a three-phase system, one set of electrodes must be low during integration to divide the registers into separate charge collecting areas. Two of the three possibilities are taken in alternate fields to produce an interlaced output. It should be noted that in all of the systems using integration within the registers, a true interlaced system is not obtained. In each field of a frame, the full integration area is used but the charge is divided differently in the two cases. In effect the pair of fields show a one, or two thirds for

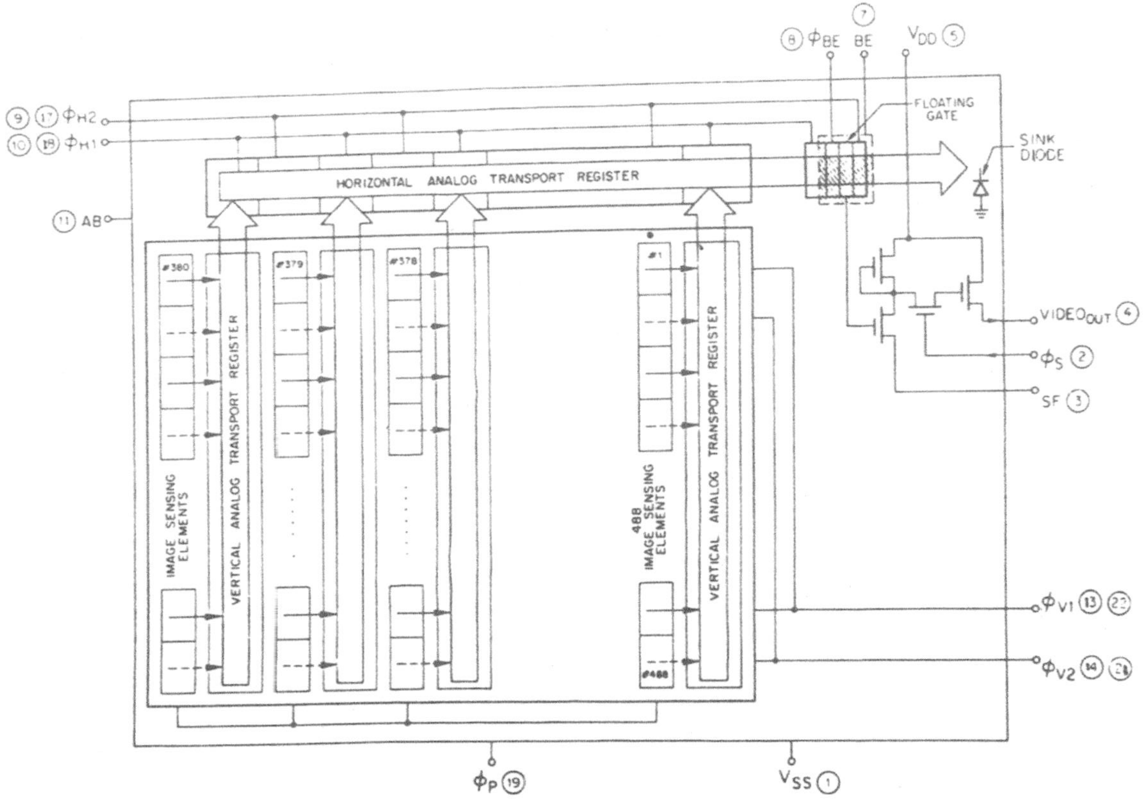

Figure 4.20. A CCD area array with interline transfer. (Courtesy of Fairchild Camera and Instrument (UK) Ltd.)

a three phase clock, frame line shift but have less resolution, in the frame scan direction, than a true interlaced system.

The array of Figure 4.20 shows an alternative method of detecting the output of a CCD shift register. An extra electrode, the floating gate, is provided in the register. The charge is moved under this electrode by the clock signals and induces in the electrode a voltage change proportional to the charge in the packet. This voltage output is buffered by a source follower and, in the circuit shown, passes through a sampling gate to the output amplifier. In some systems a separate recharge system is provided to set the initial potential of the gate between each output.

4.3.7. *Area Resistive Gate Sensor*

The area resistive gate sensor is related to the CCD and also uses a conventional CCD system for its output circuit.[31] Light detection occurs in separate photosites under positive polysilicon electrodes (Figure 4.21). Each line of photosites has a separate photogate electrode drive in by the stage of an MOS shift register. Orthogonal to these electrodes are the resistive gates also of polysilicon. A potential is applied along the length of these electrodes creating a drift field in the underlying silicon. The charge packets from one line of photosites

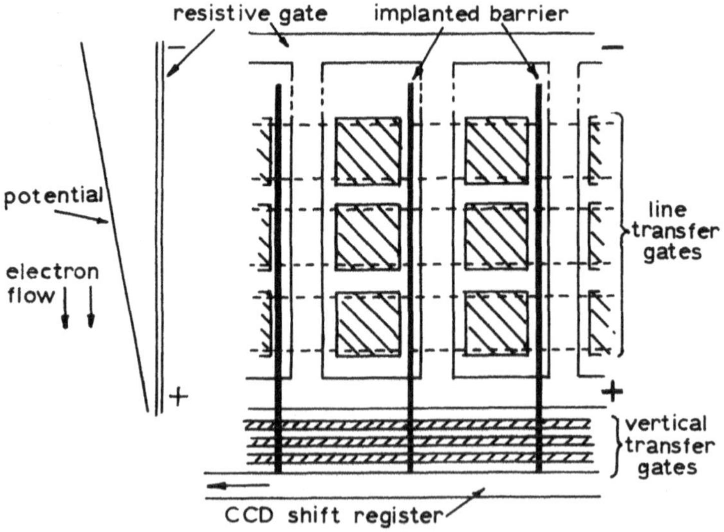

Figure 4.21. An area sensor using the resistive gate process.

are transferred to the adjacent resistive gates by reducing the photogate potential to zero. The drift field drives the charge into the first stage of a short CCD register. All of the charges from one line of photosites are shifted simultaneously into the output shift register. This is repeated for the other lines. A new line is shifted under the resistive gates as soon as the previous line has been taken into the output shift register. As a result the charge has the whole of a line scan period to move along under the resistive gate. Except for the photosites, the chip is covered by a metal layer to exclude the light.

4.3.8. *Area CID Array*

In the Charge Injection Device (CID), charge packets are not moved across the array but are moved locally or injected into the substrate, the action from which the name is derived.[3,17,19,26,32] A layer of *n*-type silicon is formed on a *p*-type substrate to produce an epitaxial junction and is covered by a silicon oxide layer. Each photosite consists of two areas, each under a transparent electrode of polysilicon or metal oxide (Figure 4.22). The electrodes from the array of photosites provide the system for *X-Y* addressing the array. The epitaxial junction

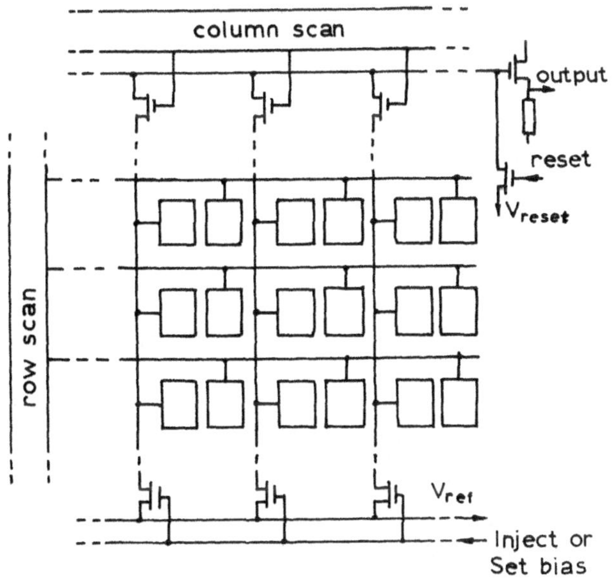

Figure 4.22. A CID array using the charge injection process.

is reversed biased, *n*-layer negative, to remove the majority carriers. The application of a negative voltage to the electrodes produces a depletion layer below the electrodes, in the same way as in the CCD, but the voltages are reversed because *n*-type silicon has been used. In principle *p*-type silicon could be used with positive voltages.

Normally the row electrodes are at a large negative potential, about −20 volts, and the holes from the electron-hole pairs, separated by the photons, collect under the electrodes forming an inversion layer. The column electrodes are charged to a potential about half that of the row electrodes by connection to the reference voltage line, V_{ref}. To read the content of a row of photosites, the column electrodes are allowed to float and the appropriate row electrode is raised to substrate potential. The charge packets move under the floating electrodes causing a change in the potentials of these electrodes, by capacitive coupling, proportional to the size of the charge. The column lines are connected, in turn, to the output conductor by MOS gates (see Section 4.3.2). The output line is reset to a fixed potential before each connection. After charge sharing has been completed between the column and output conductors, the result is buffered by a source follower. While the selected row electrode is at substrate potential, the column electrodes are connected to the V_{ref} line which is also at substrate potential. This removes the depletion layer in these sites and the charge is injected across the epitaxial junction into the substrate. The potential of the V_{ref} line is changed to about a half of the row drive potential, and the selected row returns to −20 volts, allowing integration of the light to recommence. A new row is selected and the process continues. Interlacing can be obtained by accessing alternate rows in the first scan and the others in the second scan. The column and row scan drivers are usually MOS digital shift registers although decoders could be used.

The CID is very flexible in the way that it can be operated. For example the whole array can be cleared by bringing all of the electrodes to zero. Integration will start from the time that the biases are reestablished. If the illumination is removed after a suitable integration period, e.g., after a flash of light, and only the bias voltage (−10v) is applied to V_{ref}, the picture may be scanned many times without loss. A usable picture has remained in an array at 200 °K after three hours of scanning at the normal television rate. With very low levels of lighting it is possible to use long integration times and to monitor the buildup of the picture by normal scanning.[14] Also, if decoders are used to address the array, random addressing, without clearing, is possible.

The CID is not afflicted by blooming as excess charge is conducted across the junction.

4.3.9. *The DRAM as a Sensor*

The dynamic MOS RAM (DRAM) can become a cheap binary sensor if placed in a package with a window.[33,34] Each bit of data is stored as a charge packet. Normally the charge is lost slowly due to thermal conduction and it is necessary to refresh the data at regular intervals, e.g., 2 ms, to restore the charge to its original level. The effect of light is to increase the rate at which the charge is lost, and the time at which the charge falls below the level signifying a change in value of the stored bit is dependent upon the light level. The charge state can be set by a "write" action; after a specified time the bit can be read by a "read" action. Only a binary result can be obtained and the value will have changed if the integrated light at that cell has exceeded the threshold value. By cyclically reading all addresses and resetting with a write action the device can be scanned. Alternatively, only the cells being used need be read and reset. Naturally the refresh process is not used. A 32k device can have a 128×256 array and the pitch in the two directions is unlikely to be equal. The maximum scanning rate is that of the device in its normal memory application. There may be a little distortion due to the limited positional accuracy of the cells.

Since the device has not been designed for optical input, the electrode structure may obscure much of the array giving a low sensitivity. Also, the storage space may be divided into several separated areas and, if the device operates in a byte access mode, the bits of the byte may not be adjacent. The quality of the silicon may not be to optical device standards and this can produce high dark currents and nonuniform sensitivity, affecting the inherent thresholds at which the individual cells will change state. Even so, it may function well with high contrast, binary scenes.

4.3.10. *Position Sensitive Diode*

Although a position sensitive diode is not a scanning device, it can be used to determine the position of a spot of light with a reso-

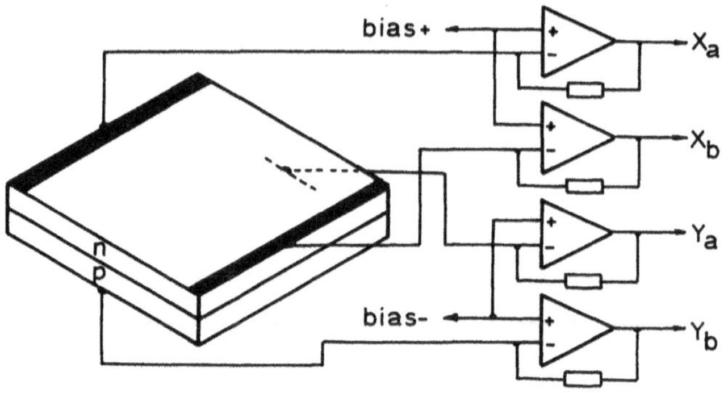

Figure 4.23. The two-dimensional position-sensitive diode. The outputs depend upon the position of a light spot on the diode.

lution up to 1 in 10,000, in X and Y. One-dimensional devices also exist. The structure is a thin, large area junction (Figure 4.23). The junction is reverse biased and conduction will occur through the junction where the spot of light falls on the device. One surface has electrodes on the X edges and the other on the Y edges. For each surface, the two electrodes are at the same potential and, due to the resistance of the material, the currents flowing to the electrodes are inversely proportional to the distances of the electrodes from the light spot. In practice the values of $(X_a - X_b)/(X_a + X_b)$ and $(Y_a - Y_b)/(Y_a + Y_b)$ are derived from the four currents and from these the position can be calculated. The response time is of the order of 10 μs and the linearity better than 1%. By calibration a higher accuracy can be achieved. It is not necessary for the spot to be very small if the light distribution in the spot remains constant. The device can be used in the presence of ambient light by using a modulated light spot. Only the modulated component of the current is used in signal processing.

4.4. Laser Scanners

The laser light source is described in Section 5.2.3.12. Due to the small angle of divergence of a laser beam, it is possible to project it as a very small spot into a scene. By the addition of a means for moving the spot across the scene and of detecting some of the light scattered from the spot, an image acquisition system can be produced.[35] Although the light that can be collected for the detector may be only a small part of that emitted, an acceptable signal-to-noise ratio is possible because relatively high spot energies can be used and, due to the monochromacity of the light, narrow band filters can be used to exclude from the detector most of the ambient light. Even so, sensitive detectors and good light collection are required. Photomultipliers are frequently used for this purpose but phototransistors, and in particular avalanche transistors, can be used. The photomultiplier has the advantages of high gain, high speed, and low noise, and can have a large collection area. The photocathode should be selected to match the laser wavelength, the S20 cathode being suitable for the He-Ne laser.

The interaction between the microstructure of a reflecting surface and the high coherence in a laser beam results in an effect called speckle. The polar diagram describing the scattering of the light is a complex and variable function. This will add a significant noise component to the detected signal, particularly when the light is collected from a relatively small angle.

In most systems a single lens cannot be used to produce a small spot. The effective lens aperture is determined by the beam diameter and, since this is small, the resulting diffraction pattern at the scene would be large and give a large spot. Using the Rayleigh criterion, that the peaks of two spots should coincide with the first minimum of the other, the minimum angular separation of the spots is $1.22\lambda/D$ for a uniform circular beam of diameter D at a wavelength λ, and $1.27\lambda/D$ for a beam with a Gaussian distribution, the normal case for a laser, of diameter D at the half-intensity level. This corresponds to an MTF (see Section 5.1.1.4) of about 15%. Usually an MTF of 50% is required and the minimum separation must be doubled. Basically, a first lens is used to produce a rapid divergence of the beam and a second lens, with a much larger aperture, focuses this beam onto the scene as in Figure 4.24, but in practice more complex systems may be required due to the operation of the deflector itself. In acousto-optic and electro-optic systems, the light interacts with the bulk of the material and the size of the projected spot is dependent upon the optical qualities of the material and the finish of its surfaces.

The variation of the spot size will pass through a minimum at a certain distance from the lens and its diameter near this point will depend upon the effect of diffraction at the lens, the convergence of the beam together with the distance from the minimum, and aberrations in the optics. To obtain the maximum range, i.e., depth of field, in which the spot is less than some maximum size, determined by the required resolution of the scanner, a compromise must be made in the size of the aperture of the lens. Reducing the aperture will increase the spot size due to the first effect but reduce it due to the other two. From the convergence aspect the depth of field is $\pm vd/D$, where v is the distance from the lens of diameter D and d is the maximum spot size. The radius of the first dark ring of the diffraction pattern is $1.22\lambda v/D$. For the optimum result for a specific system, this will be related to d, e.g., by a constant k, the radius becoming kd. The depth of field becomes $\pm kd^2/1.22\lambda$. With the combined effects d will have to be slightly smaller than for the convergence effect alone. The optimum lens aperture is $1.22\lambda v/kd$ and the system is diffraction limited. It is seen that the depth of field in such a system is determined only by the square of the acceptable spot size and the wavelength of the light.

Many techniques have been devised for the deflection of laser beams and the following are some of the deflection systems that might be applied to vision processing.[(36)] Similar or different systems can be combined to produce two-dimensional scans or to combine a small angle high-speed scan with a large angle slower scan. Some of the deflection systems cannot handle wide beams, but this problem can be overcome by a more complex optical system. The first lens focuses the beam onto the first deflector. If a second deflector is used, a relay lens is placed between the deflectors to refocus the beam onto the second deflector. Finally, a lens with a large aperture focuses the beam onto the scene. The beam will sweep radially from the deflector and this may not be what is required. The sweep pattern can be modified by the use of lenses or mirrors. For example, if the deflector is placed at the focal point of a large parabolic mirror, the beam paths will be parallel. Unfortunately, the sweep velocity will not be uniform for a constant angular velocity but will increase as a secant squared function away from the center of the scan. A correction can be made either by varying the angular scan rate or by varying the sampling rate. Another possible modification is to use a telescopic optical system to magnify the deflection angle.

In continuously scanning systems the position of the beam may not be sufficiently accurately known for the purposes of sampling the video signal. This can be overcome by deflecting a small part of the scanned beam onto a grating and collecting the transmitted light onto a detector to generate a timing signal. The grating can be generated by a computer, using parameters of the system, so that a varying line spacing is produced to compensate for nonlinearity in the scan of the scene. In systems allowing random movements, the scanning path can be computer controlled to scan as required, e.g., to trace the edge of an object.

4.4.1. Galvanometer Mirror

A mirror mounted on a galvanometer movement provides a means for continuous, stepped, or random deflection of a beam for up to 10,000 resolvable spots.(37) Deflection rates of 15°/ms continuous and 100 μs for a small step are possible. By the use of a suitable current waveform, linear scans of the scene can be made. Sweeps of 40° with a repeatability of 0.01% are claimed, but for this accuracy an angle measuring system on the movement is required for servo feedback. This system, using twin deflectors, is used for laser range finders of the triangulation type (6.2.3).

4.4.2. Reciprocating Mirror

The mirror can be driven by a cam or other mechanical or electromechanical system and the spot caused to scan in a preset pattern.(38) Frequently this is useful as the frame scan for a two-dimensional system in that a large area mirror can be driven. Small mirrors may be mounted on vibrating devices such as tuning forks or piezoelectric bimorphs (a bar consisting of two layers of piezoelectric material which bends when a voltage is applied and vibrates when an ac drive of the correct frequency is used). The deflection of the laser beam may be increased by using multiple reflections between a fixed mirror and the moving mirror.

4.4.3. Rotating Mirror Drum

This is the most common and important technique for laser scanning.(39–42) A polygonal prism with polished faces, rotating upon its axis, can produce a continuous scan up to very high speeds (Figure 4.24). The sweep angle depends upon the number of faces and the

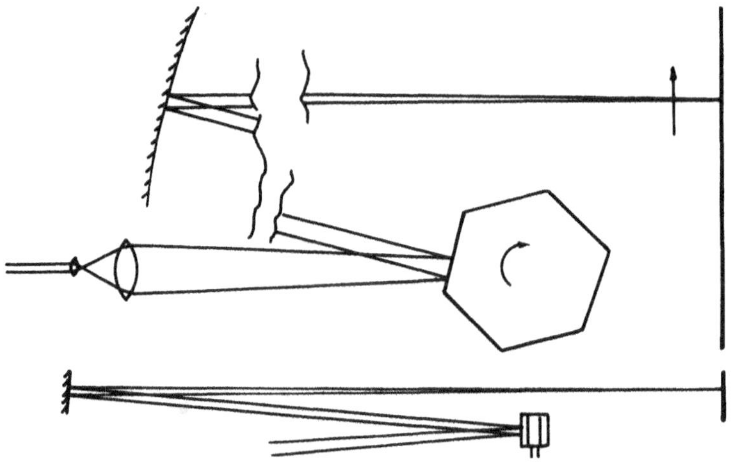

Figure 4.24. Orthogonal views of a laser scanner using a rotating polygonal mirror. A fixed parabolic mirror produces a line scan on a flat surface.

diameter of the beam at the face. At the ends of the scan, energy is lost as the beam passes over the edges of the face. The energy lost will be reflected to the start position of the next scan. Thus, either less signal is received from the ends of the scan or there is a time interval between the scans. If this is not accepable, the original beam can be split into two so that as one beam is brought to the end of the scan the other is starting the next scan.

The prisms are made from metal or glass. Prisms for use at high speeds may run on air bearings, to reduce friction, and be driven by gas turbines. To reduce air friction, the prism may be in a low-pressure enclosure or in helium. The faces of high-speed prisms are cut to be convex when at rest so that they are flattened by the centrifugal forces when in use. Standard equipment can resolve 20,000 points, with spot diameters down to a few microns, at 5000 scans per second; the high-speed scanners can reach 100,000 scans per second. The drum may not be the simple prism as shown. The faces may be at an angle to the axis, i.e., pyramidal, and complex patterns have been used to produce specific scan patterns.

The path of the scanning spot may be modified by the use of mirrors or lenses. A parabolic mirror is used in the system in Figure 4.24 to produce a straight line scan. For the high resolution systems the shapes of the correcting mirrors may be complex. Also, extra plane mirrors may be introduced to fold the system into a more convenient structure. In the simple parabolic system the linear scan is not made with a constant velocity.

Light reflected specularly from flat surfaces may be collected by allowing it to travel almost the same path in reverse and directing the light onto a detector which is effectively at the focus of the mirror. Frequently light guides are used for collection (see Section 5.1.4). This may be a transparent rod coated with a reflective layer except for a narrow gap along its length where the light enters from the scene. Photomultipliers can be placed at one or both ends of the rod. The side of the rod opposite to the gap may be shaped, e.g., roughened or given a serrated pattern, to reduce direct reflection back through the gap and to direct the light towards the ends of the rod.

4.4.4. Acousto-Optic

A repetitive pressure wave in a transparent medium produces periodic changes in the refractive index and can act as a diffraction grating to a beam of light. The sine of the deflection angle is inversely proportional to the wavelength of the sound wave. It is also proportional to the light wavelength, hence is suitable for scanning with monochromatic light only. The light pattern in the scene is that generated by the acoustic diffraction grating and therefore consists of an undeflected beam and pairs of first, second, etc., order beams. It is usual to use a first order beam since this is the brightest of the deflected beams. Each spot is itself a diffraction pattern produced by the limited width of the light beam in the grating. The width of the spot is inversely proportional to the width of the beam and therefore it is necessary to make this as large as possible. On the other hand, the deflection angle is changed by varying the acoustic frequency. Energy will move from the old to the new position of the spot as the change in frequency moves through the beam. Thus a wider beam gives a longer switching time. For continuous scans, for which linearly swept acoustic frequencies are used, an angle exists between the two sides of the beam, i.e., it is tapered, and it is possible to arrange the optical system so that a single point is formed at the convergence.

The concentration of energy into the first order of the diffraction pattern can be increased if high acoustic levels and wide acoustic beams are used and the incident beam enters the acoustic wave at approximately the correct angle.[35,43,44] This is known as Bragg diffraction, due to its similarity to Bragg diffraction in crystals, and the angles of the incident and emergent beams, within the medium, are equal to the Bragg angle (Figure 4.25).

$$\text{Bragg angle} = \arcsin\,(0.5 \times \lambda/W)$$

where λ and W are the wavelengths of the light and sound in the medium. The change in the angle of emergence in air with respect to the frequency change is given by

$$\delta B/\delta f = \mu\lambda/v$$

For large deflection angles it is advisable to maintain the Bragg condition. This has been achieved by changing the angle of the sound waves to match the acoustic wavelength.[44] The input can be made through a phased array transducer with the phase being frequency dependent or by a staggered transducer formed on a stepped profile at the end of the acousto-optic line. The Bragg angle is usually less than 1° and the sweep about 0.2°.

Typically, acoustic deflectors using diffraction can produce a few hundred resolvable points in a scan. The greater the number of points the wider the light path and therefore the slower the access rate. A few microseconds is typical for the beam to be gradually switched from one point to another. This is less of a problem for full line scanning than it is for random scanning. Bragg systems can place 70% of the original light in the deflected beam; the other systems can deflect about 20% into each of the first order spots.

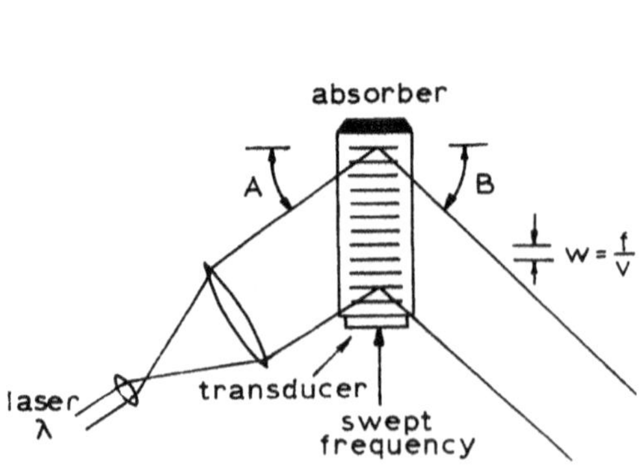

Figure 4.25. The deflection of a light beam by a travelling acoustic wave. The system operates in the Bragg mode.

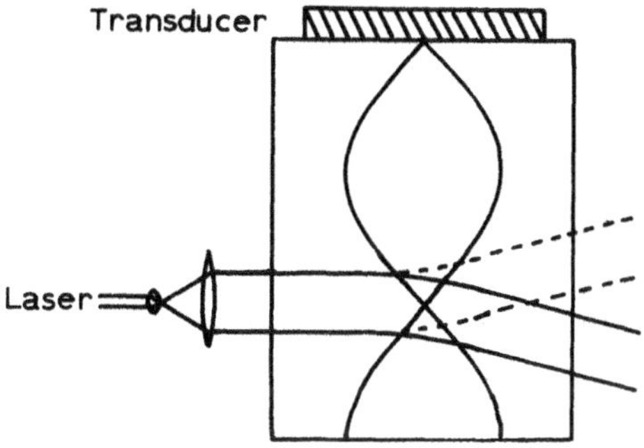

Figure 4.26. An acousto-optic deflector showing the standing wave profile.

A light beam may also be deflected by a refractive system.[45,46] A standing acoustic wave is set up in a transparent medium by resonance (Figure 4.26). The light beam is directed to pass through a pressure node of the standing wave where a varying pressure gradient, and hence a varying refractive index gradient, exists. The beam is deflected by the refractive index gradient in an oscillating manner. In general the angle of deflection is approximately a sinusoidal function of time, and compensation can be made in the processing of the scanned picture. Scan times less than a microsecond can be obtained. The resolution depends upon the light beamwidth and the acoustic wavelength, since as the beam in the medium bends into regions where the refractive index gradient is no longer linear, different parts of the beam will receive different deflections and the beam will spread. Typical deflection angles are 1 degree in solids and a few degrees in liquids at 0.5 MHz. At low frequencies the beamwidth can be large and the resolution high, several hundred points, but at high frequencies the acoustic wavelength is short which restricts the beamwidth and the resolution.

4.4.5. *Electro-Optic*

The refractive index of a medium can be varied by the application of an electrostatic field.[15,43,46–48] There will be a differential change for light polarized in planes parallel to and orthogonal to the direction of the applied field. Typical fractional changes in the refractive index are of the order of 10^{-3} to 10^{-4}, although some materials exhibit higher figures but are difficult to produce in optical quality. A triangular prism of such a material will deflect a light beam, polarized parallel to or perpendicular to the applied field depending upon the material, and sweep angles of the order of 0.1° are possible. The effect can be increased by using multiple paths in each prism and multiple prisms, which can be driven in analog or digital mode, i.e., each prism on or off. Switching times much shorter than a microsecond can be achieved, limited by the speed of the drivers and the heating of the prisms. The number of resolvable points is a few hundred. Optical systems can be used to magnify the deflection angles (see Section 5.1.5).

Also, it is possible to produce, by the application of high voltages, refractive index gradients in media. These will bend the light beam as in the acoustic system. The performance achievable is similar to that for the prism.

The differential change in the refractive indices in the orthogonal planes can be used in a different way, as illustrated in Figure 4.27. The incident light beam is polarized at 45° to these planes. By considering the beam as having two equal components polarized in the two planes, the applied field can change the phasing of these components to introduce a differential phase shift of a half-wavelength. When recombined, the resulting beam is plane polarized orthogonally to the plane of the original beam. Thus it is possible by switching with the appropriate voltage to change the plane of polarization through a right angle. The emergent beam passes through a birefringent prism that has different refractive indices for the two planes of polarization and hence gives two angles of deflection.

A system based on this technique has been built, using calcite prisms and nitrobenzene between the electrodes in the Kerr cells.[49] It has ten X and ten Y stages, to give 1024 × 1024 points. To encompass the partially deflected beam, the later stages are larger. Switching times are 300 to 900 ns, the longer times applying to the larger cells, and

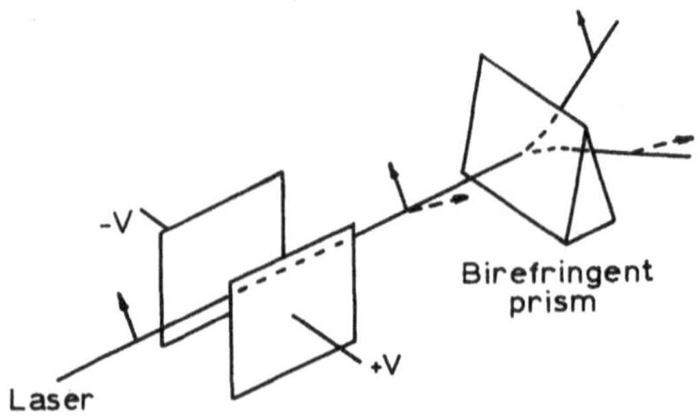

Figure 4.27. One stage of an electro-optic deflection system. A suitable electro-optically active liquid is placed between the electrodes.

are limited by the speed of the drivers. The deflection angles are magnified by telescopic lens systems.

The electro-optic systems are inherently fast, but due to the high voltages and the cell capacities, are limited by the response times of the drivers. The deflection angles are small but can be magnified optically. The number of resolvable points in one dimension is approximately 100–1000. The deflection angle is dependent upon the wavelength of the light and would separate mixed wavelength beams. In some cases it may be possible to produce corrections by the use of appropriate phase plates and dispersive prisms.

4.5. Other Scanners

4.5.1. Flying Spot Scanner

The spot on the phosphor of a cathode ray tube can be projected onto a scene by an optical system and the light scattered by the scene received by a detector. By driving the deflection in a suitable manner the spot may be moved across the scene as required. Raster, point-to-point, and other scans are possible. The light level cannot be high and a sensitive detector, such as a photomultiplier, must be used. The light has a wide bandwidth and therefore the spot cannot be isolated from the ambient light by the use of filters. Consequently the ambient lighting must be at a low level. The light produced is dependent upon the efficiency of the phosphor, the maximum permissible current density for the phosphor, the spot size, and the scanning speed. The impact of electrons on the phosphor produces heat which if excessive can damage the phosphor. For this reason it may be necessary to use less current if the spot is likely to have long dwell times. Many phosphors and guns, with dispenser cathodes, used in flying spot scanners are designed for high current operation. A special green phosphor, P53, has a maximum luminance in excess of 100 kcd/m^2.

Electrostatic focusing and deflection can be used but magnetic systems are used for the best performance. To simplify the optical system many scanner tubes have flat screens, although this complicates the electron optical system. To maintain a small spot it may be necessary to use dynamic focusing, in which the focusing is adjusted in relation to the radial position on the screen. To satisfy simultaneously the requirements of the optical and electron optical sides of the screen, a fiber-optic face plate may be used, with fibers of 3 μm diameter (see Section 5.1.4). An MTF of more than 2000 points per diameter at 50% can be achieved. Due to the mutual repulsion of the electrons within the beam and the size of the gun aperture, it is difficult to produce a small spot at high currents. The granularity of the phosphor can limit the resolution and also produce a fixed spatial noise pattern of brightness variations. Various colors are available from single or mixed phosphors. A color scanning system can be built by using a white phosphor and three detectors with the appropriate filters. Delays in the system, producing the effects of lag, are caused by the turn-on and decay times of the phosphor, the latter being the greater. The fastest give a decay to 10% in 150 ns.

This scanner is most suited to flat surfaces such as paper and photographs, especially transparencies. Its main advantage is the freedom available in the scanning pattern in that the spot may be moved as required, or held stationary, in the latter case with a restriction on the beam current. Small parts of a scene can be scanned and edges or other features followed.

4.5.2. *Flying Field Scanner*

Clearly, moving a line scanner, e.g., a linear detector array, in a direction orthogonal to its length, in the image plane of an optical system will produce a two-dimensional scan. A similar effect occurs if the sensor is stationary and the optical system is moved at right angles to its optic axis and to the scan direction of the sensor. This is the principle of the flying field scanner. A repetitive field scan can be produced by an oscillating displacement of the imaging lens or by using multiple lenses mounted in a cylinder which rotates around the sensor. Stops must be provided so that light from two lenses cannot reach the sensor at the same time. As a result there must be a gap between the field scans. A similar effect can be produced by placing an oscillating mirror or a mirror drum in the optical system.

4.6. Comparison of Scanning Systems

The performance of each type of scanner depends upon the way in which the various parameters are balanced, including the cost parameter.[3,50] This makes it difficult to make a clear comparison be-

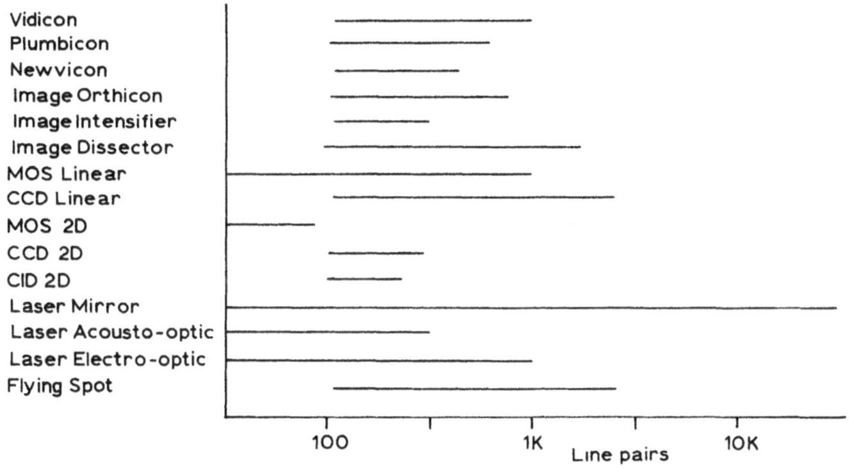

Figure 4.28. The limiting resolutions of scanners in line pairs/scan for line scanners and line pairs/picture height for area scanners.

tween the various types of scanner. Also, scanners may be found whose performance in one or more respects is significantly different from the performance generally expected from that type. Many devices, too, have yet to reach the limits of their development. A comparative indication of the various properties is attempted in Figures 4.28–4.33.

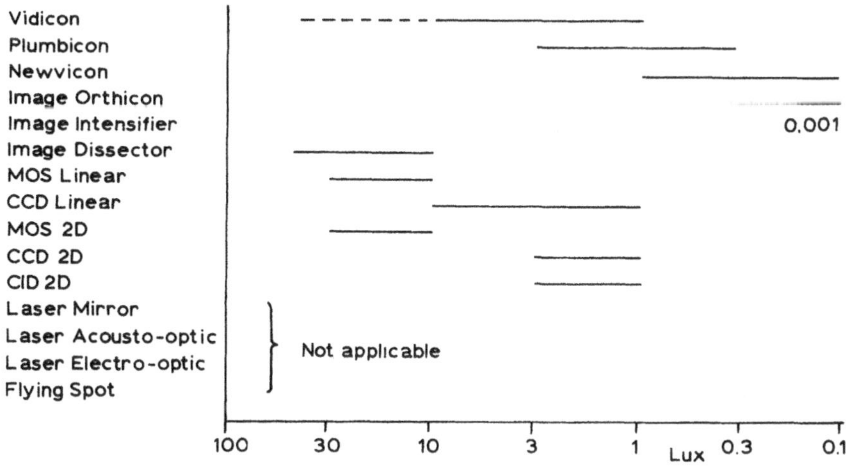

Figure 4.29. The sensitivity of scanning systems in terms of the illuminance required for a 25 Hz scan repetition rate.

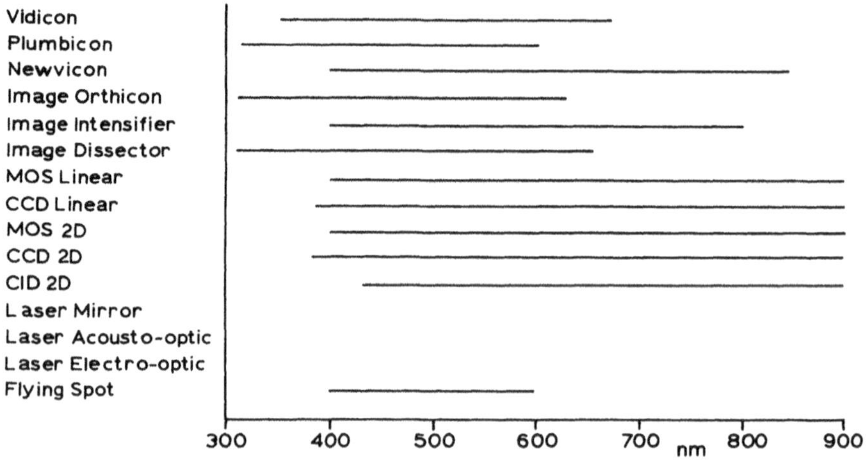

Figure 4.30. The optical bandwidth of scanners, indicating the range in which the sensitivity is greater than 40% of the peak sensitivity. In general, laser scanners are available at several wavelengths, in each case the bandwidth is narrow.

Figure 4.28 indicates the order of resolution to be expected. The limiting resolution criterion implies an MTF of the order of 20% (see Section 5.1.1.4). The sensitivity of the scanners is expressed as the light level on the sensor required to give a high-level signal when the repetition rate is 25 Hz. This aspect does not apply to scanners which

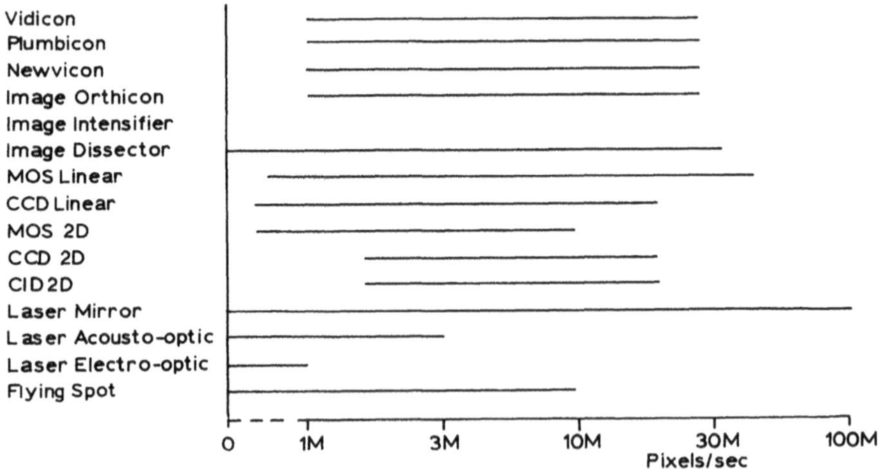

Figure 4.31. The scanning rate of scanners, the size of a pixel being the resolution limit.

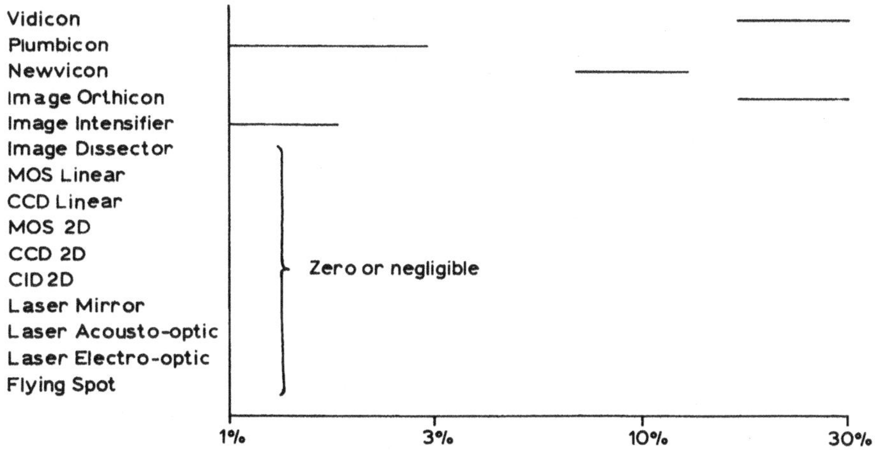

Figure 4.32. The decay lag of scanners expressed as the remanent image after 60 ms.

are their own light source. The optical bandwidth, Figure 4.30, applies to the basic sensor. For resolution and spatial noise reasons it may not be possible to use the infrared response of the solid state devices. The laser systems usually use narrow bandwidths but multiple wavelength systems are possible. The decay lag, Figure 4.32, of some devices may be increased at higher pixel rates if the speed does not allow the

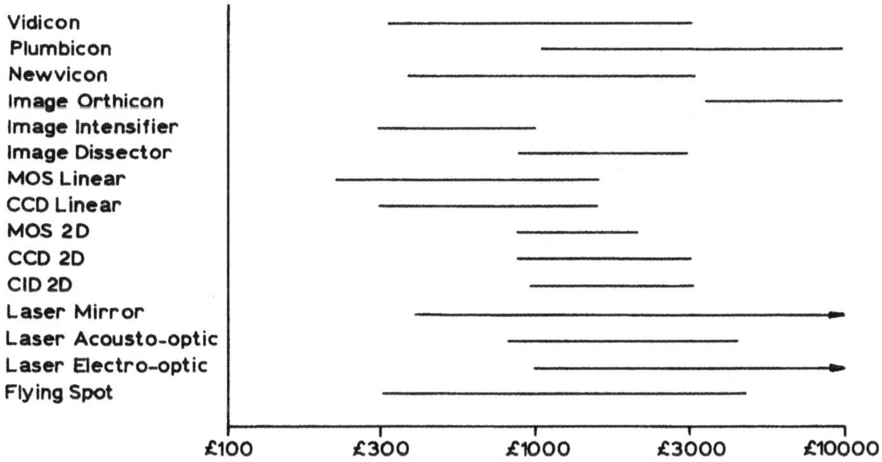

Figure 4.33. These figures give a rough indication of the cost of a basic scanning system (1985 prices).

complete resetting of the sensor element. Some solid state devices can exhibit lag under these conditions (see Section 4.3.2). The figures for cost, Figure 4.33, give only a very rough indication; the price will depend upon the overall quality of the system.

4.7. Color Scanners

The application of color sensitive scanning systems is beginning to receive attention (11.8.3).[51–54] The concept of color as related to human and to picture processing is discussed in Section 8.2.2. In the majority of color scanning systems, the same basic idea of three primary colors, red, green, and blue, is used. When the human does not have to be considered it is not necessary that these color bands be selected or that the division of the range into three should be made. There is greater freedom to select the ranges on the basis of the requirements of the colors present in the scene. If possible, the choice should be so that the maximum discrimination is obtained between the important parts of the scene.

It is not usually possible to apply the separation of the bands through the medium of the lighting, including the beam of a laser scanner. The sensor would have to have a bandwidth adequate to cover the total band, and separate pictures would be scanned for each of the bands, the lighting color being switched between the bands. The required lighting band could be selected by suitable filters and/or types of light source. The more usual process is to use a broadband light source and filters, associated with the sensors, to select the required color bands. The light may be divided into separate beams, by partially reflecting mirrors or prisms, and each filtered to pass the required band. This results in a significant light loss, more than 67% for a three-color system, and it is better to use dichroic mirrors (see Section 5.1.3.1) to separate the several colors.

This light separation system will be within the imaging optics of the scanning systems. It is most conveniently placed between the lenses and the sensors. In such a system the color splitting system and the sensors, one for each band, are arranged so that each sensor is in an imaging plane of the lens. It is then possible to change the focusing of the lens or to change lenses without producing a differential change in the focusing on each sensor. A typical system in a camera using electron beam scanning tubes is to place a complex prism between the lens and the tubes.[3] The light, having entered perpendicularly through one transparent face, falls on one dichroic mirror where one band is separated and then onto a second where the remaining two bands are separated. The three separated beams leave the prism and form images on the three tubes. The three exits from the prism are at the same optical distances from the entrance. The same system can be used with solid state cameras.

Single sensor systems can be made incorporating a tube or array.[3,19,55,56] For the latter, interference or absorption filters, appropriate to the bands required, are deposited as thin strips across the array, with each strip covering one column of sensors, for an area array. The output from the array alternates between values for the different color bands. Naturally, this results in some loss of spatial resolution in nonwhite scenes and, as resolution is already less in these arrays than is desirable, techniques are used to retain as much resolution as is possible. One method is to repeat the cycle "white–color 1–white–color 2." With the higher white content, more resolution is retained and adequate information is present to deduce the magnitudes of the components of a three-color system. In an alternative system, green, yellow, and cyan are used. Yellow covers the green and red sections of the spectrum and cyan the green and blue. Thus all of the primaries can be measured. Green is present in the output from each sensor and there is no loss of resolution in the green, the color for which the eye is most sensitive. Other systems use various chequer patterns of colors, with each photodetector having its own filter, to improve the resolution.

Because the position of the scan lines in a tube in not stable, the only arrangement possible is one in which the color bars are orthogonal to the scan lines. The video output alternates between the colors, and timing information is required to identify the colors. One method is to place black lines between each group of color bars.[11] The drop to black can be used to synchronize a clocking system used to route the color sections of the output.

Two sensor systems exist in which one receives the full light band and the other has two filters to provide the color information.

Color measurement systems use diode or photomultiplier sensors preceded by interference or absorption filters. Calibration involves setting the relative gains of the sensors. Ideally, a broadband light source giving uniform output across the band is required, but it is normally sufficient to use a source, such as an incandescent lamp, whose power output variations with wavelength are known.

References

1. A special issue on cameras, *RCA Review, 36*(3), 383–651, Sept. (1975).
2. L.M. Biberman, S. Nudelman, eds., *Photo Electronic Imaging Devices* (2 volumes), Plenum Press, New York (1971).
3. R.E. Flory, Image acquisition technology, *Proc. IEEE, 73*(4), 613–637, April (1985).
4. E.L. Dereniak and D.G. Crowe, *Optical Radiation Detectors,* John Wiley, New York (1984).
5. Resolution in camera tubes, Technical publication M80-0027, Mullard Ltd., London (1980).
6. *Mullard Technical Handbook, Book 2, Part 2a,* Camera tubes and image intensifiers, Mullard Ltd., London.
7. *Philips Data Handbook, Electron Tubes, Book T10,* Camera tubes and image intensifiers.
8. J.H.T. Roosmalen, A new concept for television camera tubes, *Philips Tech. Rev. 39*(8), 201–210 (1980).
9. Plumbicon camera tubes—recent developments, Technical publication M80-0028, Mullard Ltd., London (1980).
10. *CCIR XIIIth Plenary Assembly 1974 XI* (Report 624), Characteristics of television systems, International Telecommunication Union, Geneva (1975).
11. B. Grob, *Basic Television and Video Systems,* McGraw-Hill, New York (1984).
12. A.A.J. Franken, 2/3 inch Plumbicon tube with enhanced performance, *Electronic Components and Applications 1*(2), 71–76 (Feb. 1979).
13. R. G. Neuhauser, The Saticon photoconductor, *RCA Engineer 27*(3), 81–85, (May/June 1982).
14. Y. Talmi, ed., *Multichannel Image Detectors,* Vol. 2, American Chemical Soc., Washington, D.C. (1983).
15. L. Levi, *Applied Optics,* Vol. 2, John Wiley, New York (1980).
16. J. Mays, Precision image isocon TV camera, *Proc. SPIE 182, Imaging Applications for Automatic Industrial Inspection and Assembly,* 83–93 (April 1979).
17. K.N. Prettyjohns, ed., *Proc. SPIE 501, State-of-the-Art Imaging Arrays and their Application,* San Diego (Aug. 1984).
18. W.F. Kosonocky, Visible and infra-red solid-state sensors, IEEE Conf. Proc. Intl. Electron Devices Meeting, 1–7, Washington (Dec. 1983).
19. B. Kazen (ed.), *Advances in Image Pickup and Display, Volumes 2–6,* Academic Press, New York (1975–1985).
20. W.C. Dash and R. Newman, Intrinsic optical absorption in single-crystal germanium and silicon at 77°K and 300°K, *Physical Rev. 99*(4), 1151–1155 (Aug. 1955).
21. D. Lake, Solid state cameras, *Proc. 4th Intl. Conf. on Robot Vision and Sensory Controls,* 75–83, London (Oct. 1984).

22. F.O. Huck, C.L. Fales, S.K. Park, D.E. Speray, and M.O. Self, Application of information theory to the design of line-scan and sensor-array imaging systems, *Optics and Laser Technology 15*(1), 21–34 (Feb. 1983).
23. J.D.E. Benyon and D.R. Lamb, *Charge Coupled Devices and Their Application,* McGraw-Hill, New York (1980).
24. H-F. Tseng, G.P. Weckler and S. S. Li, Charge transfer and blooming suppression of a charge transfer photodiode area array, *IEEE Jnl. Solid-State Circuits SC-15*(2), 206–213 (Apr. 1980).
25. R. Melen and D. Buss, *Charge-Coupled Devices: Technology and Applications,* IEEE Press, New York (1977).
26. D.F. Barbe, ed., *Topics in Applied Physics,* Volume 38, Charge-Coupled Devices, Springer-Verlag, Berlin (1980).
27. N. Kadekodi, A. Claproth, T. Vo., A. Anyiwo, L. Sheu, and A. Ibrahim, A 5732-element 1.2″ linear CCD imager, IEEE Intl. Solid-State Circuits Conf., 36–37 (Feb. 1984).
28. G. Boucharlat, J. Chabbal, and J. Chautemps, 256 × 256 pixel CCD solid state image sensor, *Proc. 4th Intl. Conf. on Robot Vision and Sensory Controls,* 85–90, London (Oct. 1984).
29. U. Feddern and S. Zur Verth, The frame-transfer sensor, *Electronic Components and Applications, 6*(4), 223–229 (1984).
30. E. Oda, I. Akiyama, T. Kamata, Y. Ishihara, A. Kohno, K. Arai, and T. Kitagawa, A CCD image sensor with 768 × 490 pixels, *IEEE Intl. Solid-State Circuits Conf.,* 264–265 (Feb. 1983).
31. H. Heyns and J.G. van Santen, The resistive gate CCD area-image sensor, *IEEE Jnl. of Solid State Circuits SC-13*(1), 61–65 (Feb. 1978).
32. D.M. Brown, M. Ghezzo, and P.L. Sargent, High density CID imagers, *IEEE Trans. Electron Devices ED25*(2), 79–84 (Feb. 1978).
33. W. Donnelly, Low cost imaging system uses DRAM as photosensor, *Electronics Industry 9*(11), 9–11 (Nov. 1983).
34. D.G. Whitehead, I. Mitchell, and P.V. Mellor, A low-resolution vision sensor, *Jnl. Physics E (GB) 17*(8), 653–656 (Aug. 1984).
35. *Proc. SPIE Conf.,* Laser Scanning and Recording, *498,* San Diego (Aug. 1984).
36. J.D. Zook, Light beam deflector performance: a comparative analysis, *Applied Optics 13*(4), 875–887 (April 1974).
37. K. Pelsue, Precision, post-objective, two-axis, galvanometer scanning, *Proc. SPIE Conf., High Speed Read/Write Techniques for Advanced Printing and Data Handling, 390,* 70–78, Los Angeles (Jan. 1983).
38. W. Reimels, Low wobble resonant scanners, *Proc. SPIE Conf., High Speed Read/Write Techniques for Advanced Printing and Data Handling, 390,* 58–63, Los Angeles (Jan. 1983).
39. T.S. Fisli, Multifunction document processor, *Proc. SPIE Conf.,* Advances in Laser Scanning and Recording, *396,* 20–27, Geneva (April 1983).
40. R.N. West, Three laser scanning instruments for automatic surface inspection, *Proc. SPIE Conf., Advances in Laser Scanning and Recording, 396,* 102–110, Geneva (April 1983).

41. L.M. Hubby, Optical system design for a laser printing system, *Proc. SPIE Conf., High Speed Read/Write Techniques for Advanced Printing and Data Handling, 390,* 79–84, Los Angeles (Jan. 1983).
42. J.C. Urbach, T.S. Fisli, and G.K. Starkweather, Laser scanning for electronic printing, *Proc. IEEE 70*(6), 597–618 (June 1982).
43. *Proc. SPIE Conf., Advances in Laser Scanning and Recording, 396,* Geneva (April 1983).
44. A. Korpel, R. Adler, P. Desmares, and W. Watson, A television display using acoustic deflection and modulation of coherent light, *Proc. IEEE 54*(10), 1429–1437 (Oct. 1966).
45. H.G. Aas and R.K. Erf, Application of ultrasonic standing waves to the generation of optical-beam scanning, *J. Acoust. Soc. Am. 36*(10), 1906–1913 (Oct. 1964).
46. V.J. Fowler and J. Schlafer, A survey of laser beam deflection techniques, *Proc. IEEE 54*(10), 1437–1444 (Oct. 1966).
47. I.P. Kaminow, *An Introduction to Electro-Optic Devices,* Academic Press, New York (1974).
48. W. Kulcke, K. Kosanke, E. Max, M.A. Habegger, T.J. Harris, and H. Fleisher, Digital light deflectors, *Proc. IEEE 54*(10), 1419–1429 (Oct. 1966).
49. H. Meyer, D. Riekmann, K.P. Schmidt, U.J. Schmidt, M. Rahlff, E. Schroder, and W. Thust, Design and performance of a 20-stage digital light beam deflector, *Appl. Opt. 11*(8), 1732–1736 (Aug. 1972).
50. L.J. Pinson, Robot vision: an evaluation of imaging sensors, *Proc. SPIE Intl. Soc. Optical Engineering* (USA) *442,* 15–26 (1983).
51. M. Ueda, F. Matsuda, and S. Sako, Color sensing system for an industrial robot, *Proc. 10th Intl. Symp. on Industrial Robots and 5th Intl. Conf. on Industrial Robot Technology,* 153–162, Milan (March 1980).
52. M. J. Chen and D. L. Milgram, Binary color vision, *Proc. 2nd Intl. Conf. on Robot Vision and Sensory Controls,* 293–306, Stuttgart (Nov. 1982).
53. R.A. Jarvis, Expedient 3-D robot colour vision, *Proc. 2nd Intl. Conf. on Robot Vision and Sensory Controls,* 327–328, Stuttgart (Nov. 1982).
54. D.M. Connah and C.A. Fishbourne, The use of colour information in industrial scene analysis, *Proc. 1st Intl. Conf. on Robot Vision and Sensory Controls,* 340–347, Stratford-upon-Avon (April 1981).
55. K. Ishikawa, S. Hashimoto, Y. Sone, and T. Kunii, Color reproduction of a single chip color camera with a frame transfer CCD, *IEEE Jnl. of Solid State Circuits SC-16*(2), 101–103 (April 1981).
56. T. Watanabe, K. Hashiguchi, T. Yamano, and J-I Nakai, A CCD color signal separation IC for single-chip color imagers, *IEEE Jnl. Solid-State Circuits SC-19*(1), 49–54 (Feb. 1984).

5

Optics and Lighting

5.1. Optical Systems

In this section we introduce optical systems used in intelligent automation. Further information is available in the many standard books on optics.

5.1.1. Lenses

5.1.1.1. Symbols and Equations. The following list comprises symbols and equations relevant to our discussion of lenses:

F – The focal length of a lens system
D – The diameter of the effective lens aperture
M – The linear magnification ratio, image size to object – size
U,V – The distances of the object and image from the corresponding principal planes
N – The numerical aperture:

$$1/U + 1/V = 1/F$$
$$M = V/U$$
$$U = F(1 + 1/M), \quad V = F(1 + M)$$
$$F^2 = (U - F)(V - F)$$
$$N = F/D$$

Depth of field = $\pm N(1 + M)d \,/\, M^2(1 \mp d/MD)$ from perfect focus

Total depth of field = $2N(1 + M)d \,/\, M^2(1 - (d/MD)^2)$ where d is the maximum diameter of the circle of confusion

Required radiance of a surface = $4\,N^2(1 + M)^2 I/\pi$ assuming no lens losses, where I is the required illuminance at the image

5.1.1.2 Terminology. The following terms refer to lenses, but are also applicable to mirror systems and to mixed systems.

Optic axis—The axis of symmetry through a lens system is called the optic axis.

Circle of least confusion—Due to distortions in a lens system and diffraction at the edges of lens apertures, it is not possible to bring rays which have passed through a point on one side of a lens to a point on the other side. The rays will pass from convergence to divergence with the beam having a minimum diameter in between. This minimum point is called the circle of least confusion.

Focal plane—When a parallel beam of light enters a lens system the rays will converge to a point, actually a circle of least confusion, or will appear to diverge from a point after passing through the lens. The locus of this point, as the angle with respect to the optic axis is varied, is the focal plane. There is a focal plane for each direction of traverse through the lens. Due to distortion in the normal lens, it is necessary to consider only those rays which pass close to the center of the lens. Also, due to the lens distortions the plane may not be flat and the point may not be very small, in which case the circle of least confusion is used. The focal plane meets the optic axis at the focal point.

Principal plane—The locus of the point at which the paths of a ray entering a lens, parallel to the optic axis, and of the ray leaving the lens would intersect if projected, is the principal plane. There is a plane for each direction of traverse. In a thin lens the planes are roughly coincident with the lens. Generally, in composite lenses the planes are separated but each may or may not be the nearer to its corresponding focus.

Focal length—This is the distance between a focal point and the corresponding principal plane, measured along the optic axis. The focal lengths for the two directions are equal. Occasionally the power of a lens is given in diopters, and its value is equal to the reciprocal of the focal length measured in meters. The focal length can be positive or negative.

Aperture—The effective aperture of a lens system may be defined by the outer dimension of a component lens or by a fixed or variable iris in the lens system. It is usual to express the size of the aperture as the numerical aperture or the F number of the lens. It is equal to the focal length divided by the aperture diameter and is written F/N, e.g., *F*/4 for a numerical aperture of 4. The aperture size affects the light collecting power of the lens and also its resolution. These aspects are dealt with below. The term "stop" is often used with respect to a change of aperture size. A change of one stop changes the light transmission by a factor of two.

5.1.1.3. Aberrations. Most lenses are formed from spherical surfaces and these are not ideal, even for rays parallel to the optic axis (see Circle of least confusion, in Section 5.1.1.2). By using many surfaces of different curvatures and between materials of various refractive indices, an acceptable compromise is reached. The distortion effects increase as a power of the off-axis distance.

Astigmatism occurs when the power of the lens is different in two planes in which the light beam is travelling. This may be due to lens aberrations or to the inclusion of a cylindrical lens in the system. Lens systems usually exhibit some astigmatism for light beams travelling at an angle to the optic axis. The rays from a point source produce a pattern similar to that of Figure 5.1 in which two focused lines are formed at different distances from the lens. The lines will be in and orthogonal to the astigmatic plane of the system. The presence of astigmatism can be shown by viewing a grid pattern. The two sets of lines will be in focus at different distances from the lens.

The refractive index of a glass is a function of the wavelength of the light, i.e., the glass is dispersive. Almost all materials have a refractive index which decreases with increasing wavelength. By a choice of glasses for the components of a complex lens, the combined dispersion can be minimized and the resulting lens is achromatic. A lens is designed for a specific range of wavelengths and its performance

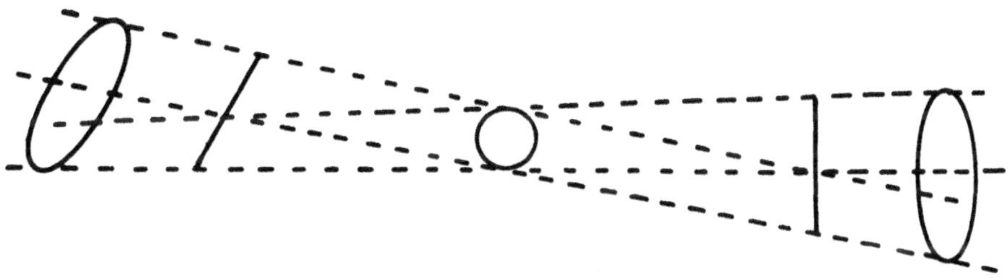

Figure 5.1. The beam shape at the "focus" of an astigmatic system, showing the two focal lines and the circle of least confusion.

may be less satisfactory outside that range. The compensation of various aberrations leads to a compromise and by designing for narrow optical band operation greater compensation can be given to the other aberrations.

Perfect imaging of an object is not possible. Not only does the imaging plane have a different shape, i.e., a flat object may not produce a flat image, but also the ratio of the distance of an image point from the optic axis to that of the object may be a function of that distance. This leads to barrel or pincushion distortion, so called from the shape of an image of a square object.

To minimize the effects of distortion it is necessary that lenses be used for the object and image distances, and within the maximum off-axis angle, for which they were designed. Most lenses are not symmetrical and are designed for the distance to be short on one side and long on the other. Provided that this is satisfied it is not important which is the object and which the image. Photographic and television camera lenses are examples of these. Copying lenses are designed for distance ratios nearer to unity. Lenses of one type can be combined to produce the effect of the other. For example, lenses designed for large imaging ratios can be placed with their "long distance" sides facing, to produce an imaging ratio near to unity. The penalty may be that the apertures of the lenses will interact to produce vignetting (See Section 5.1.1.6), and at least one lens should be used at full aperture. In the cases of simple lenses, the rule which gives the least distortion is that the bending of each of the light rays should be shared by the different surfaces, e.g., the short distance for a plano-convex lens is on the flat side.

5.1.1.4. Resolution. The resolution given by a lens is primarily a function of its aberrations, both geometric and chromatic. It is also affected by diffraction from the edges of the aperture, reflection at the lens surfaces, and scattering within the glass, on the surfaces, or in the lens housing. It is expressed completely in terms of the modulation transfer function (MTF) for each point in space with respect to the lens. When a grating having a sinewave variation of its transmission is imaged through an optical system, a similar sinewave grating image is produced. The ratio of the intensity modulation of the image to that of the object grating is related to the pitch of the grating. This relationship between the modulation depth ratio and the spatial frequency is the modulation transfer function and is equivalent to the frequency response of an electronic circuit. (The intensity modulation is equal to a half of the peak-to-peak variation divided by the mean level.)

The effective MTF of a system comprising an optical system, a sensor, and signal electronics can be calculated by convolving the MTFs of the component parts, i.e., taking the products of the values for each spatial frequency. For this purpose the equivalent MTF of the electronics is the ratio of its frequency response and the scanning rate. The MTFs of a series of optical systems cannot be combined in this way unless images formed between them are fixed in space by a diffusing screen. This is apparent when it is realized that a composite lens with a good MTF can be formed from components having less ideal MTFs, but having aberrations which are compensatory.

On the optic axis, the MTF is usually independent of the direction of the grating, but off-axis it will normally vary with the angle. The radial MTF corresponds to a grating with its lines parallel to the line from the optic axis to the test point, and the tangential MTF for a grating orthogonal to this. In consequence, the related resolutions are along lines orthogonal to these gratings. The MTF of a lens usually falls with an increase in the off-axis angle of the ray path and therefore it is inadvisable to use lenses with focal lengths which are unnecessarily short and result in large off-axis angles. If possible, this angle should be kept below 30°.

The shapes of the MTF curves are important. In the lens design a compromise is made between response at low frequencies and that at high frequencies. For example, in a television system it is usual for the electronics and transmission system to apply a limit to the highest frequencies handled. Therefore, there is no point in having spatial frequencies in the image that would result in electrical frequencies higher than that limit. The line scan pattern provides a similar limitation in the frame direction. Consequently, television lenses are designed to give the best performance up to the limit. On the other hand, a photographic emulsion is capable of recording very high spatial frequencies and it makes sense for the lens to handle such frequencies, although this will result in a performance at low frequencies which is less than that of a television lens. Typical MTF curves illustrating this are shown in Figure 5.2. Each lens will give a less satisfactory

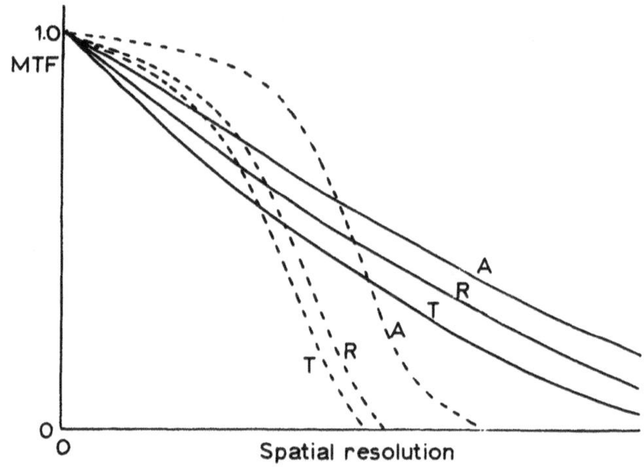

Figure 5.2. The resolutions of a typical photographic lens (solid lines) and a television lens (broken lines), on the optic axis (A) and for radial (R) and tangential (T) directions off-axis.

performance when used in the wrong application. For most vision systems the television lenses will be best, although they may be inadequate for high resolution systems.

The resolution of a lens may be improved by reducing its aperture, because most of the aberrations are caused by the use of sections of the lens remote from its axis, but there is a limit to the minimum aperture that can be used without again reducing the resolution. An aperture produces a diffraction pattern whose size is inversely proportional to the size of the aperture; thus a focused spot cannot be infinitely small but is a central diffuse spot surrounded by rings and is usually blurred by the lens aberrations. The angle subtended at the aperture by the radius of the first dark ring of the pattern is equal to $1.22\lambda/D$, where λ is the wavelength of the light and D the diameter of the aperture.

In the photographic criterion for resolution, or when viewing by eye, two points are just resolvable when the central peak of the diffraction pattern of one image falls on the first dark ring of the other. This gives an MTF of 20% for a pitch of $1.22\lambda/D$ radians. In the image plane of the lens this corresponds to a grating of $1500/N(1 + M)$ cycles/mm for a wavelength of 560 nm. The figure will be less if a greater modulation depth is required in the image. By comparison, the spatial frequency, measured near the axis in the focal plane, resulting from the aberrations in a well designed lens, should be between 3000 and 15,000 cycles/focal length, for an MTF of 20%.

5.1.1.5. Depth of Field. Away from the image point, the circle of confusion becomes larger and is a section of a double cone with its base at the effective aperture of the lens. The depth of field applies to the object space and is the range through which the object can move while producing at a fixed imaging plane, e.g., the position of a sensor, a circle of confusion whose diameter is less than some specified value, d (Figure 5.3). The limits of this range are not symmetrical about the position giving the sharpest image, the two distances being given by

$$\pm N(1 + M)d \,/\, M^2(1 \mp d/MD)$$

with the positive direction being away from the lens. For small apertures, diffraction will increase the spot size and the acceptable radius of the first dark ring will be related to d, e.g., by the constant k, and equal to:

$$kd = 1.22\lambda V/D = 1.22\lambda N(1 + M)$$

For the maximum value of N so given, the depth of field becomes

$$\pm k(d/M)^2/1.22\lambda(1 \mp d/MD)$$

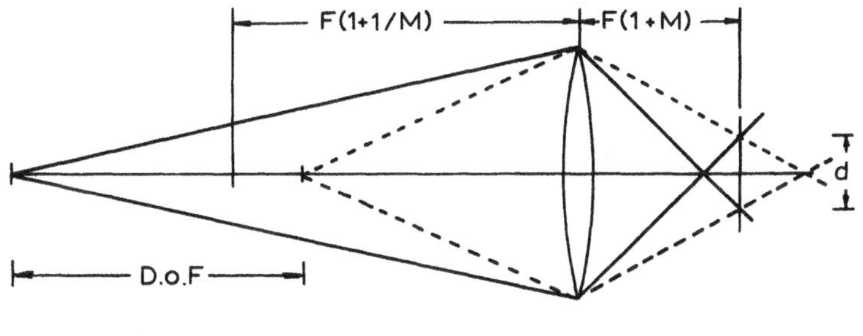

Figure 5.3. The figure for calculating the depth of field of a lens system. The image plane is shown to the right of the lens and the corresponding position for an object to be in focus, to the left.

where d/M is the equivalent size of the circle of confusion in the object field and its maximum acceptable value will be related to the pixel size in the object field. To give a similar loss of resolution to that without diffraction, a slightly smaller value of d will have to be selected. When D is large with respect to d/M, and hence the pixel size in the object space, the depth of field is proportional to $(d/M)^2$ and is independent of the camera system, except in wavelength.

Similar considerations apply to the image region for a fixed object, the permissible variation being called the depth of focus.

The maximum acceptable diameter for the circle of confusion depends upon the sensor and the application. In photography two criteria are commonly used: that the diameter should be less than 1/1000th either of the focal length of the lens or of the width of the picture. Other systems specify the 50% MTF or the limiting resolution, i.e., 20% MTF. In systems with solid-state sensors the decision is dependent upon the sensor and the subsequent processing of the video. It has been found that there is only a slight degradation of the image when the maximum diameter of the circle is equal to the pitch of the array.

5.1.1.6. Light Collection. The light forming a point of the image is that leaving the corresponding point of the object within a cone defined by the effective aperture of the lens. Considering the simple case of an object on or near the optic axis and radiating J watts/m^2 per steradian in the direction of the lens, i.e., a flux density of J watts from a square meter of surface into a solid angle of a steradian, the light collected by the lens is, if D is small with respect to U,

$$J\pi D^2 / 4U^2$$

$$\text{i.e.,} \quad J\pi D^2 M^2 / 4F^2(1 + M)^2 \quad \text{watts}/m^2$$

The error is less than 1% if D is less than a quarter of U. The irradiance of the image is this value divided by M^2:

$$J\pi / 4N^2(1 + M)^2 \quad \text{watts}/m^2$$

Taking the system in reverse, if the irradiance of the image is to be I watts/m^2, the required flux from the object must be

$$4IN^2(1 + M)^2/\pi \qquad \text{watts}/m^2/\text{steradian}$$

This value together with the characteristics of the surface of the object enables the required irradiance of the object to be calculated. If the object has a Lambertian surface with no absorption (see Section 5.2.1), which is receiving Q watts/m^2, then $J = Q/\pi$. The irradiance of the image becomes

$$Q / 4N^2(1 + M)^2 \qquad \text{watts}/m^2$$

and the required irradiance of the object is

$$4IN^2(1 + M)^2 \qquad \text{watts}/m^2$$

The above assumes that there is no light loss in the lens. In practice there will be a reduction of the light by a factor K, about 0.8 for a normal compound lens, and the required flux from the object becomes

$$4IN^2(1 + M)^2/\pi K \qquad \text{watts}/m^2/\text{steradian}$$

and the irradiance of a Lambertian object becomes

$$4IN^2(1 + M)^2/K \qquad \text{watts}/m^2$$

The transmission of a lens is affected by absorption in the glass, reflection from the refracting surfaces, and scattering from particles within the glass or at the surfaces, e.g., by dust. Reflection at the refracting surfaces is reduced significantly by the process called blooming or coating. In the simple form of this process a layer is placed on a surface which divides media of different refractive indices. The refractive index of the layer is equal to the geometric mean of these refractive indices and the thickness of the layer is equal to a quarter of the wavelength at the center of the spectral band. Reflections from the two surfaces of the layer cancel because they are equal and in antiphase. Common coating materials are magnesium fluoride and silicon oxide. The light loss due to surface reflection, and also to absorption, may be greater for a lens used outside its design spectral range.

These formulae apply close to the optic axis. For points away from the axis there are several causes of a reduction of the light flux. Let A be the angle subtended by the off-axis distance at the effective aperture of the lens.

1. From the off-axis point the aperture appears to be elliptical and its area is reduced by $\cos A$.
2. The distance of the object from the lens is increased by $\sec A$ and the area of the aperture subtended at the object is reduced by $\cos^2 A$.
3. The light falls on the image plane at an angle and consequently covers a larger area. This applies another cosine factor derived from the angle of the ray on the image side of the lens. In general this angle is close to A and it can be taken as another factor $\cos A$.

This gives a combined reduction factor of $\cos^4 A$. For a point which is off-axis by a distance equal to half the distance from the effective aperture, A is arc tan 0.5 and $\cos^4 A$ is 0.64; thus the light is reduced to less than two thirds of the axial flux.

The above applies when there is a single effective aperture. In a compound lens there are many other apertures and for large off-axis angles the edges of these apertures may block the light from the effective aperture. This produces a more rapid reduction in the light level and is called vignetting. (Vignetting is a photographic technique in which the edge of a portrait is faded by printing through a second aperture held in front of the enlarging lens. The name comes from a related technique in which vine leaves (vigne) were printed around the portrait.)

5.1.1.7. Focus and Zoom Controls. Often parts of a lens system can be moved with respect to the others. By this means, the focal length of a lens or the positions of its principle planes may be changed without moving the bulk of the lens. This may be used to bring an image into focus. In a zoom system the position of the image is not changed but the focal length, and hence the magnification, is varied. Zoom ranges of 10 to 1 are typical. A zoom system will have an independent focusing control. Naturally, lenses with these facilities involve compromises and will have a lower performance/cost ratio than that of a fixed lens.

5.1.1.8. Lens Measurements. The following applies to positive lenses. Negative lenses produce virtual images from real objects, i.e., the image is beyond the surface of the optical system as it is viewed. For these it is necessary to use a supplementary positive lens, whose properties are known, to produce a real, and therefore accessible, image.

To find the position of one of the focal planes at the optic axis, place a plane mirror against the opposite side of the lens. Place a pin, a grid, or a small light source in the region of the focal plane, near to the axis, and, viewing along the axis, compare the object with the reflected image. The image will be inverted. Move the object along the axis until there is no parallax. The object is now in the focal plane. This process can be repeated with the lens reversed.

The product of the distances of the object and image from the focal planes on their side of the lens, measured away from the lens, is equal to the square of the focal length:

$$(U - F)(V - F) = F^2$$

To find the focal length of a positive lens, place the object, as above, on one side of the lens and find the position of the image by observing the lack of parallax with a similar object on the image side. The focal length can be calculated from the distances from the known focal planes.

The principal planes are at a distance of a focal length from the focal points measured towards the lens.

5.1.1.9. Lens Construction. Most lenses are molded from glass, and ground and polished to the required profile. The latter processes require that the surfaces be spherical. The quality of the bulk material is important. Doublets and triplets are formed by gluing together simple lenses having complimentary profiles on the mating surfaces, but made of materials having different refractive indices and dispersion so that geometric and chromatic aberrations can be minimized. Lens systems are formed from these compound lenses or simple lenses, with air between them, and aperture stops suitably positioned. The surfaces of the elements may be coated to reduce reflections which can degrade the image. Many lens units are provided with standard threads or mounting units to match the fitting on the camera. For example, the "C" mount is used for most of the medium range of television cameras and maintains the correct distance to the sensor when lenses are changed.

Materials other than glass can be used. Plastics, such as polycarbonate, can be molded to give a good surface finish and high quality bulk material. Plastic lenses are lighter and generally cheaper than glass lenses. Molded lenses can be made with aspheric surfaces by which geometric aberrations can be corrected. It is difficult to produce aspheric surfaces when grinding and polishing are involved. Molded lenses are of adequate quality for light collection and direction where imaging is not required. If the glass of the lenses will withstand high temperatures they may be used close to high-power lamps.

5.1.1.10 Fresnel Lenses. There is a requirement for large lenses in lighting systems and in light collection, in which a lower quality of performance is acceptable. If constructed conventionally they are heavy and may produce significant light loss in the thick glass nearer the optic axis. For light travelling parallel to the optic axis inside the lens the light is deflected by the surfaces at the same distance from the axis and the thickness of the lens serves no useful purpose. In the Fresnel lens, the bulk of the glass is removed and the surface shape retained (Figure 5.4). These lenses can be used only for a limited range of object-image positions about those for which they were designed, and near to the axis. For other positions the rays in the lens will not be nearly parallel to the axis and some will be affected by the discontinuities in the surface resulting in a deterioration of the image. That shown in Figure 5.4 is designed for use between infinity and the focal point. Double-sided and negative lenses can be produced. The lenses are produced by molding and can be made from glass or plastic.

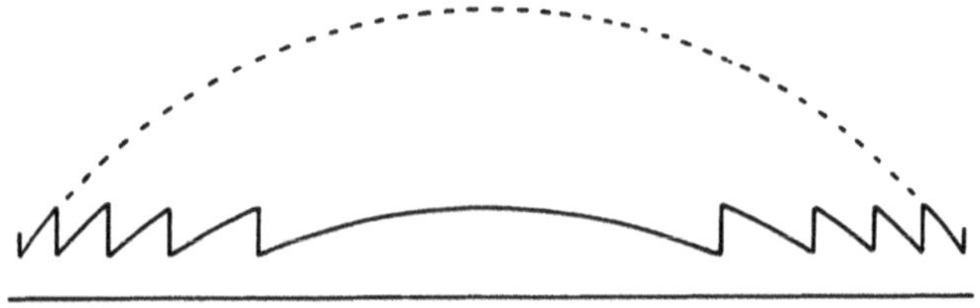

Figure 5.4. A Fresnel lens, equivalent to a plano-convex lens.

5.1.1.11. Cylindrical Lenses. In some systems it is necessary that the lens be effective in one plane only; for these cylindrical lenses are used. They introduce a controlled degree of astigmatism into the system either to correct for existing astigmatism or to spread a point object into a line image.

5.1.2. Mirrors

Other than their use in scanning systems, plane mirrors are used to fold optical systems, in order to change the general direction of the light beam. The reflecting surface may be made thin so that a proportion of the light passes through and the light is split into two beams. Cylindrical, parabolic, and elliptical mirrors, active in one plane, serve a similar function to that of the cylindrical lenses, and spherical, paraboloidal, and ellipsoidal mirrors are effective in all three dimensions.(1)

The active surface of most mirrors is metal which may be backed by bulk metal of the same or different material, or by glass or plastic. For lighting applications the reflecting surface may be on the back of a transparent sheet which provides protection for the surface. In other cases the surface is protected from contamination, and reduced reflectivity, by a thin transparent layer. For typical reflecting surfaces the reflectivity at 300 nm is 30%–80%, at 550 nm, 50%–95%, and at 1000 nm, 70%–97%, thus the reflectivity increases towards the infrared.

Because the light does not pass through bulk material it is not subjected to chromatic aberrations, but at the same time the mirror loses the advantage of the use of different refractive indices for the correction of spherical or similar aberrations. Good image formation occurs only for the object and image at the foci of an ellipsoidal mirror, becoming the same point for a spherical mirror, and at the focus and infinity for a paraboloidal mirror. The aberrations increase as the aperture of the mirror is increased. The imaging of mirrors for other object and image positions is improved by using aspheric surfaces and by combining the mirrors with correcting lenses.

The reflecting surface of some mirrors is an interference filter formed on a transparent base (see Section 5.1.3.1). These reflect within a limited wavelength band or bands and transmit the light of other wavelengths. A common application for these is as "hot" or "cold" mirrors to separate the visible radiation from the heat of the infrared.

5.1.3. Filters

5.1.3.1. Spectral Filters Spectral filters affect a light beam by changing the relative proportions of the energy at different wavelengths in the beams transmitted or reflected by the filter. The variation may be gradual, as in a compensating filter used to correct the spectral response of a light source or sensor, or sharp, as in band-pass, band-stop, high-pass, and low-pass filters. The latter can be used in the sensor for separation of white light into three color bands for color systems or for the reduction of the infrared component in solid-state cameras.

Filters operate by the processes of absorption or interference.(2,3) Absorption filters are produced by dissolving dyes in glass, plastics, or gelatine or by dissolving metal oxides or suspending colloids in glass. The light which is not transmitted is converted to heat. This heat is usually small, and can be ignored, when the filters are used in the sensors of the system, but if they are used where the light flux is high, the effect of the heat upon the filter and its surroundings must be considered. Cutoff gradients of 10 to 1 in 50 nm are common; 10 to 1 in 20 nm can be achieved.

Interference filters are capable of much sharper cutoffs. Basically these are thin dielectric films on a glass support, with partially reflecting metal films on each side of the dielectric. Interference occurs between beams reflected from the metal films and, depending upon the wavelength and the direction of the emerging beam, may reinforce or reduce the intensity. As a result, light in one part of the spetrum may be transmitted while that in the rest is reflected. The required filter characteristics are obtained by using multiple layers of dielectric of the appropriate thicknesses. By using a combination of dielectrics it is possible to construct filters without the metal reflecting layers, the reflections at the layer boundaries being produced by the change in refractive index. This avoids the light loss produced by the metal layers. Filters which split the beam into a long wave component and a short wave component, without absorption, are known as dichroic filters.

The absorption in interference filters is much less than that in absorption filters, so heating of the filter is less of a problem. When only one of the resulting beams is to be used it should be remembered that the other may contain a large amount of energy and this will be converted to heat wherever that beam goes. Because the filters use interference, the cutoff wavelengths are dependent upon the angle of incidence of the light, the wavelength increasing as the angle from the normal increases. Consequently, these filters are specified for use at particular angles, e.g., normal or 45°. In an optical system the angle of incidence of the various beams may be different and the resulting filtering effect will not be the same for different parts of an image. Another consequence of the interference process is that the cutoff edges will repeat at harmonic wavelengths. Often this is not a problem since this occurs outside the working range of the system, but when this is not so, a combination of filters may be necessary. The interference filter may be combined with an absorption filter, in which case the absorption filter should be on the side away from the light source if the light flux is high.

Two interference filters can be used to split white light into the three bands for a color system. One beam from the first filter is split by the second and the resulting three beams from the two filters pass to the three sensors.

5.1.3.2. Neutral Filters and Diffusers. Neutral filters are intended to produce a uniform reduction of the light flux across their operating band. They can be absorption filters or partially reflecting mirrors, the latter producing two beams without a significant light loss, i.e., it is a beam splitter.

Greater uniformity in the illumination of a scene or in the collection of light can be obtained by the use of diffusing screens. They may be translucent material such as plastic, or opal or ground glass, or reflecting surfaces such as white paint or a mirror with a patterned surface which scatters the light. The use of diffusers will normally result in the loss of light due to absorption in the diffuser or by the scattering of light outside the working space, and therefore the degree of diffusion should be kept to the minimum necessary for the required effect.

5.1.3.3. Polarizing and Birefringent Filters. Normally, light consists of rays, the electromagnetic field directions of which are randomly oriented. Light can be polarized, i.e., one of these directions selected, by passing the beam through birefringent crystals, by reflection at the appropriate angle from suitable surfaces, and by transmission through dichroic polarizing sheets.[2] (This use of dichroic should not be confused with that used in "dichroic mirrors," which is an entirely different process.) These sheets consist of large aligned molecules supported by a plastic or a glass sheet. For each plane of polarization in the incident light, the energy can be considered as divided between the polarizing direction of the polarizer and a direction orthogonal to this. The division of the energy will be proportional to the cosines of the angles between the original plane and these two directions. If the original beam has random polarization, the total energy of the beam will be divided equally between the two orthogonal planes. Except for the birefringent crystals which operate by separating the two component beams, the component orthogonal to the plane of the polarizer is absorbed, producing heat. The removal of the unwanted component is not usually complete and those filters giving the greater degree of polarization also absorb more of the desired component. The transmission factors for both components are functions of the wavelength and the filters are designed for use over limited bandwidths. Typically the transmission in the dichroic sheets of the in-line component is 70% and the orthogonal component is 0.01%. The sheets can be used up to 80 °C but above this temperature the material will deteriorate. It can also be affected by strong ultraviolet radiation.

In birefringent filters the velocity of a beam polarized in a specific direction, with respect to the axis of the filter, differs from that polarized orthogonally. Consequently, one component will suffer a different phase shift from the other in passing through the filter. The phase shift is a function of the wavelength. A plane-polarized beam at an angle to the axis of the birefringent filter can be considered as having component parts along and orthogonal to the axis. The components will emerge out of phase and when recombined will form, in general, elliptically polarized light. Circularly polarized light is produced if the original plane had been at 45° and the phase shift, 90°. A filter with this phase shift is called a quarter wave plate, and when combined with and following a linear polarizing filter, it becomes a circular polarizer.

5.1.4. Light Guides and Fiber Optics

Light can be conducted through a medium which has highly reflecting boundaries so that it is transferred in a controlled way from one place to another. The light can be spread out or collected together as required. Light will be lost at each reflection and there will be attenuation within the medium, although quartz and some plastics have very low attenuation factors. It is necessary to design the guides to minimize the number of reflections for each ray and the length of its path in the medium, as it passes from the input to the output. The reflecting surfaces may be produced by depositing aluminium or silver onto the surface of the guide. For these surfaces the main loss is in the reflections and this can be reduced significantly if the phenomenon of total internal reflection is used.

The equation for refraction at a smooth surface between media of different refractive indices, μ_a and μ_b, is

$$\mu_a \sin A = \mu_b \sin B$$

where A and B are the angles between the normal at the surface and the rays in the two media. There will be a third ray due to specular reflection at the surface, but usually its intensity is small. When the ray begins in the denser medium, μ_a, and the angle is greater than that given by

$$\sin A = \mu_b / \mu_a$$

the equation cannot apply. There is no transmission through the surface and all of the energy is reflected, i.e., total internal reflection occurs. This is a very efficient process. To obtain total internal reflection, the guide has to be designed so that the angles of incidence at surfaces are always greater than that given above.

Total internal reflection may not occur if the surface is contaminated or rough, and this is likely to happen when the external medium is air. The surface can be protected by coating it with a thin layer of low refractive index material, although this will increase the minimum angle of incidence (Figure 5.5). The thickness of the layer need be no more than a few wavelengths. This is the basis of the fiber-optic light guide.(2-4) A light-carrying core of high refractive index glass or plastic is surrounded by a cladding of a similar low refractive index material. Circular, square, and hexagonal cross-sections are common. The term "fiber" is restricted to those guides having a cross-section which is small with respect to its length, and this ratio and the material

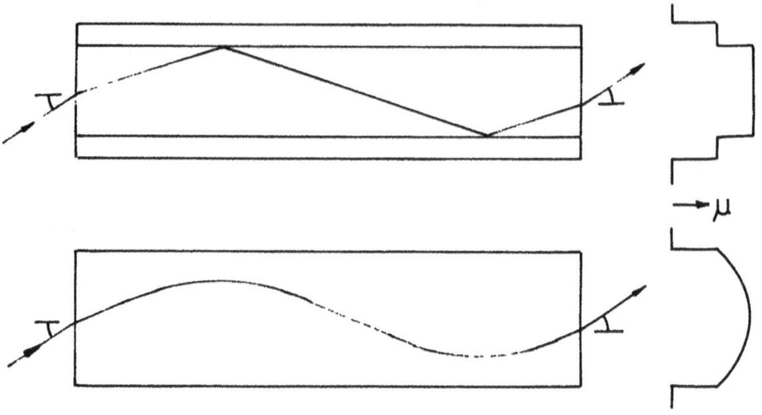

Figure 5.5. The passage of a light ray through a clad fiber and a graded index fiber. Also shown are the refractive index variations.

used determine the flexibility of the fiber. Fibers have been made with diameters of less than 5 μm. The angles at which the rays leave the end of the fiber are determined by the refractive indices of the core and cladding materials, and of the external medium, which for air is close to 1.0. The emerging beam is in the form of a solid cone. Similarly, the fiber can receive light only within that angle.

$$\text{The numerical aperture} = \sin(\text{half cone angle}) = (\text{core index}^2 - \text{cladding index}^2)^{\frac{1}{2}}/(\text{external medium index})$$

Fibers can be used individually or in bundles. The ends of the fibers in a bundle can have the same relative positions at both ends of the bundle. This is called a coherent bundle and a pattern projected onto one end of the bundle will be reproduced at the other. This is a useful device for getting a picture from a rather inaccessible position to a large camera.[(5)] The resolution is determined by the number of fibers in the bundle. Incoherent bundles are used to pipe light into areas to be scanned when the lamp may be too large or too hot. Often the ends of the fibers are deliberately randomized between the ends of the bundle so that variations in the light pattern from the lamp are not transferred to the scene. Various flexible or rigid guides are possible with different input and output shapes, e.g., line to disk, with splitting, e.g., one input to several outputs, and tapering, i.e., different sizes at the ends. A system with two ends feeding to one can be used to feed light into scene and to return, along the other guide, the reflected light. This return guide can be coherent so that a picture is obtained. Another application is the fiber-optic plate in which parallel fibers are fused together, using the cladding material, to form a solid plate. A typical application for these is in image intensifiers. Fibers have been made with no sharp transition between the core and the cladding but with an index which falls continuously with increasing radius; these are known as graded index fibers. The index gradient retains the light in the center of the fiber (Figure 5.5).

5.1.5. *Optical Systems*

A problem in vision systems is that the apparent size of an object can depend on its distance from the scanner. This may complicate a recognition system when the distance to the object is not known. The parallel optics system of Figure 5.6 solves this problem.[6] A large lens is placed in front of the camera so that the aperture of the camera lens, which is set to be small, is in the focal plane of the large lens. The only rays which can reach the camera are those which travel almost parallel to the optic axis of the system. For any value of u the position where the center of the bundle of rays from the top of the object strikes the sensor surface is the same. Thus the size of the image on the sensor is constant for all positive values of u. The value of v is not constant and the area of the bundle of rays at the sensor depends upon v and hence u. The image is, in general, out of focus but because

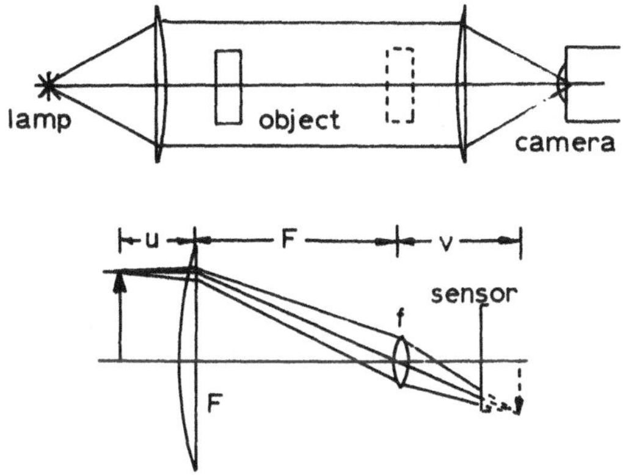

Figure 5.6. The parallel optical system. Lens f is the camera lens. $1/v = 1/f + (u - F)/F^2$. The actual imaging ration is $-v/F$, but the apparent ratio at the sensor is $(-v/F).(f/v) = -f/F$, i.e., a constant.

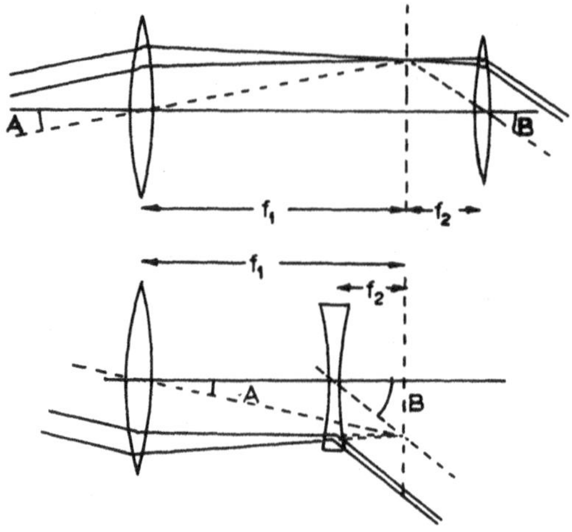

Figure 5.7. A telescopic lens system for the magnification of deflection angles ($\tan B/\tan A = f_1/f_2$).

the camera lens aperture is small this is not significant. To minimize the effect, the camera lens to sensor distance is set so that the sensor is in the center of the range of image positions. This arrangement is known as a telecentric system. Only rays parallel to the axis are used and if back lighting is required, an efficient system can be produced by placing a small lamp at the focus of another large lens. Fresnel lenses are very suitable for this application.

The telescopic, or afocal, system can be used to change the angles of beams in deflection systems. In its simplest form it consists of two lenses having a common focal plane, i.e., separated by the sum of their focal lengths (Figure 5.7). This focal plane will be between the lenses if both are positive, but beyond the negative lens if one is negative. A parallel beam entering the system reemerges at a different angle and with a different cross-section. For small angles, the ratio of the beamwidths is proportional to the ratio of the focal lengths, and the ratio of the angles is inversely proportional.

5.2. Light

In this section we will discuss the use of light in intelligent automation. Further information is available in the many standard books on light.[2,7]

5.2.1. Units

The output of a light source is usually expressed in candelas, lumens, or watts, and, as the output will vary with direction, will refer to a particular direction. Polar diagrams are used to illustrate the variation of output with direction. Until recently the candela was defined as the luminous intensity, viewed perpendicularly, of 1.667 mm^2 of a "black body," i.e., a perfect emitter, at the freezing point of platinum. This has been replaced by the luminous intensity, in a given direction, of a source emitting 1/683 watts/steradian at 555 nm. Fortunately, calibrated incandescent lamps are available which are related to these standards. A 12-volt, 48-watt car tungsten lamp overrun at 4.3 amps has a color temperature near 2856 °K and has an output of about 80 cd in a direction perpendicular to the filament.

The *lumen* is the luminous flux in a steradian from a point source of one candela (cd.sr).

The *radiant* and *luminous exitance*, or *emittance*, are the output from unit area of a surface, measured in watts or lumens, per unit area, (watts/m^2 or lumens/m^2). The lumen refers to the output integrated over a specified spectral band, with varying weighting, to correspond to the response of the human eye. The watt is an absolute unit which can be applied to a particular wavelength or for the integral of the power within a specified band.

The *emissivity* of a thermal radiator is the ratio of its radiant exitance compared to that of a black body at the same temperature.

The *irradiance* and *illuminance* of a surface are the flux incident on unit area of the surface (watts/m^2 and lumens/m^2 (= lux) respectively). An indication of lighting levels is given in Figure 5.8.

The *luminance* of a surface, in a given direction, is the luminous intensity radiated per unit area of that surface (cd/m^2 or lumen/sr/m^2).

The *radiance* is the radiant intensity for a unit area of the source (watts/sr/m^2).

The *radiant intensity* of a source, in a given direction, is the flux from the source per steradian (watts/sr).

The luminance of a surface will usually vary with the angle from the normal to the surface. If it is proportional to the cosine of the angle to the normal then it is a Lambertian surface.[8] Most matt surfaces are a good approximation of this. An interesting property of a Lam-

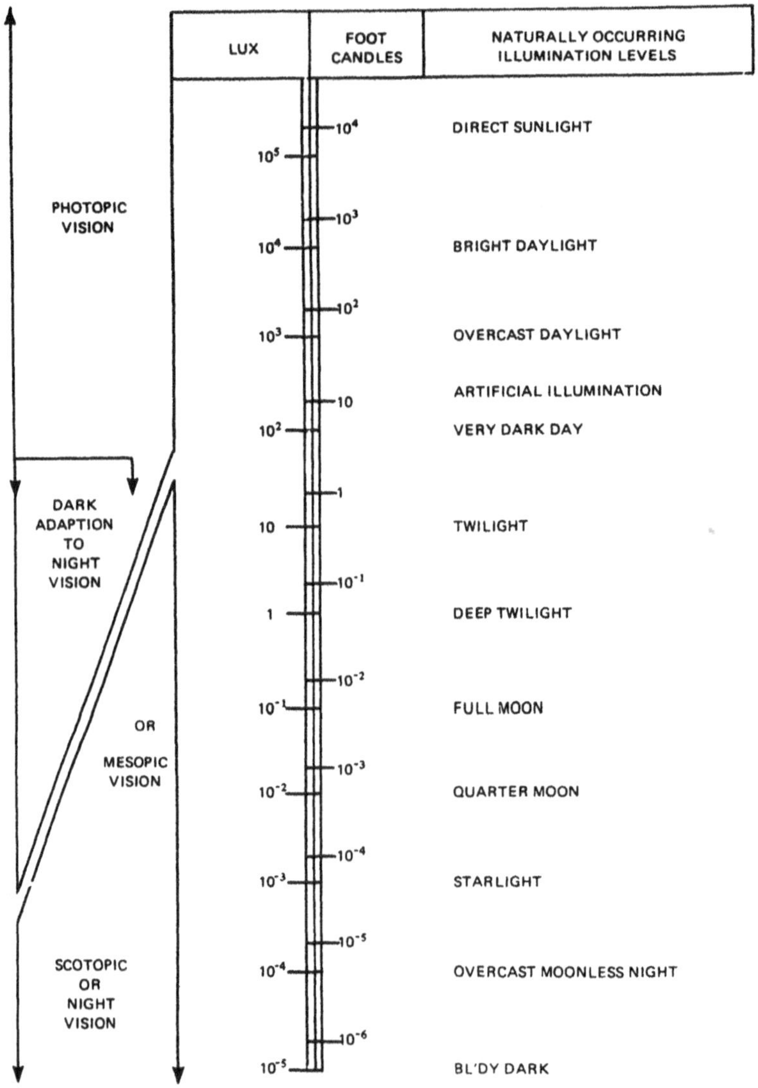

Figure 5.8. An indication of light levels. (Courtesy of Mullard Ltd.)

bertian surface is that its apparent luminance, as in a two-dimensional view, is independent of the angle from which it is viewed. For a fixed solid angle of view, the area in view increases as the secant which compensates for the cosine reduction in output. The light flux (radiance or luminance), along the normal, from a totally reflecting Lambertian surface, is $1/\pi$ watts/sr/m^2 (of surface) for an irradiance of 1 watt/m^2, or $1/\pi$ lumens/sr/m^2 (of surface) for an illuminance of 1 lux.

5.2.2. Color

The distribution of the output from an incandescent lamp is similar to that from a black body, a source whose output is characterized only by its temperature and not by its material or surface. There is an output/wavelength curve for each black body temperature, as shown in Figure 5.9[2,8] The curves are defined by Planck's radiation law. The energy density per meter waveband is

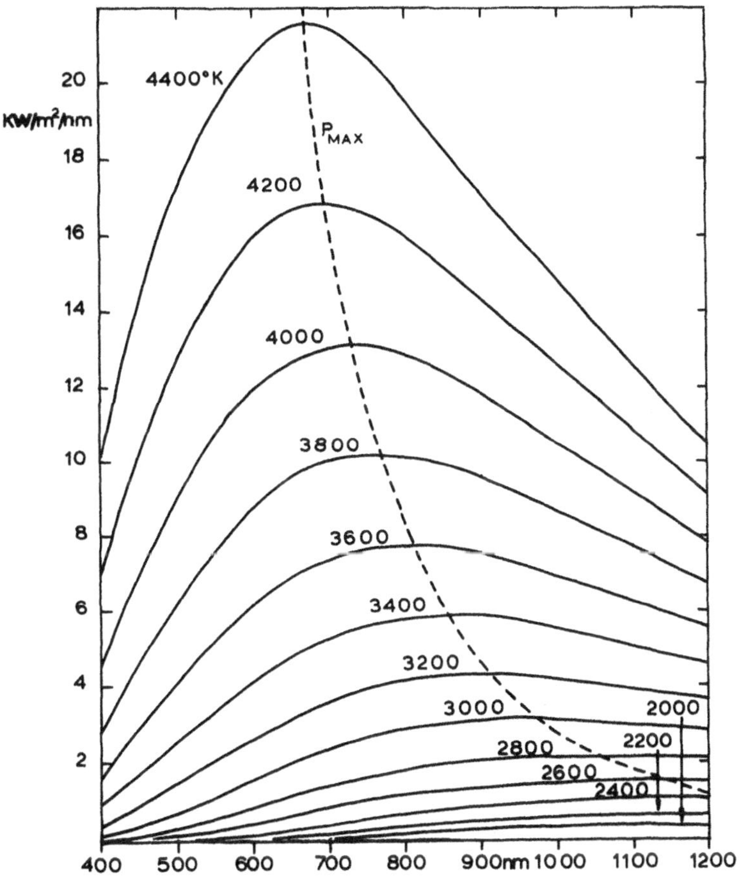

Figure 5.9. Black body radiation. The power density in a 1-nm bandwidth is shown for various black body temperatures.

$$A \; / \; \lambda^5(\exp(B/\lambda T) - 1) \text{ watts/m}^3$$

$$\text{where} \quad A = 3.7415.10^{-16} \text{ watt·m}^2$$

$$\text{and} \quad B = 1.43879.10^{-2}\, m\text{·°K}$$

The peak wavelength is inversely proportional to the absolute temperature (Wien's Displacement Law) and the peak energy is proportional to the fifth power of the temperature. The total energy, by the Stefan-Boltzmann law, is proportional to the fourth power,

$$E = 5.6697 \text{ T}^4\text{·}10^{-8} \text{ °K·watts/m}^2$$

The temperature corresponding to the curve most closely fitting, in the visible range, that of a given lamp, at a given current, is called the color temperature of that lamp. It can also be applied to other light sources and to illuminated surfaces. When incandescent lamps are used for photometric purposes, e.g., as standard light sources, it is normal for them to be used at 2856 °K, although sometimes 2870 °K is used.

Other light sources have spectra which, in general, are a combination of line spectra and a continuum. The line spectra are the result of emissions from excited atoms, in discharges or phosphors, and can be spread or split by gas pressure or magnetic or electrostatic effects. The continuum can be produced by thermal effects and by phosphors. The effects of color in the light source, the objects to be scanned, and the sensitivity of the detector need careful consideration. These affect the overall signal-to-noise ratio of the system. Also, they can affect the contrast obtained in the scene and the correct choice of colors can be advantageous.

5.2.3. *Light Sources*

Light can be generated by many processes including incandescence, excitation of atoms by conduction in solids and gases, and excitation of atoms by irradiation with electrons and electromagnetic radiation.[7,9,10]

When choosing a light source for a specific application several aspects need consideration:

1. The light output/wavelength as related to the sensitivity/wavelength of the sensor to be used. For example, mercury vapor lamps are relatively inefficient when used with silicon sensors because most of the output has a short wavelength, whereas the sensor is most sensitive in the near infrared.
2. The effect of colors used in the scene. Colors can be detected only when the appropriate wavelengths exist in the lighting.
3. Obtaining the required light distribution within the scene.
4. Removal of heat.
5. The presence of high voltages, the generation of electrical or magnetic interference, the production of dangerous radiation, e.g., ultraviolet light, the generation of ozone, by ultraviolet light or high voltages, or the risk of explosion.
6. The stability of the output and ripple on the output.

In connection with the third aspect, it should be remembered that the support structure in a lamp can cast shadows and that lens effects can occur in the glass of the lamp and produce bright and dark areas in the scene. A comparison of various light sources is shown in Figures 5.10–5.13.

5.2.3.1. Incandescent Lamps. Incandescent lamps are the most commonly used sources and are, in general, the most adaptable and

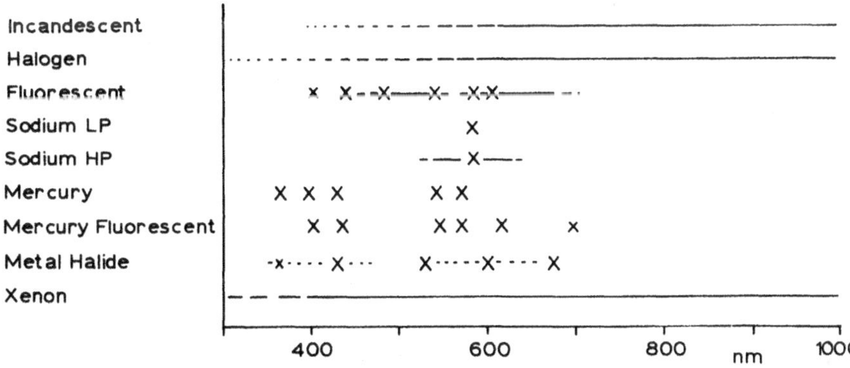

Figure 5.10. The spectral output of light sources. The broken lines indicate a lower output level. Principal spectral lines are indicated by an "X."

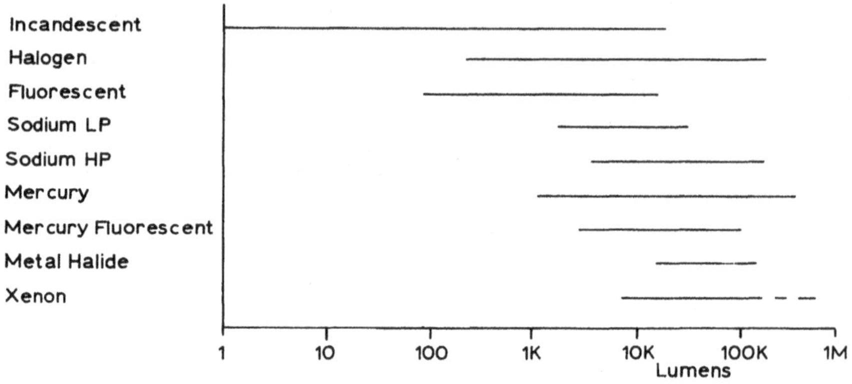

Figure 5.11. The total visible light output from a single lamp.

controllable. By the nature of the process, a considerable fraction of the energy, about 80%, is converted to heat, some being radiated in the infrared and the rest lost by conduction and convection. The efficiency of the lamp in the visible band increases with the increase of the filament temperature, i.e., the lumens/watt figure is higher, but this is accompanied by a reduction in the life of the lamp (Figures 5.14 and 5.15). The longer life-efficiency products are obtained with the thicker filaments, which consequently require larger currents. As a result, low-voltage lamps are often the best even if a transformer has to be used. Also, their filaments are more robust than the thin filaments of low-current lamps.

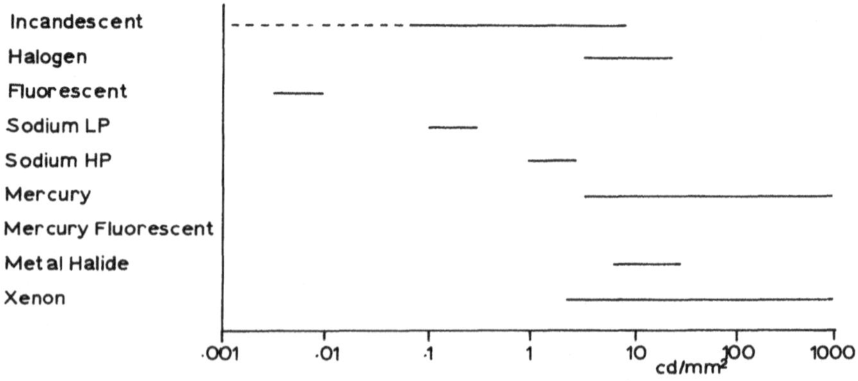

Figure 5.12. The luminances of light sources.

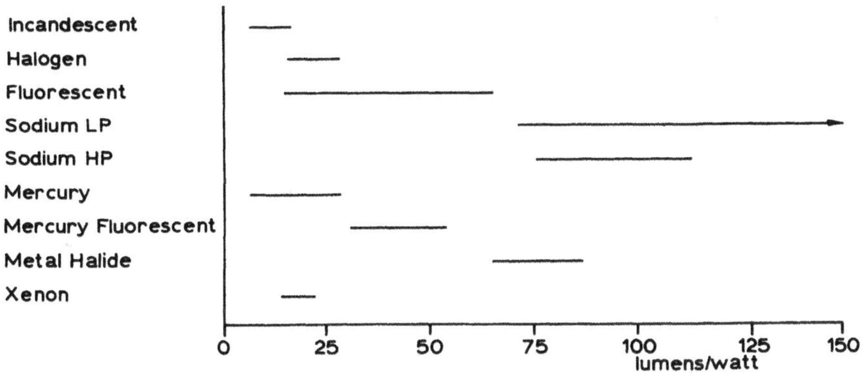

Figure 5.13. The efficiencies of light sources. The figures include an allowance for the ballast or other current control device.

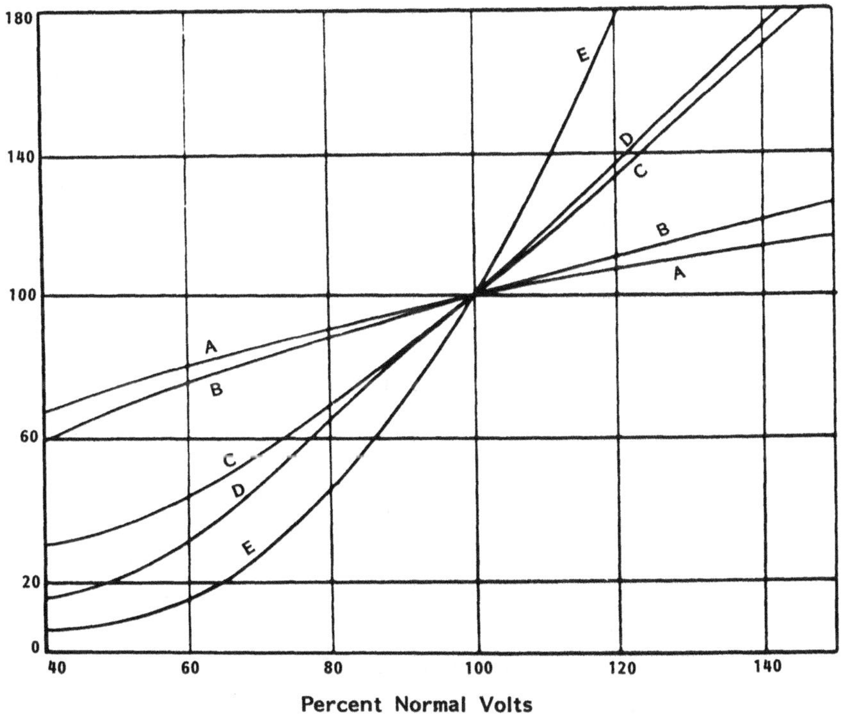

Figure 5.14. The effect of the operating voltage upon parameters of an incandescent lamp. A = resistance, B = current, C = wattage, D = efficiency (lumens/watt), and E = output (lumens). (Courtesy of Philips Lighting Division, Philips Electronics, and Associated Industries Ltd.)

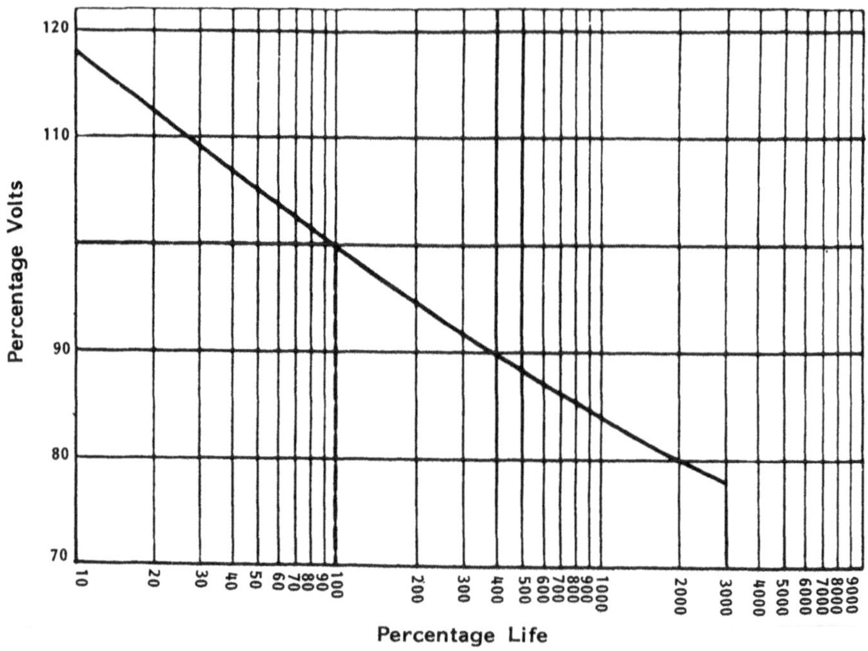

Figure 5.15. The effect of the operating voltage upon the life of an incandescent lamp. (Courtesy of Philips Lighting Division, Philips Electronics and Associated Industries Ltd.)

The filament may break due to fatigue or shock, but the normal cause of failure is the evaporation of tungsten from the filament. If the evaporation is uneven, the thinner parts of the wire become relatively hotter and the evaporation rate increases. Gas mixtures such as nitrogen with argon or krypton are used, at atmospheric pressure, to reduce evaporation. In lamps with small envelopes the evaporated metal condenses onto the envelope and reduces the output of the lamp. Long-life lamps may have special provisions for reducing evaporation but usually they are run at a lower temperature and consequently are less efficient. The highest color temperatures obtainable, consistent with reasonable life, are about 2900 °K. Typical efficiencies of high-current lamps are 20 lumens per watt. Loss of output during the life of these lamps is typically 15%.

The lamp output can be controlled from zero to maximum by varying the supply. When driven from ac supplies, a ripple of twice the supply frequency will be imposed on the output. The modulation depth is less for the thicker, higher-current filaments, for higher frequencies, and for under-run lamps. A typical figure for lamps at full voltage, 50 Hz, and running at 1 amp is 7%, peak to peak, of the light output.

5.2.3.2. Tungsten Halogen Lamps. Tungsten reacts with the halogens at temperatures between 250°C and 900°C to form the halides.[7,9] These halides dissociate at temperatures above 900°C. The tungsten halogen lamp contains these halogens and has a small envelope so that the envelope temperature is maintained between 250°C and 900°C. As a result the envelope is made from quartz or borosilicate glass. Tungsten reaching the wall is converted to the volatile halide and the wall stays clear. At the filament, dissociation occurs and the tungsten released collects on or near the filament, reducing the loss of tungsten. Using this process it is possible to run the filament at a high temperature, about 3100 °K, while obtaining a long life, typically 1000 hours. Efficiencies are of the order of 22 lumens per watt.

There are various compact forms of these lamps, some fitted with an integral reflector.[10] The most common applications of these are as vehicle headlamps and for projectors. The integral reflectors may have a metallized reflector but some have a dichroic surface (see Section 5.1.3.1) that acts as a "cold mirror" and reflects only the visible light. The result is that a large quantity of heat can be generated behind the lamp and damage cables or other items if not adequately shielded. The other common form of tungsten halogen lamp is the long cylindrical lamp used for flood lighting and copying equipment. Typical output ratings are 200 lumens per mm. These must be mounted with their axes not more than 4° from the horizontal. At greater angles the heavy tungsten halogen gas will migrate to the lower end of the lamp, disrupting the protective cycle, and the lamp will fail. The high filament temperatures result in the production of ultraviolet light which is transmitted by the envelope. This may be a danger with the high-power lamps.

It is essential that the lamp wall be maintained within the specified limits for the tungsten return process to function correctly. The temperature limitations mean that the lamp cannot be underrun by more than about 20% in voltage, depending upon the cooling conditions. If the filament temperature falls below 900 °C it will be attacked by the halogens. Also, temperature limits are specified for the lamp connections, usually a maximum of 350 °C. The efficiency of the lamp can be increased by coating the wall of the tube with a "hot mirror," (see Section 5.1.3.1) which retains the energy of the infrared within the lamp. Ripple on the output is similar to that for ordinary incandescent lamps. Care should be taken in handling the lamps because sodium salts, from the hand, left on the envelope will diffuse into the quartz during use and render the envelope opaque. Lamps should be cleaned with alcohol.

5.2.3.3. Discharge Lamps. This section is an introduction to Sections 5.2.3.4.–5.2.3.10.[7,9,10] Discharge lamps operate by the electrical excitation of the atoms of a gas, i.e., the raising of electrons to higher orbits around the nucleus, resulting in the release of radiation as they return to their original state. In consequence, the spectrum of the radiation is characterized by the gas used. This is particularly so when the gas is at a low pressure and the current density is low. At high pressures and/or high current densities, the spectral lines are broadened and also, due to the small volume in which the energy is released, the discharge is at a high temperature and radiates by incandescence. In some lamps the basic discharge is not the source of the visible radiation. Instead, the radiation has a high ultraviolet content which is used to irradiate phosphors which reradiate at the longer visible wavelengths. The spectrum of discharge lamps is, in general, a combination of a line spectrum and a continuum.

Most discharge lamps have a low or negative resistance and the current must be limited by a ballast or appropriate circuit. The discharge is started by a high-voltage pulse through either the main or auxiliary electrodes. The initial discharge starts in a gas such as argon or xenon and the heat of this discharge vaporizes the main medium. Preheated electrodes may be used to assist the discharge. During normal ac operation the output of a discharge lamp is not continuous or at a constant level. This variation will affect the output of a sensor unless its integration period is close to a multiple of the half-period of the supply. Usually exact synchronization is not necessary, in which case the effect upon the sensor output will depend upon the size of this error, the integral ratio of the periods, and the lamp output waveform. Some lamps can be operated from dc or at frequencies higher than the normal supply by the use of special drive circuits. This may result in an improved efficiency of the lamp.

5.2.3.4. Fluorescent Lamps. In fluorescent lamps, the basic discharge is in low-pressure mercury vapor, although some argon or krypton may be included to aid the start of the discharge. The discharge occurs in a long tube coated in the inner surface with a phosphor. The tubes are usually straight but lamps are available in which the tubes are bent into various shapes. The initial breakdown is produced by a pulse of a voltage much higher than that used in normal operation, which usually is assisted by preheating electrodes at the ends of the lamp to produce a supply fo electrons. A ballast unit is used to limit the current. Instant start lamps exist that will start from the 240-volt ac supply, without a high-voltage pulse or preheated electrodes, and that use a resistive ballast, e.g., an incandescent lamp.

The ultraviolet light generated by the discharge is absorbed by the phosphor and reemitted as longer wavelength light. The spectrum from the phosphor is a continuum whose shape depends upon the phosphor (Figure 5.16). It is usual to quote a color temperature for the phosphor which, to the eye, is that of an equivalent incandescent source. Also present in the output will be some of the line spectrum from the mercury vapour. Tubes are available without phosphors so that the ultraviolet light is available and are useful for scenes containing fluorescent materials.

The output from a fluorescent lamp is temperature dependent due to changes in the mercury vapour pressure.[9] The wall temperature depends upon the running current, the ambient temperature, the enclosure, and draughts. The maximum output from most types is obtained when the wall temperature is about 40 °C, and is about a half of the maximum at 15 °C and 75 °C. Typical efficiencies are 50 to 75 lumens/tube watt and outputs 2 to 4 lumens/mm for 26- and 38-mm diameters and 1 to 2 lumens/mm for 16-mm diameter tubes.

The life of a lamp is determined by the loss of emitting material from the electrodes and reduced efficiency of the phosphor. There is an accelerated loss of emitting material during warm-up. The phosphor efficiency falls to 85% in about 5000 hours and 80% in 10,000 hours.

If a fluorescent lamp is operated with a dc supply, the mercury will migrate to one end of the tube and the discharge will be affected. This effect can be avoided by frequent reversals of the polarity. When operating from dc supplies, a current limiting system must be used. The lamps can be used in the flash mode by pulsing with dc but the polarity should be reversed frequently. Gear exists for driving most lamps at frequencies higher than 50 Hz. This may result in higher

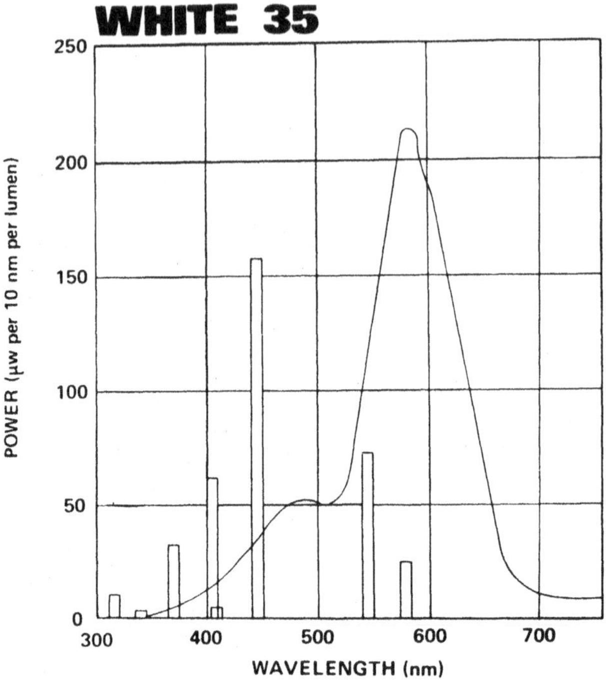

Figure 5.16. The line structure and the continuum of a white fluorescent lamp. (Courtesy of Philips Lighting Divison, Philips Electronics, and Associated Industries Ltd.)

efficiencies, e.g., by a factor of 1.15 at 20 kHz. Transistor circuits operating at 20 kHz are available. The decay time of the phosphor smoothes the fluctuations in the output and, naturally, is more effective at the higher frequencies. Most lamps show only a small variation, <10%, in their output along the tube except for the regions near the electrodes.

5.2.3.5. Mercury Vapor Discharge Lamps. In mercury vapor discharge lamps, the mercury is contained in a small quartz capsule within an envelope, containing inert gas at low pressure, and provides a fairly small source. The output is mainly in the form of a line spectrum and there can be a high ultraviolet content which may be dangerous near the lamp. Normally the glass of the outer envelope is selected to absorb some of the ultraviolet. The lamp is operated with special current limiting gear. The warm-up time is about 5 minutes. The efficiency is usually between 10% and 30%. Lamps exist with a current limiting resistor within the outer envelope, which is an incandescent filament and provides longer wavelength radiation to improve the color of the lamp.

Also available are long lamps, e.g., 1400 mm long by 20 mm diameter, consuming about 3 watts/mm. These must be used horizontally and, having no outer envelope, radiate strongly in the ultraviolet.

5.2.3.6. Mercury Fluorescent Discharge Lamps. A mercury fluorescent discharge lamp is a mercury vapor lamp with the outer envelope coated internally with a phosphor. This converts the ultraviolet light into longer wavelength light and improves the color and the efficiency of the lamp in the visible band. Typically, the efficiency is 50 lumens/watt.

5.2.3.7. Sodium Discharge Lamps. In sodium discharge lamps, the discharge medium is sodium but the discharge starts in neon. The output, characteristically yellow, is almost monochromatic, 589 nm, but there is some radiation from the neon. The lamps are operated with current limiting gear. It is not usual to preheat the electrodes, the neon discharge does this, and the lamp reaches maximum output in about 10 minutes. The discharge takes place in a long narrow tube usually folded into a U shape, with the tube in an envelope containing gas at a low pressure. The inner surface of the envelope has a "hot mirror" coating to retain the heat. As with the fluorescent lamp, it is not possible to operate with a single direction dc supply. Also, the orientation of the longer lamps must not be more than 20° from the horizontal as this will cause migration of the sodium. This lamp is the most efficient of lamps in the visible band and can produce up to 200 lumens/watt, 0.2 watts/watt, equivalent to 170 lumens/watt when gear losses are added.[11] The neon radiation can contribute a significant fraction to the light detected by sensors, such as silicon sensors, which have their highest sensitivity in the red.

5.2.3.8. High-Pressure Sodium Discharge Lamps. A high-pressure sodium discharge lamp is similar to the mercury vapor lamp. The discharge medium, sodium with some mercury, is contained in a sintered aluminum oxide tube. The high pressure broadens the spectrum to give a pale yellow light with most of the energy spread over the range 550 nm–630 nm. The efficiency is about 110 lumens/watt, warm-up time is 10 minutes, and the lamps may be mounted in any orientation.

5.2.3.9. Metal Halide Discharge Lamps. A metal halide discharge lamp is a mercury vapor lamp with the halides of thallium, indium, and sodium added to the discharge material. Their effect is to spread the light across the band, although the spectrum retains a significant, but different, line structure. The efficiency is about 80 lumens/watt.

5.2.3.10. Xenon and Krypton Discharge Lamps. Xenon and krypton discharge lamps exist as long and short discharge lamps. In both cases, the output is a continuum from 300 to 1000 nm, with some spectral lines, and includes some ultraviolet. The long lamps are available straight or coiled, consuming about 5 watts/mm at an efficiency of 20–25 lumens/watt. They can be operated from dc or ac and in pulse mode. In pulse mode the peak powers can be much higher than in continuous mode. Shorter lamps, 20–100 mm, are available which consume 80 watts/mm at the same efficiency in continuous mode. These require water cooling. There are lamps with very small arcs, 0.5–4 mm, which are useful when a "point source" is required. They operate from dc only, are rated up to 2000 watts and, with an efficiency of 15–35 lumens/watt, give up to 60,000 lumens with a luminous intensity of 6000 candela. The average luminance of these sources is 500 cd/mm^2.

5.2.3.11. Light Emitting Diodes. In light emitting diodes, current flowing through a forward biased junction in materials such as GaAs, GaAsP, and GaP excites the atoms and light is emitted as the valence band electrons return to their stable state.(7,12) Due to the limited number of possible energy transitions, the light emitted has wavelengths characteristic of the material. For the materials stated, the color moves from infrared to green as the phosphorus to aresenic ratio is increased. Silicon carbide is used to produce blue light, at 480 nm. The outputs are low and are equivalent to the smallest incandescent lamps. Typical values are ten millicandela for continuous operation (3 volts at 50 mA) and 0.1 lumens/watt, although the overall efficiency is reduced by the loss in the current limiting circuit. Much higher outputs, about 100 times greater, can be achieved by pulsing with currents of several amperes, but the duty cycle must be low to limit heating of the junction. Switching times are less than 1 μs. Operation at high junction temperature reduces the life of the devices and at normal temperatures the life should not be considered as infinite. Usually the light output is directional with a cone half-angle of 15° to 30° due to the lens effect of the encapsulation, but the radiation from the bare chip covers a larger angle.

5.2.3.12. Lasers. The laser can be used for illumination or to produce a light beam of high intensity and with a very narrow angle of divergence, less than one milliradian, which can be scanned across a scene (Section 4.4). Very narrow spectral bandwidths are possible.

Laser action can be produced in gases, liquids, semiconductors, and other solids.(7,13) The atoms of the material are excited by some means and, in reverting to their stable state, release energy having a wavelength characteristic of the material. The laser is formed by placing reflectors about the material so that the radiant energy is retained in the medium as a standing wave. The excited atoms are stimulated to release their energy by the standing wave and add it in the same phase and direction as the wave. Thus a high-power, coherent source is produced. One of the mirrors is partially reflecting and some of the energy can be extracted as a beam with very low divergence and narrow bandwidth. The energy can be introduced to the system by a discharge in a gas (by applied voltage or r.f. field), by radiation from an adjacent high-power light source, e.g., as in the ruby laser, or by current in a semiconductor junction as in the light emitting diode.

For a given spectral line and distances short compared with the coherence length, the output can be considered as having a single wavelength. The output can be partially or totally plane polarized, and several radiation modes are possible in the output beam. These aspects may affect the distribution of the energy in the beam and the passage of the beam through an optical system. For the simple modes, e.g., TEM_{00}, the energy distribution across the beam is roughly Gaussian. Other modes have more complex distributions, e.g., with a central null, and the beam divergence is greater. The efficiency of a laser is very low; an argon laser having an output of 5 watts consumes 10 kwatts and requires water cooling. Lasers can produce higher peak powers when pulsed.

The helium-neon laser, the cheapest of the gas lasers, produces a beam of a single wavelength, 633 nm, and power up to 50 mW. Its life can be of the order of 10,000 hours. Some lasers are capable of producing a beam containing several isolated wavelengths, e.g., the argon/krypton/xenon ion lasers produce radiation between 350 and 800 nm, with total powers of 20 watts and 0.5 watts at each wavelength. Typical beam divergence angles for the gas lasers are 0.5 milliradians. Also in the visible range are the dye lasers, which are tunable to produce a range of wavelengths, and the metal vapor lasers. The dye lasers use a dye in a solvent, such as ethylene glycol, and the energy is introduced by an argon ion or nitrogen laser operating in the near ultraviolet. The main application for the CO_2 laser is in machining due to the high powers that are possible, several hundred watts CW, but low power versions are suitable for scanners where the long wavelength, about 10 μm, has some advantages (see Section 5.2.4).

Of the solid lasers, the ruby laser radiates at 694 nm but is used only in the pulsed mode, giving up to 3 ms pulses at up to 200 Joules. The alexandrite laser is tunable from 700 to 815 nm and can provide a continuous output of many tens of watts. The lasers are pumped by other very high power sources.

In contrast to the other lasers, the semiconductor diode laser is very small and has a high efficiency, typically 40%. The device is basically a light emitting diode (Section 5.2.3.11) but with a structure that provides the reflectors to contain the energy. Using various LED materials, wavelengths from 400 nm into the infrared can be generated.[14,15] Outputs are several tens of milliwatts CW and higher pulsed powers are possible. Higher powers, 100 mW CW and 1 W pulsed, are available by using a phased array structure having several diodes.[16] The light is generally plane polarized and emitted in a cone of angle equal to several tens of degrees. This large angle is not a disadvantage because the source size is of the order of a micron and a lens can be used to produce a well collimated beam. The output can be modulated or pulsed by controlling the drive current. The wavelength of the output is current and temperature dependent and if a stable source is required the power source requires careful design. A stable wavelength will not be required for most automation applications.

The high beam energies and low divergence of lasers can produce a dangerous intensity at a distance.[17] This can cause eye damage and, at the highest powers, be a fire hazard. The appropriate safety standard for the use of lasers should be consulted, especially when lasers above 1 mW are used. The possible paths of beams reflected from the system or the scene should be considered. In scanning systems using the higher power lasers it is usual to switch off the laser or to block its beam if the scanning system stops.

5.2.3.13. Electroluminescent Lighting. Electroluminescent lighting depends on the principle that conduction through, or the generation of displacement currents in, certain solids can generate light.[7,18] Zinc sulphide, doped with manganese, is formed, by thin or thick film techniques, on an electrode and is covered by a transparent electrode. The device may be driven by ac or dc supplies. They provide a large area and uniform source, useful when shadows are to be minimized. Luminances of tens of cd/m^2 up to about a hundred are available with an efficiency of a few lumens per watt. Typical lifetimes are 10 khours.

5.2.4. Lighting Systems

The method of illuminating a scene is an important consideration in designing a vision system. It can have a significant effect upon the processing of the video signal. For example, a cube showing three faces to a camera can be illuminated so that the faces appear to have the same luminance. The result is that the edges dividing the faces are not visible (Figure 5.17). By unbalancing the lighting the edges appear. Generally the aim will be to obtain the maximum contrast between features which have to be detected, e.g., between an object and its background. Although specular reflection can cause a problem, in that the high light level can spoil the detected image by spreading, in the sensor, over surrounding areas and leaving a mark with a long delay or damaging the sensor; it can also be used to obtain a high contrast. An example is in the illumination of transparent materials such as glass. The position of glass tubes on a wire mesh belt was revealed by illuminating from the side so that the belt received little light but there was reflection from the glass into the camera (Figure 5.18) (see Section 11.6.1). Polarization filters on the camera may be used to reduce the effects of specular reflection when the reflecting

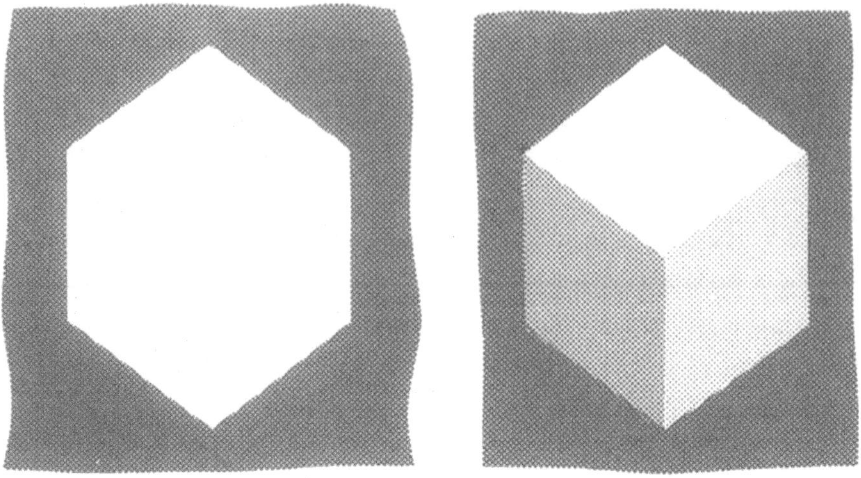

Figure 5.17. A cube, concealed by balanced lighting, is revealed by suitable lighting.

surface and the camera are in fixed positions. Specularly reflected light is usually highly polarized and may be attenuated by a polarizer set in the crossed position.

The spectral distribution of the power from the light source, the spectral variation of the sensitivity of the detector, and the presence of color in the scene must be considered with regard to the overall sensitivity and discrimination of the system. In some cases the separation of the spectral bands in the detector system, as in a color camera, may aid the separation of colored objects or areas.[19]

Advantages may be gained in using wavelengths outside the visible range. The longer wavelengths of the infrared are less affected by small particles and therefore can penetrate mist, e.g., steam, which might surround a workpiece, or dust on the object to be examined. Silicon sensors are suitable for detection in the near infrared, less than 1 μm, and vidicons are available which have special infrared-sensitive targets. Ultraviolet light can be used to reveal the presence of fluorescent materials. One application is in the detection of cracks in objects. An object made from a magnetic material can be placed in a magnetic field and sprayed with a fine magnetic material, e.g., ferrite, coated with a phosphor.[20-22] The powder will collect in the high field areas at the cracks and be visible as bright lines under ultraviolet lighting. For all materials, the object can be sprayed with or immersed

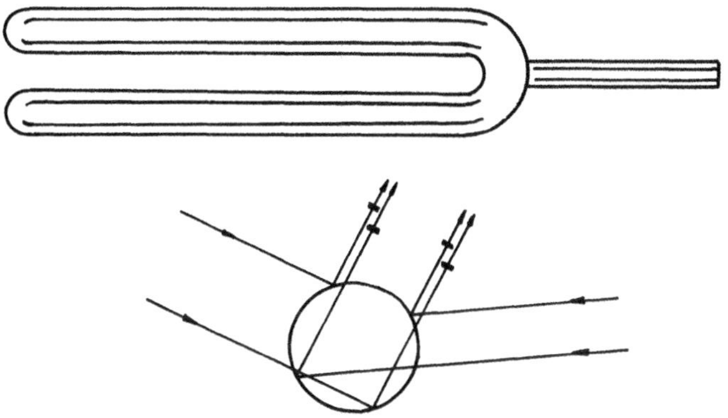

Figure 5.18. The position of the transparent glass tube is revealed by specular reflection from its walls.

in a solution of fluorescent dye. It will be retained in the cracks and be made visible as before. The effect can be enhanced by blowing fine dust onto the object to bring the dye to the surface. A dilute solution of dye can be used to reveal leaks, for example in washing machines.

Also affecting the choice of light source, when it is not itself the scanning system, is the question of whether a small source or a diffuse source is required. Small sources will produce sharp shadows but the illumination of objects will be affected less by their position and orientation when a diffuse source is used. Optical systems, particularly mirrors, Fresnel lenses, and fiber optics, can be used to produce a source of the type required. A small source can be expanded in this way and many of the most powerful sources are relatively small. A source at the focal point of a paraboloidal mirror will produce a parallel beam and at one focal point of an ellipsoidal mirror will be focused at the other. Similar effects occur in two dimensions for long lamps in parabolic and elliptical reflectors. Many light sources driven from ac, as some have to be, produce a fluctuating output. This may be acceptable if the integration period of the detector is an integral multiple of the half-cycle period, or the whole cycle period if the lamp output is spatially asymmetric and related to the supply polarity.

Ambient conditions affect the choice of lighting. Normally ambient lighting, from daylight or artificial sources, is present which the system may be able to use or at least accept as part of the lighting, but usually this is not possible. Such lighting is often variable and out of the control of the designer. Either the local lighting must be bright enough to make the ambient insignificant, perhaps with the aid of screening, or narrow bandwidths must be used, e.g., with laser or sodium light, so that corresponding narrow-band interference filters can be placed in front of the detectors. Care should be taken that bright lights do not come into the view of the sensor as this would affect the detected image, the brightness spreading over adjacent areas, or in some cases, as with television tubes, produce a long decay after-image or permanently damage the sensor. Other ambient considerations are dust or other dirty air conditions, and cooling draughts or high temperatures which can affect the operation of some discharge and fluorescent lamps. The effects of dust can be reduced by enclosures and by the use of air curtains.

5.2.5. *Structured Light*

The three-dimensional nature of scenes can be revealed to a one- or two-dimensional scanning system by the use of structured light. Instead of using uniform lighting for the scene, a coarse, high-contrast pattern can be imaged onto the scene. Figure 5.19 shows the effects of using an array of holes and a coarse grating. The apparent direction of the grating lines and the elongation of the holes gives information on the orientation of the surfaces when the positions of the light source and scanner are known.[23] Also, the illumination reveals the location of edges by the discontinuities in the edges of the shadows. It may be

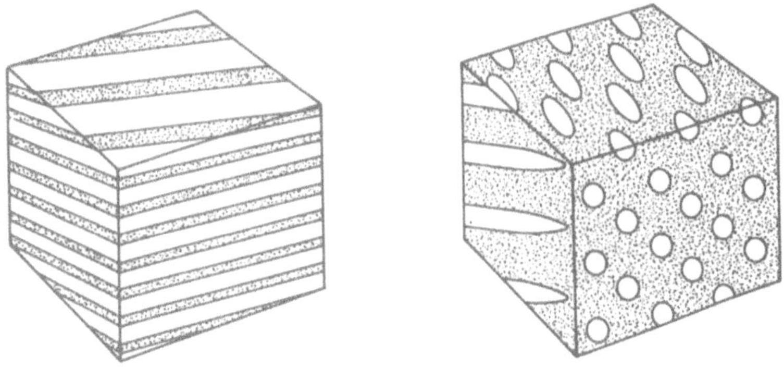

Figure 5.19. Illumination by a patterned light beam can reveal the shape of an object and the orientation of its surfaces.

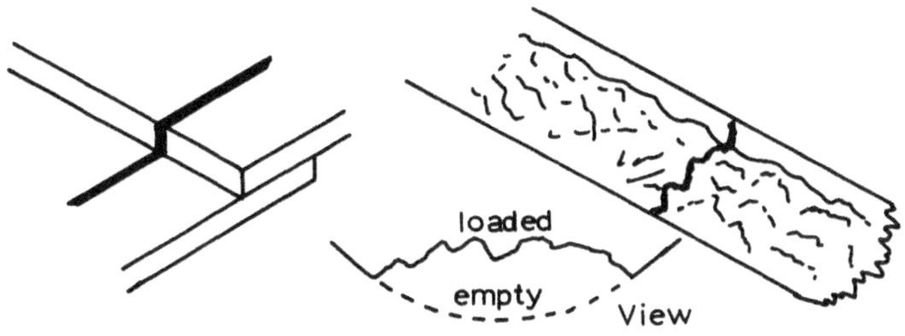

Figure 5.20. The application of a thin line of light to the location of a join in plates and to the measurement of the flow of material on a conveyor.

necessary to have two lighting systems, switched by the vision processor, to provide normal and structured lighting.

A scanned laser beam can be used as a light source for this purpose and can trace patterns as required or a laser beam may be spread into a thin plane of light by a cylindrical lens.[24] The system is basically that of a laser/camera range finder except that the whole pattern is traced by the laser during the integration period of the camera. Two applications of a straight line laser scan or a thin plane of light projected into the scene are the joining of plates, e.g., in a welding system, and the measurement of bulk material on a conveyor (Figure 5.20). In this conveyor application the area between the apparent position of the trace for an empty conveyor and that when loaded is proportional to the cross-sectional area of the material and therefore to the feed rate.

For objects having submillimeter dimensions, a grating pattern can be produced by interference between two broad, overlapping beams, from the same laser, travelling along almost parallel paths.

A technique similar to structured lighting can be applied to systems in which the picture scan is performed by a light spot, i.e., a laser or flying spot scanner cathode ray tube. In this case, the light-collecting sensor has a patterned mask against the sensor surface and the scene is imaged onto this. The effect is the same but the light path is reversed.

When the sensor is a linear array camera, objects on a conveyor belt can be revealed by the method used in the Consight system of General Motors.[25] A line of light is projected at 45° onto the belt to be in the view of the camera (Figure 5.21). Light striking a passing object will be out of the view of the camera; thus the shape of the

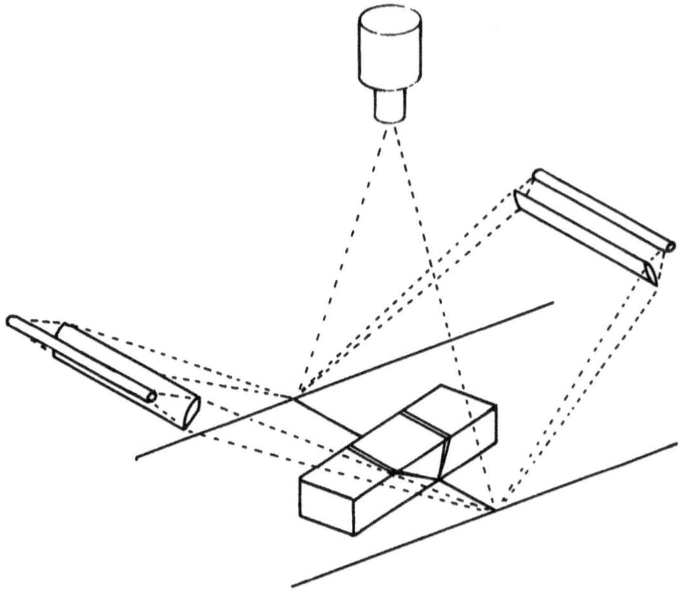

Figure 5.21. Two thin light beams are used in the Consight system of General Motors to reveal the position of an object on the conveyor.

object is revealed as a silhouette by the loss of signal. The silhouette will also contain the shadow of the object but this effect can be reduced by using a second light source illuminating from the other side. A similar system can be used to detect parts of an object which are more than a given distance above the belt.[26] The scene is illuminated at an angle by a broad beam with a sharp cutoff, the cutoff line being at the appropriate distance from the conveyor along the axis of the camera (Figure 5.22). The camera is focused on this position. Only those parts of the object above the given height from the belt will be seen by the camera as illuminated. Again, shadows can adversely affect the performance.

More than two systems can operate together when each uses its own restricted band of lighting wavelengths. Appropriate light sources are selected, e.g., sodium or laser, or filtered to illuminate in separated spectral bands, and the cameras are fitted with the corresponding filters (see Section 11.8.3). In some situations, two lighting and camera systems may operate simultaneously and independently on the same

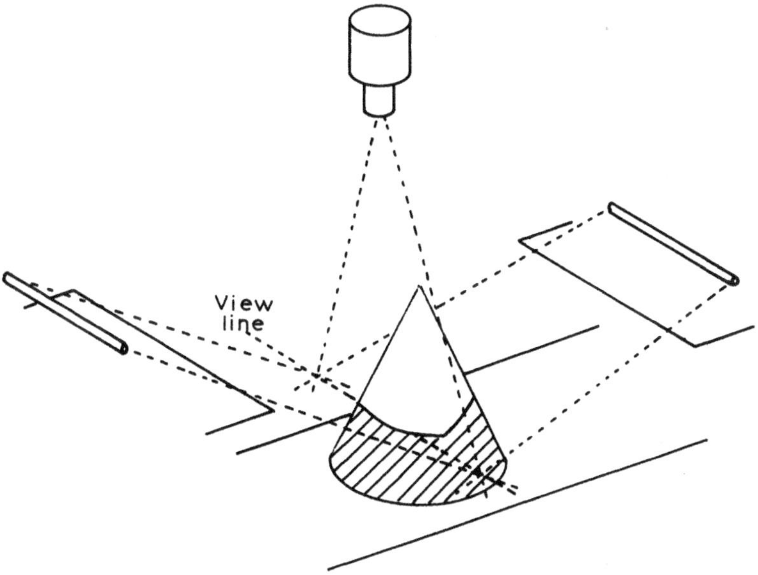

Figure 5.22. Height thresholding of an object by light beams with a sharp cutoff.

scene by using polarized light in each, with the planes of polarization orthogonal. It may not be satisfactory when objects in the scene change the polarization.

Polarized light also has an application in viewing strain in transparent materials. The strain may be due to stress applied externally, to stress produced by an included particle, or to incomplete annealing of glass. The strain produces birefringence (see Section 5.1.3.3), with the axes along and orthogonal to the direction of the strain. When viewed with plane polarizers, placed on both sides of the material, the stress is revealed as a change in transmission, particularly when the polarizers are crossed to produce a black ground. In some cases a more useful result is obtained by placing a quarterwave birefringent filter between the polarizers, the increased and decreased transmission indicating the direction of the stress.

When it is necessary to reveal the structure or roughness of a surface, a collimated beam of light at grazing incidence should be used. High points will be well lit and cast shadows, and pits will be in shadow.

References

1. W.B. Elmer, *The Optical Design of Reflectors,* Wiley, New York (1980).
2. W.G. Driscoll and W. Vaughan, eds., *Handbook of Optics,* McGraw-Hill, New York.
3. L. Levi, *Applied Optics,* Volume 2, John Wiley, New York (1980).
4. A.H. Cherin, *An Introduction to Optical Fibres,* McGraw-Hill, New York (1983).
5. A. Agrawal and M. Epstein, Robot eye-in-hand using fiber optics, *Proc. 3rd Intl. Conf. on Robot Vision and Sensory Controls,* 257–262, Cambridge, MA (Nov. 1983).
6. P. Saraga and B.M. Jones, Parallel projection optics in simple assembly, *Proc. 1st Intl. Conf. on Robot Vision and Sensory Controls,* 99–111, Stratford-upon-Avon (April 1981).
7. M.A. Cayless and A.M. Marsden, eds., *Lamps and Lighting,* Edward Arnold, London (1983).
8. E.L. Dereniak and D.G. Crowe, *Optical Radiation Detectors,* John Wiley, New York (1984).
9. J.E. Kaufman and H. Haynes, *IES Lighting Handbook,* Reference Volume, Illuminating Engineering Society of North America (1981).
10. Lighting Handbook, Lighting Division, Philips Electronics and Associated Industries Ltd.
11. L. Sprengers, R. Campbell, and H. Kostlin, Low pressure sodium lamps with luminous efficacy of 200 lm/W, *J. Illum. Eng. Soc. 14*(2), 607–615 (April 1985).
12. R.G. Seippel, *Opto-Electronics,* Reston Publishing Co., Inc., Reston, VA (1981).
13. G.K. Klauminzer Twenty years of commercial lasers—a capsule history, *Laser Focus/Electro-Optics, 20*(12), 54–79 (Dec. 1984).
14. J.C.J. Finck, H.J.M. van der Laak, and J.T. Schrama, A semiconductor laser for information read-out, *Philips Tech. Rev., 39*(2), 37–47 (1980).
15. Y. Suematsu, Advances in semiconductor lasers, *Physics Today, 38*(5), 32–39 (May 1985).
16. W. Streifer, R.D. Burnham, T.L. Paoli, and D.R. Scifres, Phased array diode lasers, *Laser Focus/Electro-Optics 20*(6), 100–109 (June, 1984).
17. A.R. Henderson, Laser radiation hazards, *Optics and Laser Technology 16*(2), 75–80 (April 1984).
18. T. Weardon, Electroluminescent display technology, *Electronic Product Design 2*(5), 57–60 (Feb. 1984).
19. C. Loughlin, Faster processing with colour, *Sensor Review 3*(3), 144–147 (July 1983).
20. D. Schmid, Industrial robot with video camera for detection of material defects, *Proc. 2nd Intl. Conf. on Robot Vision and Sensory Controls,* 19–25, Stuttgart (Nov. 1982).
21. J. Hollingham, Automating crack detection, *Sensor Review 3*(3), 130–131 (July 1983).
22. Y. F. Cheu, Automatic crack detection with computer vision and pattern recognition of magnetic particle indications, *Material Evaluation* (USA), *42*(12), 1506–1510 (Nov. 1984).
23. G.B. Porter and J.L. Mundy, A non-contact profile sensing system for visual inspection, *IEEE Computer Society Conf. Industrial Application of Machine Vision,* 119–129 (May 1982).

24. G. Betz, A general purpose 3D vision system, *Proc. 4th Intl. Conf. on Robot Vision and Sensory Controls,* 149–154, London (Oct. 1984).
25. R.D. Baumann and D.A. Wilmshurst, Vision system sorts castings at General Motors Canada, *Sensor Review 2*(3), 145–149 (July 1982).
26. J.L. Morris, The use of height thresholding in robotic vision, *Proc. 2nd European Conf. on Automated Manufacturing,* 133–140, Birmingham (May 1983).

6

Nonvision Sensors, Range Finders, and Processing

6.1. Nonvision Sensors

Vision can extend the use of a robot to a large number of tasks which are beyond that of a "blind" robot. A similar situation applies to sensors operating in simulation of the other human senses and also providing senses not available to the human. This chapter gives an insight into what is possible beyond vision.[1–8]

6.1.1. Optical Sensors

Light may be used in sensors in a manner that does not conform to the concept of vision. A simple application is the interruption of the light path between a source and a detector to detect the position of the edge of an object. The path may be the straight line between the two or it may be through an optical system using mirrors, lenses, or optical fibers. The last can be useful when it is not possible to place thc dctcctor and sensor in the appropriate position due to space constraints. Alternatively, the detector and sensor can be placed close together and the light reflected from the other end of the light path by a mirror or multicorner reflector. Such sensing devices are used to determine when an object is entering the jaws of a gripper, or some other specific location, to detect a zero position of a mechanism, e.g., in conjunction with an incremental encoder, and as the grating sensor of an optical encoder.

The combination can be used as a proximity sensor. The source and sensor are placed closed together, viewing in the same direction, but arranged so that light cannot pass directly between them. Light will be scattered from an object which comes into their view and be received by the detector. The level will rise rapidly as the range becomes short, but for very short distances the signal will fall as the light spot moves away from the view of the detector. The distance at which this change occurs depends upon the distance between the source and detector. The performance of the device depends upon the absorption and specular reflectivity of the surface of the object. By using extra detectors around the source and comparing their outputs, it may be possible to determine the orientation of the reflecting surface. The ambiguity in the orientation of hollow cylinders, closed at one end, can be resolved by using this proximity detector, placed axially, to determine the position of the closed end.[(9)]

6.1.2. *Proximity Sensors*

An optical proximity sensor has been described above but acoustic, magnetic, capacitive, and pneumatic devices can be used to detect the presence of objects close to the sensor.

In one form the acoustic detector is described in the section on range finders (Section 6.2.6), but for short ranges it is a proximity detector. An alternative method uses a single transducer whose impedance is measured while it is being driven. When the strength of the reflected signal reaching the transducer is within 40 dB of the transmitted signal, it will affect the impedance. This will occur for short ranges, up to about 100 mm, but will depend upon the reflectivity of the detected surface.

The impedance of a coil carrying an alternating current is affected by the presence of ferromagnetic and also conducting materials, the first by the change in permeability of the magnetic path, the second by eddy currents induced in the conductor.[(10)] To limit the direction in which the device is able to detect and to screen it from interference, the magnetic path is controlled as shown in Figure 6.1. The coil is wound around a ferromagnetic rod inside a cylinder, with one end closed, of similar material. The sensitivity of the device falls rapidly as the distance increases, roughly as the inverse cube. For nonmagnetic materials, when eddy currents are used, the system is sensitive to the thickness of the material. Penetration of the currents increases as the frequency is reduced and consequently the detectability of thin sheets falls at the same time.

The presence of conducting surfaces may also be revealed by

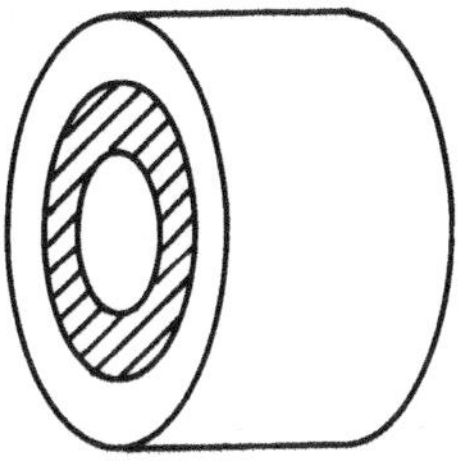

Figure 6.1. An inductive proximity sensor.

capacitive effects. When they come close to a plate electrode driven by an alternating voltage, a current is drawn. Due to the low dielectric constant of air, the device is suitable only for distances small with respect to the diameter of the electrode. It is permissible to cover the electrode with a thin layer of insulating material for protection. Also, the sensed object need not be grounded provided that it has a capacity high with respect to the capacity to the electrode.

A simple and robust sensor is provided by air leaving a small orifice.[11] The pressure around the orifice increases as it is brought towards a surface and an increase in the supply pressure is required to maintain the same flow. A common system is to feed the air through a constricting valve and measure the pressure in the pipe to the orifice. The valve and pressure sensor may be remote from the orifice if a long pipe is used, e.g., along the length of the robot's arm. Basically the sensor can be a membrane, displaced by the pressure change, operating an electrical switch, if a binary output is required, or a strain gauge, if an analog output is required. The device is suitable for ranges of a few millimeters. The passage of an object between two opposing surfaces having a small separation can be detected by placing an air jet in one surface and an orifice in the other, connected to a pressure sensor. A pneumatic contact sensor can be constructed by fitting a close wound spring to the end of a tube, with the other end of the spring closed by a short rod. Compressed air is fed to the system through a restrictor valve. When the short rod is displaced by contact with an object, air will be released and the pressure will fall.

6.1.3. Displacement and Position Sensors

Several physical properties can be used to measure, linearly or angularly, the movement between, or relative position of, two objects. These measurements may be made absolutely or only incrementally.

An absolute system has the advantage that the relative position is known soon after the equipment is switched on. In an incremental system the position information may be lost when the equipment is switched off and any movements occurring while it is off will not be detected. At switch-on, the system must move to a known calibration position, usually found with the aid of a separate detector, and the incremental sensor measures from that point. To avoid the necessity of making a large movement to a single zero position, it is possible to have a large number of accurately placed reference positions which are identified by a coarse absolute system. Range finders and some proximity sensors, suitably calibrated, can be used as position sensors.

The resistor track provides an absolute position sensitive system and is operated as a potentiometer.[6] The most common track material is conductive plastic which provides infinite resolution. Its accuracy and linearity are not high, a good device having 0.1% linearity. The linearity of the device is affected by the current drawn from the wiper and this should not be more than 1% of the track current. It is subject to wear, although lives in excess of 100 million cycles are claimed. This applies to any point on the track and this may soon be reached if there is oscillation in the drive system. Lubricants are used to reduce wear and at high speeds the wiper may be lifted from the track by the lubricant causing a loss of position information. Typical limits for circular and linear potentiometers are 10 revolutions/sec and 1 m/s.

Circular potentiometers may have a multiturn structure. The single turn design can be with or without stops at the end of the track, the latter allowing continuous rotation but with a break in the output. This difficiency is overcome in double track potentiometers in which the output from each track covers for the break in the other. Tapered tracks can be produced so that the output is not a linear function of the position. For digital systems, the output is translated to digital form by an analog-to-digital converter.

Probably the most common position sensor is the optical encoder which is available in absolute and incremental types, and rotary and linear forms.[6,8] In the incremental type, one element is a track of transparent and opaque bars, and the other element is a light source, a detector, and a narrow slit. During relative motion of the two elements the light beam is modulated. This effect is independent of the direction of motion and therefore a second sensor unit is placed at another position along the track so that its output is in quadrature with that from the first sensor. From the two outputs the amount and direction of movement can be determined. In the rotary encoder the track is formed from radial lines. Often another detector unit is provided which senses a single slit alongside the bar pattern to give a zero referencing signal.

In the absolute optical encoder there are several parallel tracks of bars, each with a single detector unit. The number of bars in each track rises as the power of two, from track to track, so that the output from the detectors represents a binary number uniquely corresponding to the position. There is a danger in the use of a simple binary code in that two outputs would have to change at precisely the same time. To avoid this, a code is used in which only one bit changes at any time. The Grey code is an example of this. Code conversion circuits are available to convert the special code into a normal binary code.

The optical encoders provide a high linearity, better than one bit, and a high resolution. In the case sizes used for robots, up to 10,000 lines can be used. Life is limited only by the bearings and the light source. Their main disadvantage is that of cost, particularly in the absolute form.

The other position sensor which is finding increasing application in NC equipment is the resolver.[8,12] This is a rotary transformer, in appearance similar to a motor, having orthogonally arranged coils on a stator and a single coil on the rotor. An alternating current is fed to the rotor, typically between 50 Hz and 50 kHz, to induce voltages into the stator windings. The ratio of these voltages is the tangent of the angular position of the rotor with respect to the stator. Standard electronic units, often constructed as small hybrid circuits, are available for driving resolvers and giving a measure of the angle in digital form. Typical resolutions are 1.3 minutes of arc, i.e., 14 bits, and accuracy ± 5 minutes. Life is limited only by the bearings and the connections to the rotor; since these are slip rings the life is very long. The use of slip rings can be avoided by building an angle independent transformer onto the unit to couple power into the rotor.

The systems can operate in the sampling mode, when the angle is measured with no prior information, or in the tracking mode, when the angle is compared with a stored value which is adjusted as necessary. The latter gives the faster response when there is no interruption of the power nor excessive angular velocities, typically less than 32 bits/ms (2 rev/sec for 14 bits resolution and 32 revs/sec for 10 bits, although rates of 100 revs/sec for 12 bits have been quoted). In the tracking mode the outputs from the stator are $A \sin wt \sin X$ and A

sin wt cos X, where X is the angle between the stator and rotor. These are multiplied by functions of the stored angle Y, and the difference taken to give

$$A \sin wt \sin X \cos Y - A \sin wt \cos X \sin Y = A \sin wt \sin (X - Y)$$

This is synchronously rectified and the resulting output is proportional to sin $(X - Y)$. The stored value, Y, is incremented by one bit, up or down, depending upon the polarity of sin $(X - Y)$. If the maximum angular velocity is exceeded there will develop an error in the measurement but this will correct itself when the velocity is reduced and the system catches up. The system contains a circuit to determine in which quadrant X lies to remove the ambiguity inherent in the measuring system.

Synchros,(6,8,12) which have three stator windings at 120°, can be used in place of resolvers and it is usual to use a Scott transformer to convert the three outputs to two, equivalent to those of the resolver. More complex multipole devices exist. The linear and rotary Inductosyns (a trade name of Farrand Controls Inc.) are related to these transducers.

Another inductive linear sensor, the linear variable differential transducer (LVDT), has three coaxial cylindrical coils, Figure 6.2. One end from each of the two secondaries abut at the center of the transducer and the primary is wound over the secondaries.(6) The moving part of the system positions a magnetic core within the coils and thereby sets the coupling between the primary and each of the secondaries. With the primary driven at frequencies from 50 Hz to over

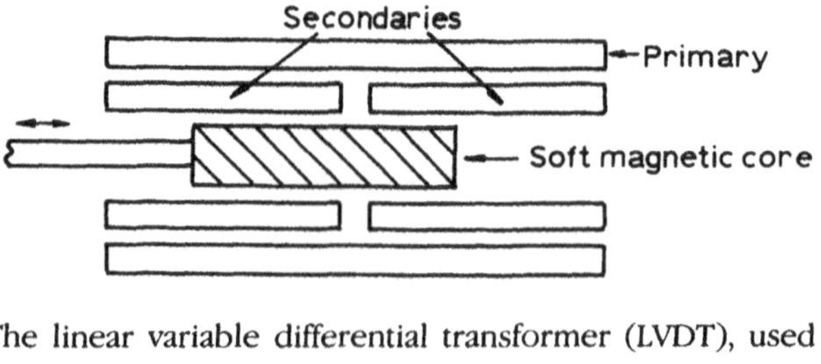

Figure 6.2. The linear variable differential transformer (LVDT), used for position sensing.

50 kHz, an output is taken from the secondaries, series connected in antiphase. The output varies linearly, to 0.1%, for a movement almost equal to the length of the core. The transducer is not subject to wear, hysteresis, or appreciable friction. The resolution is determined by the electronic circuit. Rotary versions exist. Standard circuits, often hybrid devices, are available to measure the output and provide a digital result.

A capacitive version of the LVDT exists.[6] One part of the system is two coaxial metal cylinders, with an end of each abutting, and the other a metal cylinder around the first two (Figure 6.3). An alternating voltage is applied between the first two cylinders and the third takes a potential, between these, which depends linearly upon their relative position. The measuring circuit is a Wheatstone bridge with the transducer forming two arms and equal fixed capacitors forming the others. The output is taken differentially between the third cylinder and the junction of the fixed capacitors. In an alternative version, a layer of dielectric is moved between two fixed plates and the change in capacitance is measured. The capacitance is a linear function of position. Small movements can be detected by measuring the capacitance between parallel plates. The capacity is inversely proportional to distance and rapidly becomes negligible, with respect to stray capacitances, as the separation is increased. The effect of the stray capacitance can be controlled by the use of guard rings. If the capacitance is a frequency determining component in an oscillator then the frequency is roughly

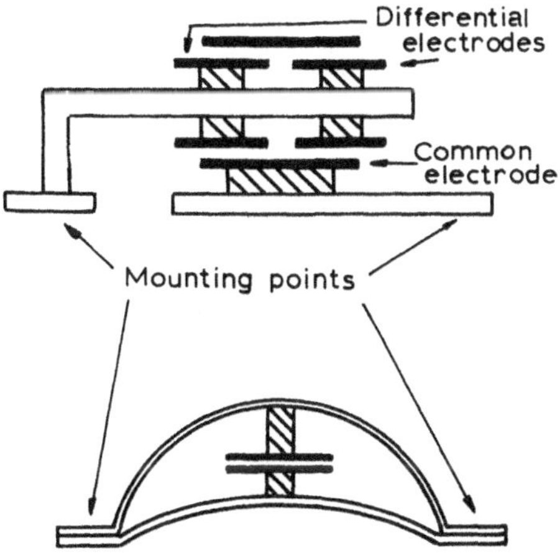

Figure 6.3. Two position sensors in which the relative movement of the mounting points is measured capacitively.

proportional to the separation. Figure 6.3 shows another capacitive system in which relative movement of the mounting points changes the separation between the insulated plates.

Sensors of the same type may be used in combination to achieve very high resolutions. A reduction drive can be placed between two rotary units so that one counts the full revolutions of the other. Also, sensors of different types can be combined. For example, an incremental optical encoder, measuring at high resolution, can be combined with a resistive encoder. Stored in the controller are the values of the outputs from the resistive encoder, which correspond to the positions at which the zero index of the optical encoder occurs, together with the corresponding count of complete revolutions of the optical encoder. These values will, due to the resistor nonlinearity, be specific to a given combination. To find the absolute position at switch-on, or if the absolute position is lost, only a small movement is required to reach a zero index. The output from the resistive encoder is read and, from the best match with the stored values, the correct absolute position is obtained.

When values of velocity, acceleration, and higher differentials are required it is normal to obtain them by the differentiation of the position/time function. There are encoders which can be used to obtain these directly. For example, the dc generator can be used to measure angular velocity. The need to suppress commutator noise restricts the bandwidth that can be used. Acceleration can be measured directly by measuring the forces applied to a small mass. The forces are measured with force sensors usually of the strain gauge type.

6.1.4. Force Sensors

Many materials exhibit useful phenomena when subject to stress. The simplest is a change in dimensions or shape, as in a spring, and if the change can be measured by a suitable displacement transducer, the force can be deduced. Another is a change within the material as in the piezoelectric and piezoresistive effects. In many structures large movements are not acceptable and sensors for small displacements are required. The displacement in a structure necessary to produce a measurable change in a sensor constitutes a compliance, which in a robotic system can increase the positioning errors and the additional flexibility can allow oscillation to occur during movements. A unit commonly used with stress transducers is the microstrain. It refers to a dimensional change and is equal to a change of a millionth of the measured dimension, i.e., 1 μm/m.

For systems in which large movements can be tolerated a compliant structure, such as a coil or leaf spring, may be used. The moving part of the structure may be connected to a linear potentiometer to detect the movement.[13] For smaller movements optical gratings, as in the incremental position encoders, are useful. Also a light source or illuminated edge on the moving component can be imaged onto a linear scanning array or a position sensitive diode (see Section 4.3). Small movements may be magnified by a lens system or by the pivoting of a mirror mounted on closely positioned fulcra on the moving component and a reference surface. Other displacement trandsucers for small movements (see Section 6.1.3) may also be used.

The most commonly used force sensor is the strain dependent resistor, either using the change in dimensions of the resistor or using the piezoresistive effect.[6,7,14–16] For the robotics applications these are supplied as thin film components. The resistor does not take the main force itself but is used to measure the deformation of mechanical structures, by being bonded to the side of the structure. For the satisfactory performance of strain gauges, care must be taken in their mounting and protection. It may be used to measure the elongation of a bar but in many robotic applications the structural requirements and choice of materials do not result in an elongation of sufficient magnitude. The most usual mode of use is for the measurement of the bending of bars, the direction of the bend being such that the length of the force sensor is changed.

Applying longitudinal strain to a resistor increases its length and reduces its cross-sectional area with the result that its resistance increases. Alloys that are used include copper-nickel, nickel-chromium and platinum-tungsten. The thickness is a few microns and the devices are usually supplied on a plastic backing film which acts as the insulator when it is fixed to the member to be measured. The performance of the device is very dependent upon the way that it is fixed to the stressed member. The gauge factor is the ratio of the fractional change in resistance to the fractional change in length. Poisson's ratio, p, for most gauge metals is about 0.3, and the gauge factor is given by[14]

$$\Delta R/R \;/\; \Delta L/L = 1 + 2p = 1.6 \text{ approx.}$$

In practice the measured values for many materials are about 2, and some are much higher. The values for the alloys given above are approximately 2.1, 2.2, and 4.0. This indicates that some other effect is occurring and this is the piezoresistive effect. The gauge factor is increased by the product of the piezoresistive coefficient and the Young's modulus of the material.

Although some metals have piezoresistive coefficients giving gauge factors higher than a pure resistor, the most significant coefficients are exhibited by semiconductors.(14) For *p*-type silicon the gauge factors range from +100 to +175 and for *n*-type, from −100 to −140. The value depends upon the doping level. Thicknesses are in the range ten to one hundred microns. Semiconductor gauges are light sensitive and should be shielded from strong light.

For a range of a few thousand microstrains, semiconductor gauges have a worse linearity than the metal gauges, typically 5% compared with less than 1%. Also, the semiconductor gauges are affected more by temperature. The temperature coefficient of resistance is from about 0.03% to 0.4%/°C compared with 0.01%/°C or less for metal gauges and the temperature coefficient of the gauge factor is typically 0.2%/°C compared with 0.01%/°C. Fortunately, compensation can be made for the nonlinearity and temperature effects by the correct arrangement of resistors on the stressed member and in the design of the measuring circuit. Temperature changes may also produce an output when the linear expansion coefficients of the gauge and the stressed member are not equal but some compensation can be achieved by placing gauges on opposite sides of the stressed member and using their outputs differentially.

The resistance changes are small and a common circuit configuration has the resistors in a Wheatstone bridge network.(14) The effect of temperature changes can be reduced by including in the bridge compensating resistors, possibly unstressed sensors, and/or placing together resistors with opposite polarity gauge factors, or placing similar resistors in positions where they will be stressed in opposite directions, i.e., on opposite sides of a bar which is being flexed. The effects of temperature can be reduced to less than 0.01%/°C. Accuracies of about 1% can be achieved with care. Hysteresis in the gauges themselves is very small and the most common cause of hysteresis in a system is due to creep in the mounting material. The sensitivity of strain gauge systems can be 0.1 microstrain and measurement times less than 10 μs, the latter set by the speed of the electronic circuit.

The sensors may be in one, two, or four arms of the Wheatstone bridge. In the first case temperature changes can affect the resistance of the connecting leads and a three-lead configuration can be used to reduce the effect. Two leads carry the power to the sensor and the third provides one of the bridge output connections. In this way there is one power lead in each of the adjacent arms of the bridge. The drive to the bridge may be ac or dc, constant voltage or current. The bridge and its leads should be shielded from interference. The resistance of some gauge materials is affected by magnetic fields. The use of connectors and slip-rings in the power carrying circuits within the bridge should be avoided due to their variable resistance. It is best if the whole of the bridge can be on the same side of such connections, with only the main power leads and the output leads being brought out.

In a robotic system the most common positions at which forces are required to be measured are at the end effector[(17)] or on the table carrying the workpiece.[(18)] The forces may be along the Cartesian axes at the position of the measurement or be torques about those axes. Ideally, the mechanical structure onto which the force sensors are to be built should cause changes in only a specific set of sensors for a given force or torque and those sensors should not respond to other forces. This may be possible when only a few of the forces are to be measured but for more complex systems the output from the several networks may have to be combined to produce measures of the individual forces. By virtue of its inbuilt compliance a force sensing unit has a significantly restricted strength and may be unable to withstand the extremes of force occurring occasionally in the equipment. To protect the device provision may be made to limit the maximum movement in each direction.

A simple structure is shown in Figure 6.4 in which the sensors are bonded to the upper surfaces of the four radial bars. The forces are applied between the inner disk and the surrounding ring. The system is sensitive to an axial force, all four sensors being affected equally, and by torques about the axes defined by the directions of the bars, when, in each case, only two sensors are affected, one positively and one negatively. The device would not sense the other two

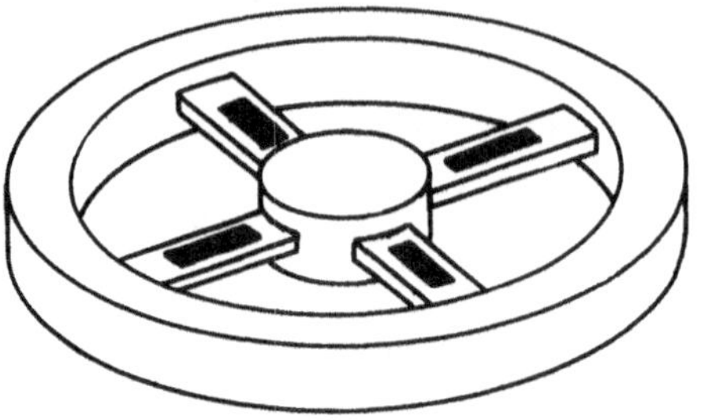

Figure 6.4. A force sensor for axial forces and torques about directions perpendicular to the axis.

forces perpendicular to the main axis or the torque about that axis. A system for the three forces and three torques is shown in Figure 6.5.

There are materials in which the applied stress affects their magnetic properties. Examples of magnetoelastic materials are alloys of iron-cobalt-boron (ICB) and of terbium-dysprosium-iron (TDI), the first in the form of amorphous ribbons and the latter as rods.[19,20] The material is processed so that the magenetic domains are aligned perpendicularly to the length of the material, but with adjacent domains in opposite directions so that there is no resultant magnetization. An external field is applied parallel to the length of the material, by a coil or permanent magnet, to bring the domains to about 45° to the field. The materials will then exhibit two magnetoelastic effects which change their internal magnetic moment, the Villari effect due to longitudinal stress and the inverse magnetostrictive effect due to changes in their length.

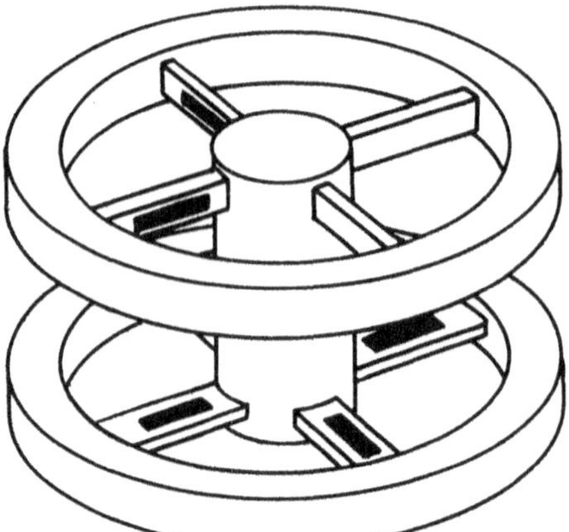

Figure 6.5. A force sensor for forces in any direction and torques about any direction.

The materials may be bonded to structural members in the same manner as the stress sensitive resistors. The value of their internal magnetic moment can be revealed in two ways. A coil is wound around the sensor and the structural member, assuming that this is not ferromagnetic, and changes in the magnetic moment induce voltages into the coil. These may be integrated and measured to obtain the total change in moment and hence in the applied force. It does not give an absolute measure. The TDI material, with low permeability but high magnetostrictive coefficient, is better suited to this mode of operation. In the second technique, more suited to the ICB material with its very high permeability, two coils are wound around the strip and the coupling between the coils is measured by applying an alternating voltage to one and observing the output from the other. Frequencies in the region of 10 kHz are used. This technique gives an absolute measure of the internal moment and hence the applied force.

The capacitive displacement transducer (see Section 6.1.3) has been used in a force measurement system in both the coaxial cylinder form and the parallel plate form.[14] A unit about 20 mm long has a capacity of about 1 pF and a gauge factor of about 0.02 pF per millistrain. Linearity is of the order of 1% for a range of 10 millistrain and the gauge factor can be temperature compensated to 50 ppm/°C. These force sensors are suitable for high temperature operation. In the parallel plate system of Figure 6.3, the insulated plates are mounted on the two arches which have common mounting points. Tension in the member on which the device is mounted will cause the plates to move apart and the capacitance to decrease. Inductive displacement transducers also have been used.[17]

Piezoelectric materials generate voltages between their polarization surfaces when subjected to stress.[6] Although the voltages can be high, the charge involved is small and is soon dissipated by leakage. In consequence they are suitable for detecting changes in stress but not for measuring static loads. The material may be crystalline, e.g., quartz, lithium niobate, or lithium sulphate, or polarized ceramic, e.g., lead zirconate titanate, barium titanate, or zinc oxide.

The magnetoelastic, capacitive, and many other force sensing systems have, in common with the resistive systems, low signal levels which require careful design to provide adequate screening and suitable processing of the signals. This may involve placing the front end of the circuits close to the sensors. The devices are temperature sensitive and compensation has to be included either by balancing the effects or by making a separate temperature measurement and applying a correction factor to the measurement. The measured parameter may not be linearly related to the force applied to the sensor and there can be a nonlinear relationship between the force on the sensor and the force to be measured. These nonlinearities may be eliminated by calibrating the system and storing the correction factors in a table.

6.1.5. *Tactile Sensors*

Tactile sensors are related to force sensors and many sensing devices may be used for both purposes.(21–23) The task of the tactile sensor includes, in some combination, the detection of the presence of an object and its shape and the forces being applied to the object, including the slipping of an object against the contact surface. The sensors may operate in a binary mode or give an analog output. When the sensor consists of a single cell it may be considered as a contact, proximity, or force sensor, as discussed in Sections 6.1.2 and 6.1.4. In this section arrays of sensors will be considered. Systems have been made based on conductive fibers,(24) conductive plastic sheet, piezoelectric sheet and total internal reflection. Resolutions are, typically, down to 1 mm.

Carbon fibers are mechanically very robust and return to their original state, with negligible hysteresis, after being subjected to very high compression. The resistance of the carbon is low but there is a high surface resistance. As a result the resistance of a pad of the material is determined by the area of contact between individual fibers and between fibers and electrodes.(25) A two-dimensional array might be constructed by placing small pads of carbon fibers between a flexible sheet carrying a ground plane and a rigid surface having an array of electrodes. The resistance of each pad would be monitored by voltages applied to the electrodes. Due to the high bulk conductivity of the fibers they must not join one pad to another. At low pressures only a few points of contact will exist in each pad, the number being very sensitive to vibration, and the resistance at each contact is variable. Consequently, the noise level is high but this drawback can be overcome by providing a small standing pressure on the array.

Conductive plastic sheet has the characteristic that its conductivity increases as the pressure applied to it is increased. A sheet may be placed between a rigid and a flexible surface each having conductors in contact with the plastic. The conductor patterns may be the same as those suggested for the carbon fiber system. Alternatively, the patterns may be parallel conductors arranged orthogonally on the two surfaces. If the thickness of the material is much smaller than the distance between the parallel conductors, conduction will be predominantly perpendicular to the surface. The array is scanned by applying a voltage between single conductors from each of the two sets, and measuring the resistance of the plastic at the crossover. To remove currents taking various routes through the other conductors, which can affect the measurement, the unused conductors on one side of the sensor should be held at the potential of the conductor being used on the other side of the array.

A system using conductive plastic sheet has been built with the sensing electronic circuits placed directly behind the sheet.[(26)] An integrated circuit is placed behind each element of the array with its surface in contact with the plastic. The circuits perform the signal processing and deliver the results to an external processor through a conventional computer bus. These circuits are protected by an oxide layer except where large electrodes make contact with the plastic. Current flows parallel to the surface of the plastic and the electrodes are arranged to minimize conduction between adjacent cells. Since each cell contains a processor and has communication with its neighbors, it is possible to perform two-dimensional processing of the data in ways similar to that of two-dimensional picture processing.

Conductive silicon rubber can be made with a low bulk resistance but a high surface resistance. A two-dimensional sensor array can be constructed by forming two layers of parallel rods of the material, one laid orthogonally across the other.[(27)] The conduction at each crossover will increase as the contact area is increased by increased pressure. The array is addressed in the same manner as that for the conductive plastic sheet. One layer of rods may be replaced by metal wires or printed conductors on a rigid surface. The whole array would be protected by a flexible insulating sheet.

A system has been made using the magnetoresistive effect in materials such as nickel-iron alloy, 81%–19% (see Section 6.1.6). The magnetoresistors are connected to form rows and are controlled by magnetic fields from parallel conductors, laid orthogonal to the rows. The conductors and resistors are separated by a compressible insulating sheet. One resistor in each row is "read" by applying a current in one of the conductors and measuring the resistance change in each row. Compressing the sheet increases the field at the resistor and increases the change in the value of the resistor. The array has 8 × 8 resistors in an area 25 × 25 mm.[28]

Some plastics are piezoelectric, e.g., the polymer polyvinylidene fluoride, and will generate charge between their surfaces when subjected to pressure. To induce this property the material has to be polarized. A high voltage is applied to the surfaces of a sheet while it is being stretched. The property can be destroyed by high voltages or temperatures above 70 °C. The sheet is fixed to a rigid surface carrying an array of electrodes and a common flexible electrode placed against the other side of the sheet.[29] Changes in the pressure in each element change the potential of each electrode. The time constant for the decay of each charge, due to leakage, is of the order of a tenth of a second; thus after each change there is a limited time available to sample each of the electrodes. Since the system makes measurements of changes, and is not an absolute sytem, it is necessary that a known condition, e.g., no forces applied, occurs frequently.

When light is shone into the edge of a sheet of plastic or glass whose main surfaces are highly polished, the light cannot escape through these surfaces due to the phenomenon of total internal reflection. Placing a pliable object, e.g., a finger, against the surface reduces the difference between the refractive indices and light can leave the surface and be scattered by the object. The angles of incidence are now such that the rays will pass through the surfaces and the fingerprint can be seen from the opposite side of the glass. With a sheet of plastic against the glass, pressure on the plastic will increase its contact with the glass and increase the amount of light that is scattered. Thus the pressure pattern is converted into a grey scale picture which may be scanned by a television camera.[30] Textures may be applied to the surface of the plastic to improve the linearity of the scattered-light-to-pressure ratio. One that has been used consists of minute cones.[31]

Two-dimensional arrays of pressure sensors give a contact picture of the object which is against the sensor. This includes its outline shape and, if an analog system is used, some information on the profile of its surface.[32] Movement of the pattern indicates that slip is occurring.

6.1.6. Other Sensors

Temperature can be measured using resistance thermometers, thermistors, and thermocouples.[6,7,33] The first two have high temperature coefficients of resistance and the third generates a temperature-dependent voltage across the junction of dissimilar metals (the Seebeck effect). The thermocouple has at least one other junction, also generating a voltage, which is kept at a known temperature. The sensitivity of the thermocouple depends upon the materials used and, at the highest, is about 70 μV/°C temperature difference. Thermistors are more sensitive than the resistors but are significantly nonlinear. They have positive or negative temperature coefficients of a few percent per degree centigrade and can have stabilities of 0.1 °C/annum. Also, by using beads of less than a millimeter diameter, very short time constants are possible, the time depending upon the conductivity of the surrounding medium. The temperature coefficient of resistance thermometers is about 0.5%/°C. Doped silicon can be used as a resistance thermometer and has a coefficient of about 1%/°C.[34]

The semiconductor diode is a temperature sensor and when current fed, the voltage across the device varies approximately linearly with the temperature. The coefficient is current- and device-dependent and, typically, is about −2 mV/°C for small currents and becomes less negative as the current is increased. In some cases it can become positive. The transistor can be used,[35] with constant current feeds to the base and collector, with an accuracy of 0.1 °C. The coefficient is of the order of +1 mV/°C, but for some devices is not so linear as the diode. It is current dependent, the coefficient increasing with increasing current.

Temperature sensors may be used to measure thermal conductivity, which can be useful in discriminating between objects. The sensor is combined with a heat source and the temperature measured.[36] The thermistor can be made to be its own heat source by passing a high current through it. The higher the thermal conductivity of the object is, the greater the drop in the temperature of the sensor. As a reference, and to eliminate the effect of the ambient temperature, a second sensor can be mounted near the first but not in contact with the object.

Inductive devices may be used to measure changes in magnetic field strength but they do not give absolute measurements. For this purpose the Hall effect probe can be used. When a current passes through a conductor, orthogonal to the direction of the magnetic field, a force is applied to the current in a direction orthogonal to the field and the current. The displacement of the current generates a potential, in this direction, which is proportional to the product of the magnitudes of the current and the field. The effect is strong in semiconductors, particularly indium antimonide and arsenide, and gallium arsenide, and fields less than 1 mT can be measured. Magnetic fields affect the electrical resistance of some materials, e.g., nickel-iron alloy, 81%–19%.[(37)] The devices using these materials can be used at frequencies above 100 MHz, and at very low frequencies, 1 Hz, fields of 1 nanotesla can be detected.

X-ray, Röntgen, radiation is detected by the use of scintillators fitted to light detectors.[(6,38)] Suitable scintillator materials, such as caesium iodide, cadmium tungstate, and sodium iodide, can be placed close to photodiodes or photomultipliers. The X-radiation is absorbed by the atoms of the scintillator and is reemitted as light as the electrons of the atoms return to their base state. A linear array of X-ray sensors has been constructed from caesium iodide on silicon photodiodes.

The microphone is an important transducer and uses either capacitance changes, the piezoelectric effect, or the induction of current into a coil moving in a magnetic field. It is used to receive sound emitted by working equipment to monitor its performance. Typical applications are testing car engines, in which the microphone may be moved around the engine by a robot, monitoring a lathe to detect tool wear, and testing shavers and other motor driven devices. The frequency range used may extend above and below that of the human. The analog signal is processed either by monitoring its voltage/time function or by measuring the relative amplitudes of its components in several narrow frequency ranges (see Section 6.4). Another application is in "schizeophonics" in which castings, etc., are struck and the sound that they produce analyzed to detect the presence or absence of cracks.[(39)] A crack will change the relative distribution of energy across the frequency band and also change the rate of decay of the sound in certain bands.

Sensing within a process may be performed by measuring a parameter within that process. A good example is the monitoring of the current and voltage in a welding arc to provide a control for the position of the torch over the weld, its velocity along the weld, and the feed of the welding wire. Another unusual example is the monitoring of the capacity between the shearing tool and the skin of a sheep during the shearing of sheep by a robot.[40]

6.2. Range Finders

The output from a scanning system is usually a two-dimensional picture, but it may be useful to have information about the third dimension. This can reveal the position and the shape of an object. The distance from a sensor to a point in a scene may be determined by methods which basically involve time of flight or triangulation. The latter is the basis for stereoscopic, structured light, and focusing systems. Also, by following the movement of an object within a scene, and using the knowledge that the object is not changing its dimensions, it is possible by comparing successive frames of the scan to deduce the position and orientation of the object in three dimensions. The procedure operates at its best with clearly defined objects. Short distances may be measured by proximity detectors which have been calibrated (see Section 6.1.2).

6.2.1. Time of Flight of Light

Light travels in air at about 300 mm/ns and by measuring the round journey time between the sensor and the point on the scene, the distance can be found. In the simplest case, a very short light pulse is generated and directed at the scene. If the distance to different points in the scene varies significantly, the source must be imaged onto the scene or, more usually, the narrow beam from a laser is used.[41] A deflection system will be required to select the required point in the scene. The returned light is collected and detected. Again the laser source provides an advantage in that the detector can be protected from the ambient lighting by a narrow-band filter. Circuit delay can be eliminated from the measurement by feeding a small part of the light from the source directly to the detector. The detector sees two pulses and the time between these is a measure of the range.

For the distances involved in most vision systems the times will be of the order of nanoseconds (1 ns = 150 mm range) and discrimination of tens of picoseconds may be required. This is not achieved easily.

An alternative method uses a sine wave modulated light source. The phase of the detected signal is compared with that of the modulating signal for the source or with that from a second detector receiving light directly from the source. In either case this will involve unknown phase shifts in the two channels and it is necessary to calibrate at a known range. Typically, modulation frequencies of the order of 5 MHz will be used. Direct phase measuring techniques can be used but high resolutions may be obtained by using the superhetrodyne method of shifting both signals to a low frequency, e.g., 100 Hz, by beating with a common oscillator, and measuring the phase between these. The phase difference is the same at the low frequency as at the high. Resolutions of 25 mm, 0.25° of phase, have been achieved.(42)

For the pulse or modulation techniques it is necessary to use a source capable of high frequency switching or modulation. Most lasers are suitable, some being directly modulated and others requiring a separate device to modulate the light beam itself. The ordinary LED may be a suitable source for a simple system. The detector also must have a high frequency response and usually a high sensitivity. Photodiodes, especially PIN diodes, may be adequate but alternatively avalanche devices or photomultipliers may be used.

6.2.2. *Focusing*

The amplitude of the higher spatial frequencies in an image reach their maximum, i.e., the image is sharper, when the system is in focus. Under such conditions if the image to lens distance is known, from a transducer coupled to the focusing mount, the object to lens distance can be calculated. The video output from a one- or two-dimensional scanner can be analyzed for frequency content, or signal rise time across a sharp edge, while the position of the imaging lens is varied. The technique can be used for a system in which the scan is performed by a light spot moving across the scene, e.g., a scanned laser, and in this case it is the projection lens that is focused. Various mechanical systems can be used to move the lens or part of the lens. A loudspeaker coil and magnet allows rapid movements to be made. Unfortunately, the direction in which the lens has to be moved cannot be deduced from a single scan and a trial and error procedure must be employed.

Examination of the depth of field equation

$$\text{Depth of field} = \pm N(1 + M)d/M^2(1 \mp d/MD)$$

shows that for long ranges, i.e., M small, and for a fixed quality of focus detection, the error increases at least as fast as the distance, so that the technique is not suited to ranges long in terms of the focal length. Also, the formula shows that the error is less if N is small, i.e., the lens aperture is large, and it may be necessary to use a large aperture for range measurement before reducing the aperture for scanning. Whether this is permissible will depend upon whether the extra light flux can be tolerated, although a neutral filter could be moved into the optical path. In this respect the aperture of the lens is equivalent to the baseline of a triangulation system.

A more complex system can be used which will indicate in which direction the system is out of focus. If, from the sensor side of the lens, the lens is viewed as two areas divided by a diameter, then, in general, the integral brightness of the areas will be different.[(43)] The direction of the difference will correspond to the brightness gradient in the scene if the image is formed behind the viewing point. A line of detectors is placed in the image plane of the system with each adjacent pair sharing a small lens, so that each receives light from only the left or right side of the main lens. When the outputs from the line of "left-hand" sensors are compared with those of the "right-hand" sensors, a similar spatial pattern will be found. If the system is "in focus," the sensors of each pair will give equal signals and the two patterns will be aligned, but if it is "out of focus" the two patterns will be displaced. The direction and degree of displacement indicates the direction and degree of "out of focus." This can be used to measure the range to the scene.[(44)]

6.2.3. *Triangulation*

The first triangulation technique in Figure 6.6 requires a scanning sensor, e.g., a television camera, and a means for projecting a spot of light into the scene at the required place, e.g., a galvanometer mirror laser scanner, with a known distance, the baseline, between the two

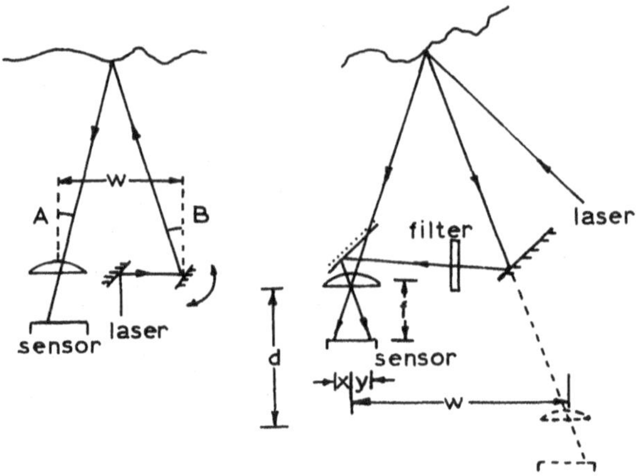

Figure 6.6. Two range finders using a laser. In the first system the range equals $w/(\tan A + \tan B)$, and in the second, $(w.f - d.y)/(x + y)$.

units. From the baseline, the angle of projection of the spot, and the angle perceived by the sensor, the range of the spot in the scene can be determined. For ranges which are several times the baseline, the error, for a given accuracy of angle measurement, is proportional to the square of the range. If a television camera is used as the sensor, compensation must be made for nonlinearities and scan drift. This may apply to some other scanners. The system can be calibrated by placing a flat surface at various distances, and at each distance moving the spot across the surface. Naturally, the spot must be bright enough, allowing for various reflectivities within the scene, to be visible to the detector. If a laser is used, a narrow-band filter can be placed in front of the detector during range measurement.

The second system shown in Figure 6.6 makes all of the measurements at the camera.[45] Deflection of the spot is used only to mark the required point in the scene. The camera views the scene through a partially reflecting mirror and also by reflection from that mirror and a totally reflecting mirror parallel to the first. In this second channel all but the light from the laser spot is blocked by a narrow-band filter. In effect, the position of the camera for the laser spot alone is shifted by distances w and d as shown. In the equation shown for the range, $w.f$ and d are constants of the system which may be determined by calibration. The distances x and y are determined by processing the picture.

In both of the above systems an electronic circuit is required to process the video and determine the spot positions. By ensuring that the apparent brightness of the spot is greater than any other part of the scene, it is possible to use a simple thresholding circuit to detect the spot. The contrast may be improved by moving a filter into the light path when the measurement is to be made. Due to low levels of returned light there may be difficulties in determining the distances of dark surfaces or of reflecting surfaces which direct the light away from the camera. Also, it is possible for the laser spot to be hidden from the camera by a feature of the scene. That the highlight observed really is the spot can be tested by moving the spot or switching off the beam.

The sensor can be the position sensitive diode (see Section 4.3.10), and with a laser diode allows a compact range sensor to be produced. No scanning system is provided for the laser and the path of the laser beam is imaged onto the one-dimensional diode. The system will detect the position of an object intersecting the path of the beam. This unit is available as a standard product. It is small enough to be handled by a robot and used as a noncontacting measurement device.[46–48] Accuracies after calibration are of the order of 0.05%. Systems combining the position sensitive diode with a scanned laser are also possible.[49]

Instead of projecting marker beams into a scene, the light sources can be within the scene, carried by the device, e.g., a robot end effector, and viewed by the camera. In one system an array of sources, e.g., four LEDs at the corners of square, are used. The sources are switched on one at a time and their positions noted. As the actual separation of the sources is known, the position in space of the array and its orientation can be determined uniquely. It is possible to use arrays which remain on at all times. Usually, irregular patterns must be used, but there is the risk of obtaining ambiguous results.

6.2.4. *Stereoscopy*

The technique of stereoscopy has been used to determine three-dimensional information from pairs of stereo-photographs, e.g., of the Earth's surface, with a human operator matching features in the two pictures. Resolutions of one in ten thousand have been achieved. The problem becomes more difficult when this is to be done automatically within a reasonable time.

By scanning the scene from two positions, a known distance apart, two similar video signals are obtained which differ in the relative positions of the images of parts of the scene which are at different distances from the scanners. In this stereoscopic system it is then necessary to identify in each image those parts of the scene which are of interest. This may be done by extracting appropriate features or by the correlation of small areas of the scene. In the past this technique has not been practicable within a reasonable time due to the amount of processing required. This limitation is becoming less with the advent of fast processors. It is necessary that nonlinearities in the scan pattern be known and compensated for in the processing. Most scanning systems can be adapted for the technique.[(50–54)]

Where two cameras are used, an element of the sensor in one defines a line in space, passing through that element and the center of the lens. Points along this line are imaged as a line on the sensor in the other camera. Therefore, when a specified feature has been identified in the two views, the position of that point in space can be found.[(55)] If each array has N by N elements, then the number of elemental volumes in the space visible to both cameras is approximately equal to N^3. The sizes of the elemental volumes will not be equal.

6.2.5. *Structured Light*

The generation of and some uses for structured light were considered in Section 5.2.5. Not only does this technique reveal the shape of an object but, when the position of the scanner and the positions and angles of the light sources are known, it is possible to compute the position of the object. Again, the basis of the system is triangulation. The light rays define lines in three-dimensional space and the meeting points of these lines and the object is imaged onto elements of the sensor in the camera. The computation can be simplified, and perhaps greater accuracy achieved, by measuring the apparent separation of the meeting points for adjacent rays. In some cases knowledge of the dimensions of the object may aid the process.

In Section 5.2.4., with reference to Figure 5.17, it was shown that the variation of the apparent brightnesses of surfaces can reveal the shape of an object. If it is known that a set of surfaces in a scene have the same reflectances or that the laws and levels of reflectance of the various surfaces are known, and also that the illuminance of the scene is nonuniform and known, then it is possible to deduce some information about the positions of surfaces from their apparent luminances.[56] The varying illumination could be provided by a small source near the camera. The illuminance would then be related to the inverse square of the distance to the surface. The angles of surfaces are revealed by changes in the apparent luminance due to the law relating luminance and angle. Due to the uncertainties in the reflectance laws and illumination levels, this technique offers only low accuracies.

6.2.6. *Ultrasonic*

The speed of sound in air, 344 mm/ms at 15 °C, is about a millionth of that of light, and consequently the use of a time-of-flight system is more practicable for a sonic system than for one using light.[57] The speed increases with temperature by about 0.6 mm/ms/°C, i.e., about 0.2%/°C, and allowance may have to be made for this. A pulse of sound may be transmitted and the echo detected by the same or another transducer, and the time of flight measured directly.[13] Alternatively, a continuous signal may be transmitted by one and the echo received by another. The phase between the transmitted and received signals is a measure of the range. So that the transducers may be small and the beam angle narrow, typically 2° to 10°, it is necessary to use a high frequency. The phase measurement results in ambiguity in the range at small separations, e.g., at 10 kHz the same phase is given at intervals of 17 mm. This may be overcome by using two signals simultaneously at slightly different frequencies. The phases of the two frequencies may be measured separately and the range computed, or the phase of the beat frequency, revealed by mixing the two signals nonlinearly, may be measured. Alternatively, a high frequency signal modulated by a low frequency may be used and the phase change in the modulation measured. Resolutions of the order of 1 mm are achievable.

Multiple ultrasonic sensors may be used to obtain a picture of a scene by using sound reflected from the scene. The number of sensors that can be used is far less than that in vision but, in compensation, extra information is derived from the time function of the sound at the receivers.[58]

6.3. Signal Processing

Rarely is the output of a scanning system in a form which can be fed directly to a computer. The output of most scanners is a serial analog signal on a single connection. There may be timing or addressing information on other connections or the timing information may be combined with the video signal as in the composite television signal (Section 4.2.1.6). In the latter case the timing and video signals must be separated in the interface. In some cases the video signal may be in digital or binary form although usually this means that a part of the preprocessing has been performed. Often the pixel rate requires that special hardware be provided for these purposes, otherwise the task would absorb all of the time of the computer. Separate hardware will free the computer for less mundane tasks.

The situation for the output from other sensors is similar to that for the vision scanner. In most cases, the most significant difference is that the frequencies involved are very much lower, usually not more than 100 kHz. Some processes that will be applied to these signals have a parallel in the processing of video signals but they may also require special processes of their own.

The effect of temperature changes on a sensor may be compensated by using the output from a temperature sensor to derive a correction. This may be performed directly by analog circuits but usually a digitized value of the temperature is used to acquire a correction value from a store, which is either used in a digital correction process or converted to analog and used for the correction.

The digital output from a nonlinear sensor may be used to access a store to obtain a correction value which is added to the original value. This is a better technique that having one corrected value for each possible input value because one correction value serves for a range of input values and the storage space is reduced.

The following are some of the basic processes that can be performed in special hardware, although some could be done in a computer. Circuits are available to perform the Fast Fourier Transform process, an efficient process for converting from the time or space domain to the frequency domain, or vice versa. Special integrated circuits are available for high speed processing and for the processing of analog signals (Section 6.4).

6.3.1. *Thresholding*

The video signal is compared with a predetermined level, the threshold, and the result output as either of two, binary, levels depending upon whether the video is greater or less than the threshold. In practice the video and the threshold level are usually presented as voltages to the inputs of an analog comparator. The threshold level need not be constant. Its value may depend upon the position in the scan and also it may vary with time. The threshold may be set at a fraction of the peak signal level, i.e., white, and a peak following circuit will be used to detect and hold the peak values. A slow decay, of suitable time constant, will be provided to follow any long term fall in the peak value. A fraction of the output from this circuit is used as the threshold. A variant of this is to set the threshold at a level which is a proportion of the range between the highest and lowest peaks of the video, Figure 6.7. This method is used in character recognition

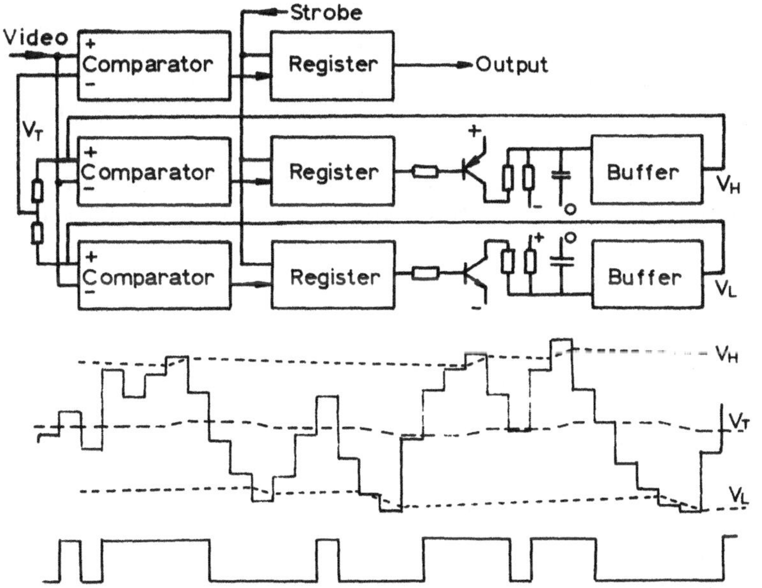

Figure 6.7. A circuit for thresholding a signal at a level V_T which is preset relative to the peak high, V_H, and peak low, V_L, levels. The lower waveform is the thresholded output.

systems to follow the white of the paper and the ink of the characters. For this system, two peak-following circuits, one positive and one negative, are required.

It is possible to store different thresholds for different pixels or different areas in a scene. This may be done using analog storage systems, e.g., CCD delay lines, for small numbers of thresholds. Alternatively, the thresholds may be stored in random access memory (RAM) as numbers and as each is accessed, corresponding to the appropriate position in the scan, it is converted to an analog value using a digital-to-analog converter (DAC) (Figure 6.8). In either case, the analog output from the threshold store is compared with the video, as described above, and, if required, the value in the store is adjusted to follow the trend in the video signal. A fraction of the stored signal is used to threshold the video in a second comparator. A system such as this will compensate for sensitivity variations or fixed offsets between separate areas of a scan due to uneven lighting or scanner nonuniformity. The process can also be performed in the digital domain after converting the video to a digital value.

The histogram of the grey values in a scene can be used in setting the level for a threshold to divide areas of differing grey level.[59] Often the relationship between the grey level and the number of pixels having that level will show a dip between the grey values corresponding to the two areas (Figure 6.9). The threshold can be set in this dip so that variations in the lighting and sensor sensitivity have less effect on the resulting binary picture.

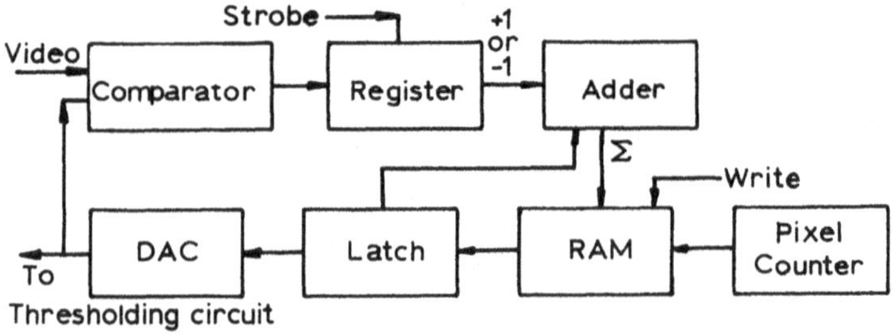

Figure 6.8. A circuit for deriving a threshold level which follows gradual changes in the signal level. For objects passing on a belt, this level would correspond to the belt.

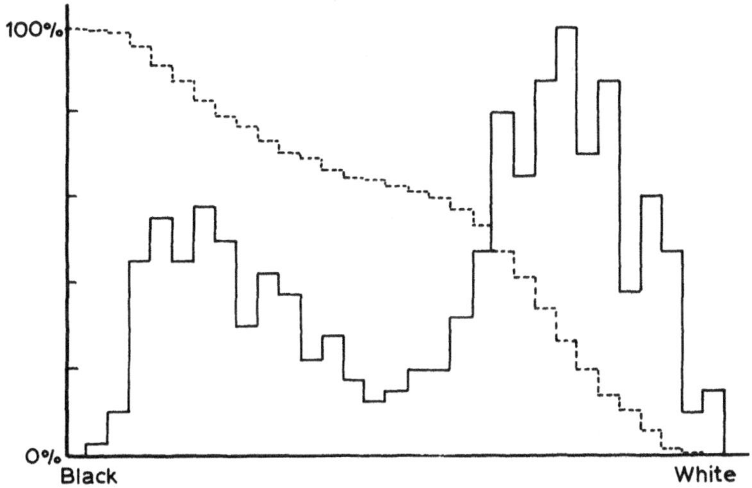

Figure 6.9. Histogram of a scene with two bands of gray level. The dotted curve shows the percentage of points which are whiter than each grey level.

6.3.2. Video Gain Correction

The gain of a system may be controlled in a similar way. In this case the video is not converted to binary but instead is attenuated in proportion to the stored value. The output analog signal is thereby corrected for gradual changes in, and spatial variations of, the illumination. When a digital storage system is used, the gains can be preset at the start of scanning by presenting the scanner with a uniform scene, e.g., a white card. In the learning mode, appropriate values are found and stored to correct for lighting and sensitivity variations.[(60)] These are not changed during subsequent scanning of the scene but are used to control the gain and perhaps correct for offsets. Analog stores cannot hold signals for long periods without variation and are not suitable for this task.

A nonuniform gain/frequency response can be used to improve the resolution of a scanning system. The effect of the size of a scanning spot or element in an array is to reduce the high frequency content of the signal; therefore, by boosting the higher frequencies some improvement can be obtained. A better result is obtained by "aperture correction." In this the signal is delayed by about a pixel period and has subtracted from it attenuated voltages from the undelayed signal and one delayed by a further pixel period. That this produces an improvement is apparent if one considers the effect on a Gaussian pulse about one pixel wide at half height. For two-dimensional aperture correction, the signal resulting from the first correction passes through two lines each having a delay of a line scan period. The corrected output is the signal from the first delay minus attenuated signals from the undelayed and doubly delayed signals.

6.3.3. Analog-to-Digital Conversion

For the computer processing of analog values, the analog input can be converted to digital form. Usually 8 bits are adequate, giving a resolution of 1 in 256. Noise in the scanning system sets a limit to the resolution that is meaningful but allowance should be made for a possible error of one least significant bit from the analog-to-digital converter (ADC). Many types of ADCs exist. The successive approximation types, in which steps, up or down, are made in decreasing binary values until a generated voltage matches the video, and ramp types, in which the time is measured for a voltage to rise linearly from zero to the level of the input signal, are usually too slow for video systems. The fastest are the flash converters. In these, a reference voltage feeds a multiresistor potentiometer, and a comparator at each output of the potentiometer compares that voltage with the video signal. The binary results of the comparison will change polarity at one point down the potentiometer and the outputs can be decoded to produce the digital value corresponding to this point.

The resolution of these systems is limited by the number of comparators required and for higher resolution systems a two-stage process is used. The output of the first stage is as above and forms the more significant part of the final result. It drives a DAC to produce an equivalent voltage which is subtracted from the video input and the difference is amplified. This difference is fed to a second comparator chain to derive the less significant part of the result. Sample rates in excess of 15 MHz are possible.

Most ADC devices, which may be integrated circuits, thick film modules, or on printed circuit boards, impose limits on the permissible rate of change, slew, of the input signal. If the video varies faster than this or there are switching transients, as in the output from some solid state cameras, it may be necessary to provide a sample and hold circuit before the ADC. Some ADC devices contain their own sample and hold circuits.

6.3.4. Correlation and Operators

Many processes (Section 8.4) require that the picture be compared with a stored pattern or operated upon by a one- or a two-dimensional matrix array corresponding to a linear or area section of the picture. It is not always required that the reference pattern be that of a solid line or area; it may be formed from scattered points. A common construction has the reference pattern held in registers and the incoming picture passes through a shift register. Because many lines of the picture may be required in the matching process, sufficient shift register space must be provided. The data in the reference registers may be loaded from the control computer prior to the processing or be permanently "wired in," in which case the register is no more than the connection to an appropriate voltage source.

The form of the comparison between corresponding stages of the shift and reference registers depends upon the process to be performed. The values to be compared may be single binary levels, i.e., black and white, multilevel or analog. In principle, the process will be the multiplication of the values at each stage and for a binary system this can be performed by an Exclusive-OR gate. The form of the data in the two registers need not be the same. For example, an analog signal may be multiplied by a multilevel binary value by the use of a multiplying DAC, the output being an analog signal. An analog signal can be multiplied by a single binary level by the use of an analog, on/off, switch. Circuits are available for the multiplication of two analog signals.

The outputs from each stage are combined to find the result of the operation by a summation process. For analog outputs in the form of currents, this is simply the summation of the currents. Adders can be used for multilevel outputs and counters for binary outputs. Frequently the value of the combined output is not required but only a signal indicating how it compared with a preset threshold or a set of thresholds. This thresholding may be performed by an analog or digital comparator. When a perfect match is required with binary outputs, the decision circuit is a multi-input AND gate.

6.4. Computer Interfacing

The term "computer" should be interpreted as including microprocessors and process controllers, which differ only in the range of peripherals and facilities that are provided. The data from the scanning system or other sensor system, in a binary form of one or many bits per pixel, can be passed to a computer directly through the computer bus (Section 10.2.4) or through a standard interface unit, e.g., of the RS 232 or the IEEE-488 type. The interfaces may accept data in a serial stream on a single line or in parallel on a multiline connection. Most such systems are byte or word orientated and it is economic in transfer time to pack single-bit pixels into bytes, using a shift register, or to pack bytes into words. Auxiliary timing signals will be required to control the transfers, perform handshake facilities, and in other ways satisfy the protocol of the interface. Also, appropriate software is required for the computer to control the interface. When the timing between the scanner and the computer cannot be synchronized, some intermediate storage of the data may be required. If the lack of synchronization extends to several bytes, than a FIFO, first in–first out, register may be required as a buffer. Other information that may be passed along the bus includes signals to control the scanner and status information from the scanner. Many standard boards are on the market for connecting sensors to computer systems.

In more complex processing systems additional circuits may be provided for storing picture data, e.g., frame stores, for executing special processes on the picture (Section 6.3), e.g., correlation with stored reference pictures, or for more generalized processing perhaps of the type possible with parallel processing arrays.[61] The processing may involve the incoming picture or be applied entirely to stored data. These provisions are made to perform the processing at speeds much higher than that possible with a digital computer and to reduce the loading on the computer and allow it to perform higher level tasks more suited to its structure. These high-speed processes and the input circuits must communicate at higher speeds than is possible under the computer bus protocol and therefore will be provided with their own picture bus with its own specialized protocol. The circuits may also use the computer bus for control purposes and for the transfer of picture data, or the results of their processing, to the computer. A typical structure is shown in Figure 6.10. Besides receiving signals, the

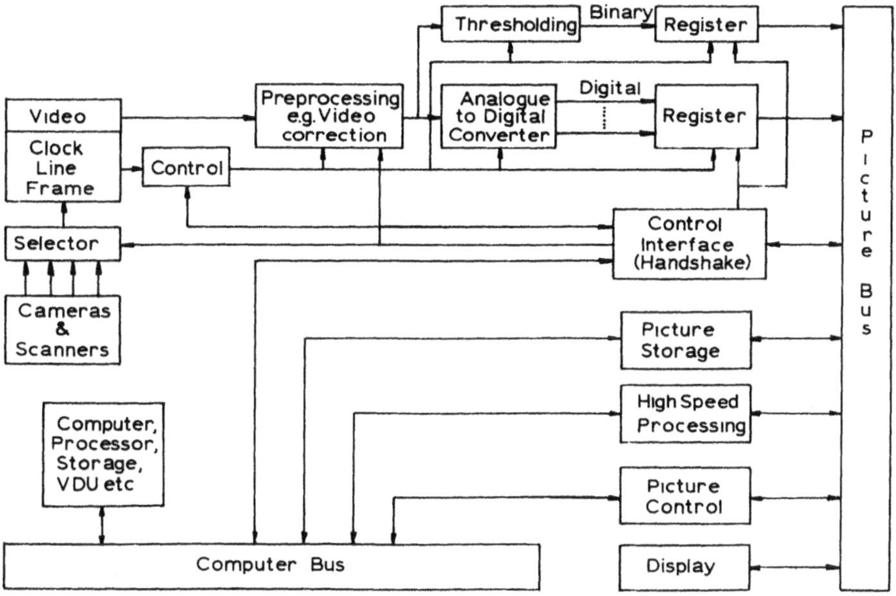

Figure 6.10. The structure of a general purpose video processing system.

system may generate signals, for example to maintain a television display. In this case it is usual, but not essential, to hold the picture in a separate store and provide circuits which independently access the store and provide the appropriate timing signals for the output. Further information on hardware for vision and signal processing is given in Chapter 10.

Two types of general purpose integrated circuits are now available for high-speed processing under software control. An example of the first is the Am29116 from AMD which is a 16-bit, high-speed digital processor capable of performing a single instruction in 100 ns.[62] A single instruction can include both of the related data input and output routines. Besides performing the simpler arithmetic and logic functions, it can operate on individual bits which is a facility often very useful in picture processing. The Texas TMS320, AMI S2811, and Signetics 8X305 are devices of this type. The second type, including the Intel 2920, are similar to the first type but have an ADC on the chip and therefore can accept an analog signal. At present these devices are limited to frequencies of a few tens of kilohertz by the ADC circuit. Many specialized integrated circuits have been produced for specific purposes, including the processing elements of parallel processing systems.[63–65]

With the introduction of the single-board computer and powerful computers on a single integrated circuit, the trend is towards the use of many processors in one system with communication between the systems. For example, individual sensors will have their own processor to correct for nonlinearities and temperature coefficients and present the data to the system in the form most suited to its subsequent use. Each drive in a robot will have its own controller, and receive from and send to the rest of the system only simple control data. Not only does this simplify the construction of the original equipment, but also it simplifies modifications and allows a higher operating speed.

The output of many sensors shows a similar form to that of video signals. Tactile devices may also produce two-dimensional information, although at a much lower resolution than that of vision systems (Section 6.1.5). Microphones, thermometers, force sensors, etc. produce an output which is a function of time, and these sensors can be interfaced in ways similar to those for the video signals. The most significant difference in many of these is that they operate at much lower frequencies and, in consequence, analog-to-digital conversion and transfer to the computer are simpler. Also, many of the analog signal processors are capable of operating on these signals.

References

1. J.A.G. Knight, Sensors for robots: the state of the art, *Proc. 2nd European Conf. on Automated Manufacturing,* 127–132, Birmingham (May 1983).
2. M.D. Stanton, H.S. Gill, and N.D. Crisp, Grippers—sensors and their control aspects, *Proc. 6th British Robot Association Annual Conf.,* 235–245, Birmingham (May 1983).
3. C. Meier, Sensor technology with robot-mounted sensors, *Proc. 4th Intl. Conf. on Robot Vision and Sensory Controls,* 231–240, London (Oct. 1984).
4. J.M. Richardson, K.A. Marsh, and J.F. Martin, Techniques of multisensor signal processing and their application to the combining of vision and acoustical data, *Proc. 4th Intl. Conf. on Robot Vision and Sensory Controls,* 395–408, London (Oct. 1984).
5. W.D. Koenigsberg, Noncontact distance sensor technology, *Proc. SPIE Intl. Soc. Optical Engineering* (USA), *449,* 519–531, Rovisec3 (1984).

6. J.A. Allacca and A. Stuart, *Transducers—Theory and Applications,* Reston Publishing Co., Inc., Virginia (1984).
7. H.N. Norton, *Sensor and Analyser Handbook,* Prentice-Hall, Inc., New Jersey (1982).
8. G.A. Woolvet, *Transducers in Digital Systems,* Peter Peregrinus Ltd., London (1977).
9. M. Schweizer and I. Schmidt, Computer controlled magazining system, *Proc. 3rd Intl. Conf. on Assembly Automation,* 379–395, Boeblingen, West Germany (May 1982).
10. H. Clergeot, D. Placko, and F. Monteil, Flexible eddy current sensors for industrial applications, *Proc. 4th Intl. Conf. on Robot Vision and Sensory Controls,* 115–122, London (Oct. 1984).
11. G. Belaforte, N. D'Alfio, F. Quagliotti, and A. Romiti, Identification through air jet sensing, *Proc. 1st Intl. Conf. on Robot Vision and Sensory Controls,* 263–272, Stratford-upon-Avon (April 1981).
12. *Synchros Conversion Handbook,* ILC Data Device Corp., New York (1982).
13. R. Dillmann, A sensor controlled gripper with tactile and non-tactile sensor environment, *Proc. 2nd Intl. Conf. on Robot Vision and Sensory Controls,* 159–170, Stuttgart (Nov. 1982).
14. A.L. Window and G.S. Holister (ed.), *Strain Gauge Technology,* Applied Science Publishers, London and New Jersey (1982).
15. J. Pople, *BSSM Strain Measurement Reference Book,* British Society for Strain Measurement, Newcastle-upon-Tyne (1979).
16. K. Bethe and D. Schon, Thin-film strain-gauge transducers, *Philips Tech. Rev. 39*(3/4), 94–101 (1980).
17. G. Piller, A compact six-degree-of-freedom force sensor for assembly robot, *Proc. 12th Intl. Symp. on Industrial Robots and 6th Intl. Conf. on Robot Technology,* 121–129, Paris (June 1982).
18. C.H. Spalding, A three-axis force sensing system for industrial robots, *Proc. 3th Intl. Conf. on Assembly Automation,* 565–576, Boeblingen, West Germany (May 1982).
19. J.M. Vranish, E.E. Mitchell, and R. Demoyer, Magnetoelastic force feedback sensors for robots and machine tools, *Proc. 12th Intl. Symp. on Industrial Robots and 6th Intl. Conf. on Robot Technology,* 131–142, Paris (June 1982).
20. J.M. Vranish, E. Mitchell, and R. DeMoyer, 'Outstanding potential' shown by magnetoelastic force feedback sensors for robots, *Sensor Review 2*(4), 200–205 (Oct. 1982).
21. R.N. Stauffer, Progress in tactile sensor development, *Robotics Today 5*(3), 43–49 (June 1983).
22. L.D. Harmon, A sense of touch begins to gather momentum, *Sensor Review, 1*(2), 82–88 (April 1981).
23. L. Harmon, Tactile sensing for robots, in: *Robotics and Artificial Intelligence* (M. Brady, L. A. Gerhardt, and H.F. Davidson, eds.), Springer-Verlag (1984).
24. B.E. Robertson and A.J. Walkden, *Tactile sensor system for robotics, Proc. SPIE Intl. Soc. Optical Engineering* (USA), *449,* 572–577, Rosivec3 (1984).
25. M.H.E. Larcombe, Carbon fibre tactile sensors, *Proc. 1st Intl. Conf. on Robot Vision and Sensory Controls,* 273–276, Stratford-upon-Avon (April 1981).

26. M.H. Raibert and J.E. Tanner, A VLSI tactile array sensor, *Proc. 12th Intl. Symp. on Industrial Robots and 6th Intl. Conf. on Robot Technology,* 417–425, Paris (June 1982).
27. J.A. Purbrick, A force transducer employing conductive silicone rubber, *Proc. 1st Intl. Conf. on Robot Vision and Sensory Controls,* 73–80, Stratford-upon-Avon (April 1981).
28. J.M. Vranish, Magnetoresistive skin for robots, *Proc. 4th Intl. Conf. on Robot Vision and Sensory Controls,* 269–284, London (Oct. 1984).
29. P. Dario, R. Bardelli, D. de Rossi, L.R. Wang, and P.C. Pinotti, Touch-sensitive polymer skin uses piezoelectric properties to recognise orientation of objects, *Sensor Review 2*(4), 194–198 (Oct. 1982).
30. D.H. Mott, M.H. Lee, and H.R. Nicholls, An experimental very high resolution tactile sensor array, *Proc. 4th Intl. Conf. on Robot Vision and Sensory Controls,* 241–250, London (Oct. 1984).
31. K. Tanie, K. Komoriya, M. Kaneko, S. Tachi, and A. Fujikawa, A high resolution tactile sensor, *Proc. 4th Intl. Conf. on Robot Vision and Sensory Controls,* 251–260, London (Oct. 1984).
32. R.C. Luo, W.H. Tsai, and J.C. Lin, Object recognition with combined tactile and visual information, *Proc. 4th Intl. Conf. on Robot Vision and Sensory Controls,* 183–196, London (Oct. 1984).
33. E.L. Dereniak and D.G. Crowe, *Optical Radiation Detectors,* John Wiley, New York (1984).
34. A. Peterson, Silicon temperature sensors, *Electronic Components and Applications, 5*(3), 206–211 (June 1983).
35. M.P. Timko, A two-terminal IC temperature transducer, *IEEE J. Solid-State Circuits SC-11*(6), 784–788 (Dec. 1976).
36. R.A. Russell, A simple thermal touch sensor, *Robotics Age 6*(10), 19–22 (Oct. 1984).
37. U. Dibbern and A. Petersen, The magnetoresistive sensor—a sensitive device for detecting magnetic field variations, *Electronic Components and Applications, 5*(3), 148–153 (June 1983).
38. D. Mullenberg, Real-time radiology of critical engine components, *Sensor Review 4*(1), 35–38 (Jan. 1984).
39. G.J. Curtis and P.A. Lloyd, Schizeophonics, then and now, *Chart. Mechan. Eng. 27*(9), 55–60 (Oct. 1980).
40. J.P. Trevelyan, S.J. Key, and R.A. Owens, Techniques for surface representation and adaptation in automated sheep shearing, *Proc. 12th Intl. Symp. on Industrial Robots and 6th Intl. Conf. on Robot Technology,* 163–174, Paris (June 1982).
41. M. Manninen, A. Halme, R. Myllyla, An aimable laser time-of-flight range finder for rapid interactive scene description, *Proc. 7th British Robot Association Annual Conf.,* 107–114 (May 1984).
42. A. Chappell, ed., *Optoelectronics—Theory and Practice,* 391–397, Texas Instruments Ltd. (1976).
43. J.E. Orrock, J.H. Carfunkel, and B.A. Owen, An integrated vision/range sensor, *Proc. 3th Intl. Conf. on Robot Vision and Sensory Controls,* 263–269, Cambridge, Mass. (Nov. 1983).

44. Anon., Distance sensing uses automatic focusing technology, *Sensor Review* 4(4), 172–173 (Oct. 1984).
45. D.M. Connah and C.A. Fishbourne, Using a laser for scene analysis, *Proc. 2nd Intl. Conf. on Robot Vision and Sensory Controls*, 233–240, Stuttgart (Nov. 1982).
46. W. Brunk, Geometric control by industrial robots, *Proc. 2nd Intl. Conf. on Robot Vision and Sensory Controls*, 223–231, Stuttgart (Nov. 1982).
47. Anon., Optocator speeds on-line inspection, *Sensor Review* 2(2), 52–53 (April 1981).
48. A. Astrop, No need to touch with latest probe, *Machinery and Production Engineering*, *139*(3589), 47–49 (Nov. 1981).
49. N. Nimrod, A. Margalith, and H.W. Mergler, A laser-based scanning range finder for robotic applications, *Proc. 2nd Intl. Conf. on Robot Vision and Sensory Controls*, 241–252, Stuttgart (Nov. 1982).
50. J.Y.S. Luh and E.S. Yam, 3-D vision for robotic systems, *Proc. 1st Intl. Conf. on Robot Vision and Sensory Controls*, 303–312, Stratford-upon-Avon (April 1981).
51. O.D. Faugeras, F. Germain, G. Kryze, J. D. Boissonnat, M. Herbert, and J. Ponce, Toward a flexible vision system, *Proc. 12th Intl. Symp. on Industrial Robots and 6th Intl. Conf. on Robot Technology*, 67–78, Paris (June 1982).
52. R.J. Fryer, ARCHIE—an experimental 3D vision system, *Proc. 4th Intl. Conf. on Robot Vision and Sensory Controls*, 197–208, London (Oct. 1984).
53. N. Ahuja, N. Bridwell, C. Nash, and T. S. Huang, Three dimensional robot vision, *IEEE Computer Society Conf. on Industrial Applications of Machine Vision*, 206–213 (May 1982).
54. D. LaCoe and L. Seibert, 3D vision guided welding robot system, *The Industrial Robot 11*, 18–20 (March 1984).
55. M. Bernasconi, R. Delazer, and M. Gini, Precision measurement by stereo vision system, *Proc. 10th Int. Symp. on Industrial Robots and 5th Intl. Conf. on Industrial Robot Technology*, 349–360, Milan (March 1980).
56. R.A. Jarvis, Range from brightness for robotic vision, *Proc. 4th Intl. Conf. on Robot Vision and Sensory Controls*, 165–172, London (Oct. 1984).
57. M. Briot, J.C. Talou, and G. Bauzil, The multisensors which help a mobile robot find its place, *Sensor Review* 1(1), 15–19 (Jan. 1981).
58. K.A. Marsh, J.M. Richardson, J. S. Schoenwald, and J.F. Martin, Acoustic imaging in robotics using a small set of transducers, *Proc. 4th Intl. Conf. on Robot Vision and Sensory Controls*, 261–268, London (Oct. 1984).
59. P.A. Dewar, A fast histogramming chip, *VLSI Design* 5(6), 100–101 (June 1984).
60. I. Dinstein, F. Merkle, T.D. Lam, and K.Y. Wong, Imaging system response linearization and shading correction, *Intl. Conf. on Robotics, IEEE Computer Soc. Press*, 204–209, Atlanta (Mar. 1984).
61. W. Hannaway, G. Shea, and W.R. Bishop, Handling real-time images comes naturally to systolic array chip, *Electron. Des.* (USA), *32*(23), 289–300 (Nov. 1984); and W. W. Smith, P. Sullivan, Systolic array chip recognises visual patterns quicker than a wink, *Electron. Des. (USA)* *32*(24), 257–266 (Nov 1984).
62. S. Joshi, D. Mithani, and S. Stephonsen, A bipolar micro-programmable processor, *Electronic Product Design* 3(2), 34–38 (Feb. 1982).

63. King-sun Fu, ed., *VLSI for Pattern Recognition and Image Processing,* Springer-Verlag, Berlin (1984).
64. G. Edwards, Programmable signal processing chips, *New Electron. 15*(4), 74–76 (Feb. 1982).
65. S.S. Magar, E.R. Caudel, and A.W. Leigh, A microcomputer with digital signal processing capability, *IEEE International Solid-State Circuits Conf.,* 32–33 and 284–285 (Feb. 1982).

7

Robot Systems

7.1. The Robot

The combination of vision and other sensory systems with robots is increasing, enabling robots to perform tasks that would not be possible for robots equipped with no more than rudimentary sensing devices, such as microswitches. For example, an assembly robot may determine the positions and orientations of the components and their suitability for assembly, guide the assembly process, and check the resulting assembly for correctness. It may also check on failure during assembly, i.e., the inability to insert a component, and avoid or recover components which have been dropped. It is, therefore, appropriate to have some knowledge of the structure of robots and of their operation before considering complete sensor-assisted robotic systems.

7.1.1. *Basic Structure*

The robot has been defined in many ways but here we define it as a reprogrammable device designed to manipulate objects, including tools, through complex motions according to information stored in a control unit. The paths are incrementally defined and systems that move between mechanically set stops, which determine the rest positions, are excluded. A robot consists of segments which can move with respect to one another. A segment of a robot may move with respect to an adjacent segment by linear movement with respect to that segment or by rotation about a common joint, the axis of which is usually, but not necessarily, at right angles to both segments or colinear with at least one of the segments. Each of these movements is called a "degree of freedom." In order to be considered a robot, the system should have at least three degrees of freedom (the human arm has more than thirty). The concept of degrees of freedom of a robot can be compared with that of the six degrees of freedom of an object in space. To define completely the position and orientation of an object, six parameters must be set.

Some of the more common configurations with three degrees of freedom are shown in Figure 7.1.[1] In the Cartesian system no rotations are used. From the cylindrical system, through the spherical, or polar, system to the jointed, revolute, or articulated system, the number of linear movements is reduced and that of the rotations increased. Another system shown incorporates articulated and linear movements, and is often referred to as the SCARA type (selective compliance as-

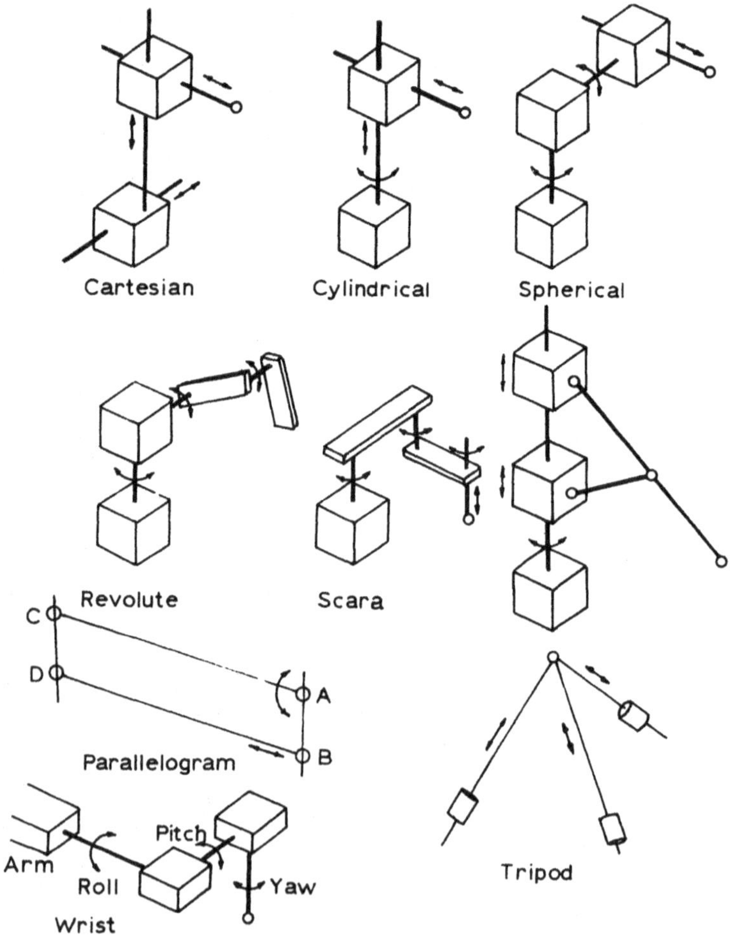

Figure 7.1. Typical structures used in the construction of robots.

sembly robot arm), although to be correct this term should be applied only to those incorporating selective compliance (see Section 7.2.1).[2] Other designs derive from the desire to increase the speed of the robot by reducing its inertia. Most of the mass lies in the drive units, so it is advantageous to place these where they require little movement. This is achieved in the tripod structure in which the motors pivot about fixed points and control the length of the segments between the motors and the wrist.[3] The parallelogram can achieve a similar result in that both drives are positioned at the base. One, at A, raises the next stage, CD, and that at B rotates CD. An advantage of this system is that the drive at A does not change the orientation at CD. This independence of the drives simplifies the control system.

Reference books giving functional details of the robots on the market are published by several organizations.[4–6]

Extra degrees of freedom may be incorporated in a wrist fitted to the end of the arm. A three degrees of freedom wrist is shown in Figure 7.1. The order of these joints may differ in specific designs. The last segment will carry a fixing surface for the end effector. Usually rotation is employed to enable this surface to be orientated, with respect to the segment carrying the wrist, over a large angular range. This may be performed by the first two motions of the wrist with the third rotating the surface about its axis.

Care should be exercised when determining the number of degrees of freedom possessed by a robot. It is possible that some motions do not increase the flexibility of the robot although they may simplify the control of the robot. Four or five degrees are adequate for many tasks and six is necessary only when a high flexibility is required. Also, it is possible that during the motion of a robot, a configuration of the joints may occur that renders impossible a desired movement of the end effector, i.e., the system becomes degenerate. If these configurations cannot be avoided it may be that extra degrees of freedom must be incorporated.

The end effector may be a device to take hold of an object, for example a gripper or sucker, or it may be a tool such as a screw driver or a paint spraying gun. It may be, or include, sensory devices such as a camera, force sensors, or a range measuring device. There can be either one effector permanently attached to the robot, a combination of devices, or an effector that is automatically exchangeable.

The robot requires power units to drive the individual joints, position sensors to determine the relative positions of the segments of the robot, and a controller to drive the robot to a sequence of positions as required by the application. The controller drives the power units to minimize the differences between the required positions and those indicated by the sensors.

7.1.2. Power Systems

The most common power media used in robots are electric, hydraulic, and pneumatic; motors or cylinders are used to convert these to mechanical energy. In hydraulic and pneumatic systems, electrically operated valves are used to control the flow of the fluid, although fluidic controllers have been used in pneumatic systems. In these two systems the fluid pressure can be generated away from the robot and relatively lightweight cylinders can apply high powers at the joints. Electric motors develop the power for a specific joint and although their power-to-weight ratio used to be inferior to that of the hydraulic and pneumatic systems, the introduction of new designs and new magnetic materials has removed this disadvantage.

The pneumatic systems have the disadvantage that since air is compressible, it is very difficult to control the valves to position the actuator accurately while operating at high speeds. Generally, the only application of these in a robot system is for the control of joints in the wrist, when these have only two or three positions, or operation of the gripper. Pneumatic drives can give a very high speed of operation and have been used in the main drives of robots to produce rapid movement to roughly the required position. At this point the piston is clamped and an electric motor provides a small adjustment to reach the exact position. A pneumatic version of the electric stepping motor exists (Figure 7.2) in which the pistons drive into the serrated edge of a bar or disk.[7] One of the three pistons is under pressure at any time and the change of pressure to one of the others, which one determining the direction of movement, moves the bar by one third of the pitch of the serrations.

Due to the very low compressibility of hydraulic fluids, accurate positioning is possible. The design of control valves for this purpose is not simple. Twin or double acting pistons are used to drive in opposite directions. Alternatively, hydraulic motors may be used, in which case rotary motion is produced. There is also a hydraulic version

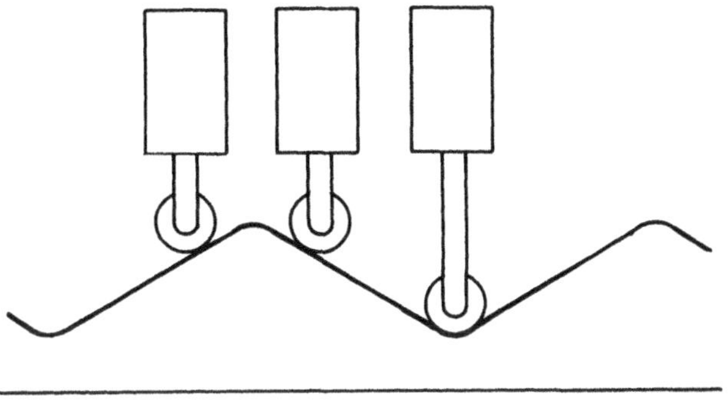

Figure 7.2. A pneumatic or hydraulic stepping motor. The right-hand cylinder is applying the drive.

of the stepping motor (Figure 7.2). Two disadvantages of hydraulic systems are that the power unit is continuously consuming a significant amount of energy and producing a significant amount of noise. Also, there is a slight seepage of oil and the risk of a greater release of oil if a pipe or seal failure should occur. This may exclude them from use where flames exist or in the presence of food or other contaminatable products. On the other hand, the hydraulic as well as the pneumatic systems can be used in explosive atmospheres, such as in paint spray booths, where electrical systems may be barred.

A new development is a close approximation to the muscle. The principle of the mechanism is that when the pressure within a rubber or soft plastic tube is increased, the increase in the diameter of the tube is accompanied by a reduction in its length. This contraction can be used to power a joint either directly or through a cable linkage. It is light and easily made in any size and so is convenient to fit to the robot arm. It has an advantage, when compared with cylinder units, in that there is no static friction. As a result, high precision is claimed for the device.

The electric motor is now the most common drive unit and the most common forms are the stepping motor and the dc motor. Recent developments in motor design have produced the brushless dc motor and a viable ac motor.[8–10] The motors can be driven directly by electronic systems, simplifying the control process. Besides the power consumed to produce the motion, some may be used to balance any load on the joint between movements. Many robots incorporate brakes to hold joints between movements and avoid this extra power drain. The brakes can be applied by springs and released when a movement is required. Also, they provide a safety feature in the event of a power failure in that motion at the joint is arrested and the robot does not droop or collapse.

The stepping motor exists in three basic forms; the permanent magnet, the variable reluctance, and the hybrid types.(11,12) Since the operation of these motors is not widely known, a brief description will be given here. The rotor is not energized and current is applied to a stator having a large number of poles. The windings on the poles are arranged and driven in a systematic way so that for each of the permissible drive patterns the rotor is held in one of several stable rotational positions. By switching the currents, the rotor can be moved on to the next stable position. Thus the motion of the rotor is basically incremental and not continuous. This feature is useful in that by counting the switching changes applied, the rotation as an integral number of steps is known without the use of a subsidiary position transducer, although a zero position sensor may still be required for absolute position control.

In the permanent magnet motor, the rotor is a magnet with several poles, alternately of opposite polarities. The stator may have many more poles but is driven to produce a field pattern with the same number of poles as the rotor. Figure 7.3 shows a simple version with a two-pole rotor. By reversing the polarity of the stator poles that are furthest from the poles of the rotor, e.g., on coil 3, the rotor will move through an angle equal to the pole pitch of the stator. The direction of rotation is determined by the pair of stator poles which are reversed. If, instead of reversing the current, it is only switched off, the rotation

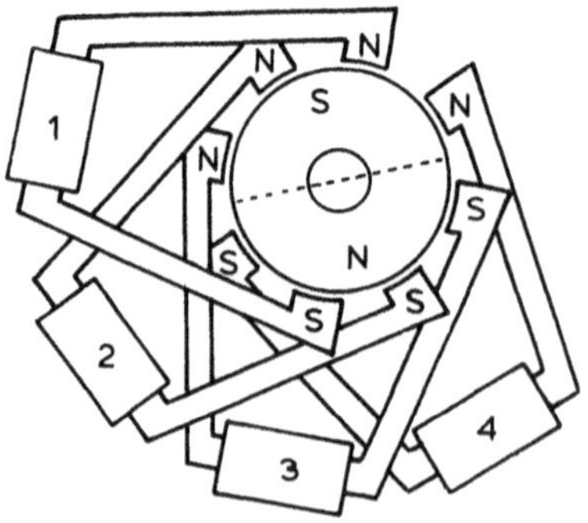

Figure 7.3. The permanent magnet stepping motor with a two-pole rotor and eight-pole stator, driven by the four coils 1 to 4.

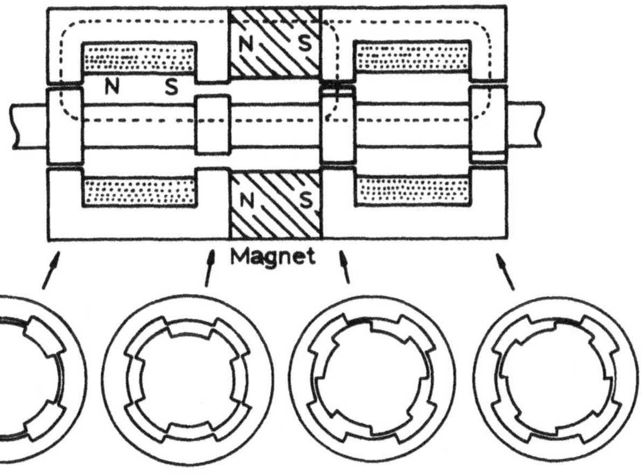

Figure 7.4. A hybrid stepping motor showing the path of the magnetic field. The left-hand stack is in the high-field state.

will be halved. Typical motors have step angles of 3.75°, 7.5°, and 15° in the full step mode, with rotors having 24 or 12 poles.[12]

Hybrid motors may be of several forms, but each has a permanent magnet and a winding system that reinforces the field along some paths and cancels it along others. Basically there are two sections with a magnet placed axially between them, in the rotor or the stator. One form (Figure 7.4) has four sets of poles with a winding between each pair and a magnet in the center of the stator.[13] Only one coil is fully energized at one time and the field that it produces will add to the field from the permanent magnet in one set of poles, cup 1, and cancel in cup 2. The permanent magnet field exists alone in the other cups. The stable position for the rotor is that shown in Figure 7.4. By switching the current to the other coil, the rotor will align with one of its cups, which one depending upon the direction of the current. Thus the direction of rotation can be controlled. Cups with 50 teeth are typical and give a step angle of 1.8°.[12]

The variable reluctance motor does not contain a magnet. Single-stack and multiple-stack types exist, the latter having three to seven stacks spaced along the axis of the motor. The principle of the motor is that the rotor aligns itself with the stator so that the circuit reluctance is at a minimum, i.e., the teeth on the rotor align with those on the stator (Figure 7.5). Figure 7.5 shows the energized stack of a multistack

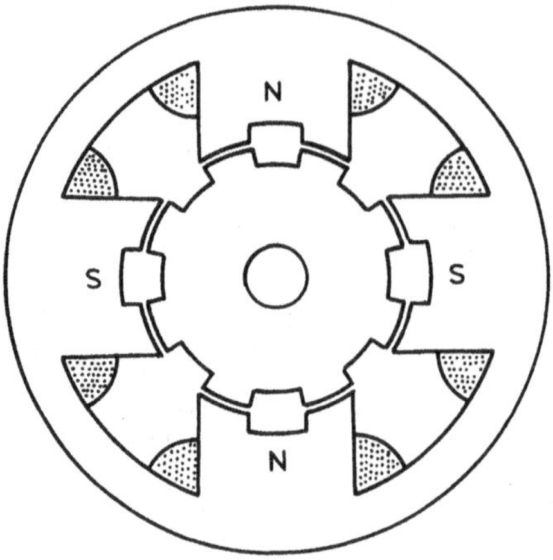

Figure 7.5. The energized stack of a multistack variable reluctance stepping motor. A rotation of 45° would be given by one complete cycle of phase changes.

motor. The other stacks, which would not be energized at this time, would have their stator-rotor angles displaced from this position by $1/n$, $2/n$, etc. of a tooth pitch, where n is the number of stacks. By switching the current from this coil to that of a stack which is $1/n$ or $(n - 1)/n$ out of position, the rotor will move in the appropriate direction. By using multiple teeth on the poles it is possible to produce small movements without having a large number of poles.

In the single-stack motor of Figure 7.6, the number of teeth on the stator and rotor are unequal. One pair of coils is energized and the rotor aligns its teeth with those of the energized stator poles. By switching the current to another winding, the corresponding teeth will come into alignment.

The relationship between the rotor movement and the applied pulses holds true provided that the motor is able to respond to the shift pulses. If the pulse frequency is too high or changes too rapidly, pulses will be missed or the rotor will run beyond the required position. This is dependent upon the inertial load on the motor. Control is lost completely when the pulse frequency is too far from that appropriate to the motor speed. Special control systems are available for generating the variable frequency pulse trains as directed by simple input data, e.g., the size of the movement that is required.[14] The available "working torque" is speed dependent, because at high frequencies the time constant of the coils does not allow the currents to reach their maxima or zero. For each motor, curves are available which relate torque and step rate (Figure 7.7). The *pull-in* curve shows the

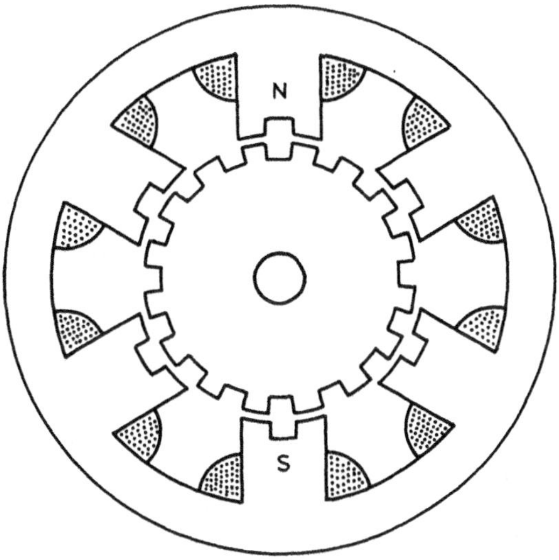

Figure 7.6. A single-stack variable reluctance stepping motor. A complete cycle of the three phase changes would give a rotation of 22.5°.

limiting combination of torque and speed below which the motor can start from rest without losing steps. For stopping, the curve is usually slightly more generous. The *pull-out* curve sets the limit below which the motor can operate without losing steps. The torque derives from a combination of the inertial and frictional loading.

Due to the pulsed nature of the currents it is possible to generate oscillations in the rotor when the mechanical properties of the load and the coupling to the load are adverse. The positional accuracy of

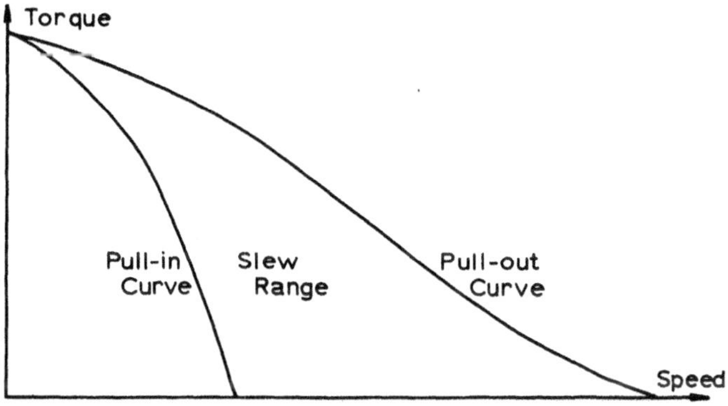

Figure 7.7. Characteristic curves for a stepping motor, relating output torque and rotational speed or drive pulse rate.

a rotor, at zero torque, is typically 5% of a step angle, but when loaded, the rotor will be displaced to allow a tangential component of force to be generated within the motor. The "holding torque" is the maximum that can be applied before the rotor jumps to the next stable position.

In view of the large number of versions of each type of stepping motor, it is difficult to compare their characteristics. Those containing a magnet have a natural detent action which may be sufficient for them to retain their position when the drive is off, although the force is much less than that with the drive on. The hybrid motor gives a higher torque for a given body size but the variable reluctance motor usually has a lower inertia and is more suited to applications requiring high acceleration. Also, it is more often available with large step angles which give higher angular speeds.

The risk of losing a step can be overcome if a position sensor is connected to the output shaft. This allows greater freedom in the control of the motor and higher torques and speeds are possible. Operated in this way, the variable reluctance motor can surpass the performance of the dc motor due to its low inertia and high efficiency, better than 80%.[15]

In the conventional dc motor, the rotor is wound and the stator is either wound or is a permanent magnet. The wound stator at one time provided the lighter motor, but developments in permanent magnetic materials have reversed the situation and also removed the need for the extra power drain of the stator coils. The rotor may be a soft magnetic material carrying the rotor windings. In this case the inertia of the system is high and reduces the acceleration that can be achieved. It is possible to separate the winding from the core and to leave the latter stationary. The rotor coil is then wound on a light former, the combination having a much lower inertia. A variant of this places the magnet stator within the coil and uses the iron casing of the motor as the return magnetic path. A low inertia can also be achieved by printing the rotor windings on a plastic disk using the same techniques as for printed circuit boards. The disk rotates between stator magnets. Wound versions of the disk motor are available. In the drive circuits for dc motors, increasing use is being made of pulsed outputs which provide a more efficient driver than the conventional analog circuits. DC motors are generally more efficient than stepping motors.

The commutator, used to distribute the current between the many sections of the winding on the rotor, limits the maximum current that can be fed to the rotor, is subject to wear, and generates sparks which may be a danger in flammable atmospheres. The development of magnetic materials with high remanence, e.g., samarium-cobalt, enables low-inertia permanent magnet rotors to be produced. Current is fed to the many sections of a stator winding but without the aid of a mechanical commutator. Instead, an electronic commutator is used which routes the currents using angle information from a transducer on the rotor shaft. These brushless dc motors are sometimes classified as ac motors.

Another development is that of ac motors, i.e., motors having a wound rotor inductively energized from the stator, with a high power-to-weight ratio. With the development of associated servo control circuits, it is now possible to use these in place of dc motors in robot systems. The control circuits are based on the inverter which generates, from dc, an alternating voltage of a variable frequency appropriate to the speed of the motor. Again, for efficient operation of the transistor circuit, the sine wave of the drive may be approximated to by a series of width-modulated pulses. To maintain a torque, current must be induced into the rotor windings. Consequently, the direction of the field in the stator must rotate faster than the rotor, and must continue to rotate when the rotor is stationary.

7.1.3. Power Transmission

The power may be delivered directly to a joint by a linear or rotary actuator but most systems use an intermediate transmission system. The most common reason is to match the force/speed characteristics of the actuator to the load. Another is to enable the actuator to be placed away from the joint so that the weight of the actuator is less of a load on itself or on other actuators. Shafts, belts, and drive chains may be incorporated. There is a danger in making the drive path too remote in that the repeatability of the system may be reduced. If the path is between the drive and the position sensor, instability may occur. In a direct drive system, a conventional motor may be used but it is also possible to design a special motor so that the rotor and stator are integral parts of the two sides of a rotating joint.

Probably, the most common torque matching technique is the use of gears or rack and pinion systems. Designs incorporate methods of reducing backlash and taking up wear. A gear system frequently used in robots is the Harmonic Drive.[16] It has the advantages of high efficiency, 75 to 90%; high ratio, up to 320 : 1; high tooth engagement, 10%; and low backlash, 3 minutes of arc. Basically, the unit consists of three parts, (Figure 7.8): a fixed outer cylinder with internal teeth, a flexible cylinder with external teeth, and an elliptical former, fixed onto the drive shaft, on which the flexible cylinder is supported by a ball race. There are two less teeth on the flexible cylinder than on the outer cylinder. Input is applied by the drive shaft to the elliptical former and the reduced speed output taken from the flexible cylinder, although it is possible to run the system in reverse. One rotation of the center shaft causes the flexible cylinder to rotate by two teeth positions. In an alternative connection, the flexible cylinder is held stationary and the output is taken from the outer cylinder.

Power transmission over a distance, e.g., from one end of a segment to the other, may be performed by rods or concentric cylinders

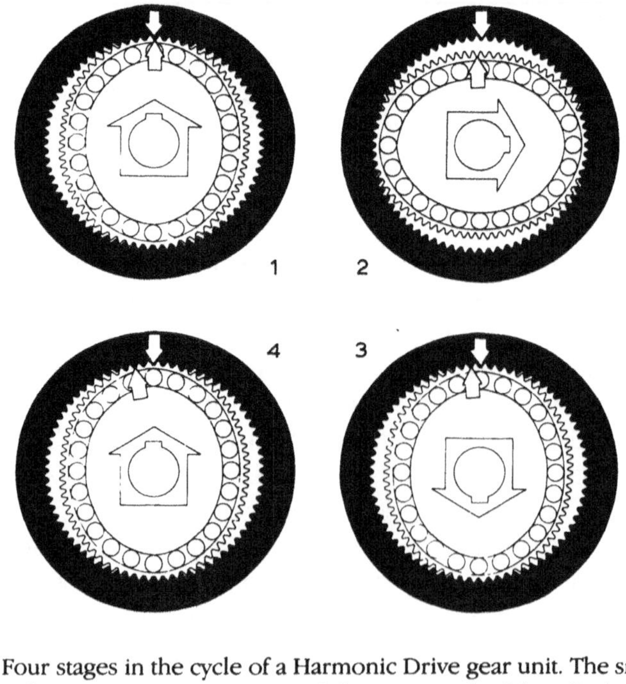

Figure 7.8. Four stages in the cycle of a Harmonic Drive gear unit. The small arrows mark teeth on the stationary outer gear and the rotating flexible cylinder. (Courtesy of Harmonic Drive Ltd.)

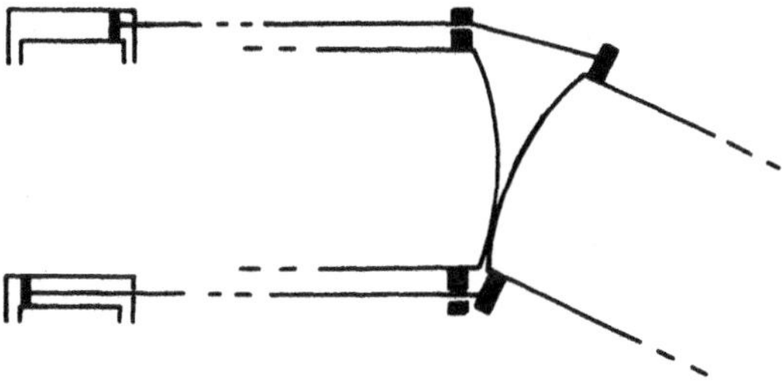

Figure 7.9. A rolling joint driven from remote cylinders by a wire system.

operating in either a push–pull or a rotational mode. The latter is a common method used to transfer power from the arm, through the first wrist joint, and into the wrist. Metal chains are used, and there is an increasing use of toothed and steel belts. These methods also allow a small change in the drive ratio. In a good design, the backlash can be made very small. Another method of transferring power is by lead screws driven by ball screw units. These incorporate a large reduction in motion between the rotary drive and the shaft, with low wear and low backlash.

Power may be transferred from power units in the base of the robot to actuators in the arm and wrist by the use of wires. These may run through the segments of the robot and pass over pulleys at the joints, or through a tube, as in the Bowden cable. The use of cables to control the motion of a joint is illustrated in Figure 7.9. The joint may be of the ball-in-cup variety or may be formed by two surfaces touching at a point. The joint has two degrees of freedom, and by the use of two orthogonally arranged pairs of wires, with the diametrically opposite wires of each pair acting differentially, the joint may be set in any position. This principle can be used to produce a multijointed, flexible arm suitable for reaching into awkward positions.(17)

7.1.4. *Position Sensors*

To control a joint, the relative position of the segments meeting at that joint must be known to the controller. Stepping motors should move one step for each pulse from the driver, and therefore the change in position is known. If a "zero position" had been detected previously and no steps have been lost, due, for example, to the pulse rate being too high, then the absolute position is known. For other systems, separate position sensors are required. They may measure linear or angular movements, either absolutely or only incrementally.

Position and displacement transducers are described in Section 6.1.3. The most frequently used transducers are the optical, synchro/resolver, and resistive. Although the obvious position for the transducer is at the joint, this may lead to servo loop stability problems if the drive is remote with flexibility in the transmission, or if there is backlash in the drive. Transducers may be used in combination, with one giving high resolution position information and the other giving an approximate position (see Section 6.1.3).

It is necessary, in most cases, to relate the data from the position sensor to the actual position of the joint prior to the use of the robot. The system must be zeroed by taking a reading when the joint is in a known position. This is sufficient if the robot is to be taught by leading it through its path. Alternatively, if the path is computed or generated without the use of the robot, and the accuracy required exceeds that available from the position sensors, it will be necessary to calibrate the sensors. The joint is placed in a series of known positions and the difference between the expected and the actual outputs recorded. During operation, values interpolated between these corrections can be used to adjust the control data.

When values of velocity, acceleration, and higher differentials are required, it is normal to obtain them by the differentiation of the position/time function. There are encoders which can be used to obtain these directly. For example, the dc generator can be used to measure angular velocity. The need to suppress commutator noise restricts the bandwidth that can be used.

7.2. End Effectors

The end effector of a robot may be a device to grip an object or tool or be the tool itself, e.g., a camera or other special device permanently attached to the robot or coupled by an automatic changeable tool unit. In the latter case, the robot may itself change the tool that it is using. Several tools or grippers may be mounted together, possibly on a turret. The robot has made possible the use of two special tools which would be hazardous if used manually. These are the high pressure water jet, used for cutting, and the high power laser beam, used for cutting and welding. In the latter case, mirrors placed at the robot joints direct the beam from the laser to the end effector. These are special systems that are not the subject of this section. Here the more general handling devices are considered. These may grip, suck, use magnetism, or in other ways hold objects, and may include sensory devices to monitor their actions. Also, some power arrangements are required to drive them.

It may seem desirable to have a general purpose unit that could perform a very wide range of tasks but so far this ideal has not been achieved, and effectors are designed for a specific range of tasks. As the effector is at the end of the robot system it is necessary that its weight be minimized; also problems arise in feeding the drive to the effector.

The most common effector is the gripper. This may have simple jaws or a complex finger structure which closes around the object.[18] The properties of the object to be handled and the way in which it is to be handled bear upon the design that is used. The grip must be adequate to support the weight of the object and the inertia during acceleration, and also to withstand any interaction forces between this object and another, e.g., during insertion or withdrawal.[19] It is possible that the size or shape of an object may change while it is being handled, for example, when the object is being worked in a forge. A simple gripper may be used for a rough robust object having convenient parallel surfaces; a more complex design is required for a smooth fragile object with only curved surfaces. The surface of the contact surfaces of the gripper may be textured to increase the friction between the gripper and the object. This texturing may be in the form of patterning or roughening of the contact surfaces, or other materials, such as rubber or padded materials, may be added.

A common jaw action has the contact surfaces remaining parallel through their motion so that they are correctly aligned for gripping objects of a range of sizes. The jaws may run in slides or be supported by a linkage mechanism. Other motions may be obtained in similar ways, for example, a circular motion to grip cylinders. Pneumatic cylinders are often used to operate grippers due to their low weight and the ease of feeding their motive power. With double-acting cylinders, the air can be used to open and to close the gripper. Alternatively, the drive in one direction can be provided by a spring. There are two advantages in using the spring to close the gripper: it is possible to predetermine the force that will be applied to the object and, in the event of a power failure, the gripper will not open. Bistable spring arrangements may be used in which the actuator switches the state of the gripper but the opening and closing are performed by springs.

Other actuators include the pneumatic motor and the hydraulic cylinder and motor, all relatively light, and providing a simple means of power transfer. Electronic solenoids are not attractive due to their weight but small electric motors may be used, either dc or stepping, and allow the distance between the gripping surfaces to the controlled. The drive to the actuator may be taken by shafts or gears through the wrist if the wrist is simple, but a three degrees of freedom wrist is usually too complex for this to be possible. Another way to remove the power source from the wrist is to use Bowden cable systems.

Another common end effector is the suction unit, which in the simplest case is a single cup but may consist of multiple cups or even a large porous surface. These are suitable for flat or curved, smooth surfaces, surfaces which may be difficult for a gripper to hold. The atmospheric pressure and the suction area limit the weight that can be handled and the inertial effects that can be withstood. The vacuum can be generated by a pump or by compressed air in a venturi unit.

When ferromagnetic materials are to be handled it is possible to use electromagnets to hold the objects. The electromagnet may contain a permanent magnet to hold the object without the dissipation of power. This provides a safety element to retain the object if the power should fail. The object is released by applying an electric pulse which temporarily cancels the field. These grippers can be used when the poles of the magnet match the surface of the object. The poles can be provided with flexible surfaces by fitting them with a plastic bag, or bags, containing a magnetic powder.[20] When the gripper is brought against an irregular surface the bag will conform to the shape of the surface. The powder provides a low impedance path for the magnetic field and will go solid when the field is applied. Alternatively, several electromagnets mounted on a flexible surface or adaptable frame may be used.

The handling of flexible objects, e.g., cloth and plastic sheet, presents special problems, so special effectors have been developed for this purpose. Part of the problem arises from the difficulty in separating the layers. As in paper-handling systems, air jets can be used to blow between the layers and then the top layer can be lifted by a suction device or a blade slid between the layers.

The end effector may consist of several devices, for example, several grippers or tools or a mixture of different types. The wrist is moved to bring the required device into the operating position. The devices comprising the end effector may have separate actuators or share the same actuators. In some applications more than one end effector may be used simultaneously. An alternative solution to the requirement for different effectors for a given task is to use changeable devices.(21) The effectors are stored in a rack within reach of the robot. The coupling between the robot and the effectors needs careful consideration since it must provide a reliable connection, must be simple for the robot to join and to release, and must carry the power connections (e.g., electrical or vacuum line) or signal lines (e.g., from a sensor) between the robot and the effector.

The changing of an end effector can be time consuming when a robot is required to perform several different processes on a workpiece. There exist the options of either fixing the workpiece and using several tools, etc. to operate on the workpiece, or moving the workpiece between several fixed tools. Another solution in batch production is to perform each operation on the whole of the batch before changing the effector and performing the next operation.

7.2.1. Compliance

In an assembly operation it may be necessary to bring parts together between which there is only a small clearance, e.g., when placing a pin into a hole. The smallest step and the repeatability of the robot are limited; it is probable that the path of the pin will not coincide with the hole and, if the robot is rigid, insertion will not be possible. This problem can be solved by providing some compliance in the robot structure if it is of the right type.(22) Two examples of the provision of controlled compliance are in the SCARA design of robot and in the remote center compliance device (RCC).

As seen in Figure 7.1, in the selective compliance assembly robot arm (SCARA), the two main segments are rigid in the vertical plane and have long bearings.(2) The first segment rotates about the fixed vertical axis of the robot and the second about a vertical axis at the end of the first. The gripper mounting rotates about a vertical axis in the end of the second segment. Vertical motion may be provided at this point or at the first axis. This construction is rigid in the vertical plane but has compliance in the horizontal plane and about the vertical axes. When this robot attempts to place a pin into a hole, a small error in the position of the pin will result in horizontal forces due to contact between the entry chamfers on the pin and the hole. The compliance will allow the pin to move and the entry will be facilitated. Rotation about a horizontal axis is resisted, an action which reduces the chance of jamming. For pins or objects which are not round, misalignment will also produce a torque about the gripper axis. The compliance about this axis will allow the necessary rotation to occur.

The remote center compliance devices are built into the end effector. They result from extensive studies of the process of insertion and the causes of jamming during insertion.[23] The principle of the device is that insertion is facilitated when the gripper is free to move laterally and to rotate about the end which is to enter the hole, i.e., a point remote from the gripper.[24–26] In effect, the direction of the force applied by the robot is through this remote point. This allows the pin to move laterally when contact is made between the tapers on the pin and the hole, and allows rotation of the pin as it enters

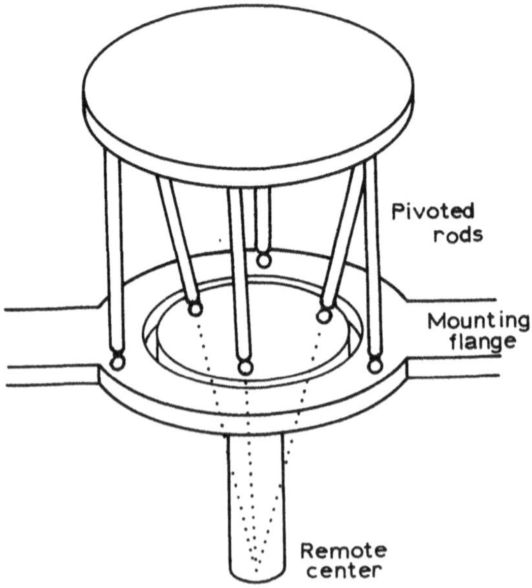

Figure 7.10. Remote center compliance. The position of the "remote center" is determined by the inner three rods. The outer rods permit tangential movement without rotation about the center.

Figure 7.11. A passive compliance device with rods having lengthways elasticity.

the hole. The structure of the device consists of two sets of linkages (Figure 7.10): one with three parallel rods, allowing lateral movement, and the other, with three rods aligned on the remote center, allowing rotation about that center.[1] Knowledge of the forces active within an RCC device can help in the control process and it can be instrumented for this purpose.[27,28] Also, the degree of compliance can be controlled.[29]

A similar device, the passive compliance device (PCD), uses six linkages which have lengthways elasticity (Figure 7.11).[1] The upper plate is fixed to the robot and the lower carries the gripper. As for the RCC device, lateral movement and rotation about a remote center are permitted.

7.3. Control Systems

Several factors are brought together in the control of the actuator of a joint, including the present relative position of the segments at a joint, their required position, and the speed, not necessarily constant, with which the movement is to be made. It may also be necessary to coordinate the operation of several joints so that the effector follows a specified path. Further, the outputs of sensors may be used in determining the path and final destination.

The path to be followed by a robot is stored within the system. In early systems this might be in the settings of a group of devices, such as potentiometers, or on paper tape, but now it is common to store this data in a computer memory system. This can be a read-only memory (ROM) store or a random access memory (RAM) store that is loaded from a magnetic tape, often a cassette, a floppy disk, or from another computer. In the potentiometer systems, the outputs of groups of potentiometers are selected by switch drums and compared with outputs from position-measuring potentiometers at each joint. Difference signals are applied to the actuator drive units to reduce the differences to less than a preset error. In modern systems, a computer or controller reads the stored values and, having combined these with data from the robot, outputs appropriate control signals to the actuator drive units. In either case, the next set of position information might be selected when all the joints have completed their operation, or when all the joints are within some preset distance of the required position. This latter facility saves time when intermediate points have been specified to define the path of the robot, but it is not required that the robot stop at them or even pass exactly through them.

The positions of the segments of a robot are, in the simplest case, defined by the settings of each individual joint as indicated by the position sensor at that joint. This may not be the most convenient method when the robot has to relate to the world of its environment or to the world around its end effector. This introduces two more coordinate systems, world and tool coordinates, to that of the joint coordinate system. The concept of tool coordinates is that, for a Cartesian system, three axes can be defined. For example, in a gripper there are the axis parallel to the fingers, the axis along which the jaws move, and the axis perpendicular to these. The choice of origin and orientation of these axes will depend on what is most appropriate to the function of the gripper and will be related to the world coordinates. Rotation can occur about all three axes.

At any time, the position of the end effector of the robot can be fully described in any of the coordinate systems. Usually there will be a unique set of coordinates in each system to fully describe the robot but ambiguities can occur. For example, a revolute robot (Figure 7.1) has two ways of presenting its end effector in a given position. It can move between these positions by rotating through 180° on its first joint, while taking the arm over the top using the second joint, and making suitable adjustments to the remaining joints. Only the joint coordinates uniquely describe this robot.

Even so, it is often more convenient to record the position data in one of the other coordinate systems and to use the computer to convert the information for the control of the individual joints. It may then be necessary to store additional data to control which of the alternative attitudes are taken by the robot.

In the computer controlled systems, the control information may be divided between the program that is being run and the path data list. In the simplest form the data list consists of sets of values defining, through one or more of the coordinate systems, the positions at the joints at each stage of the operation. It may also include information on velocities and on dwell times at a position, and instructions to wait for a signal from another device or to read position information from an input port. Alternatively some or all of this information may be in the program itself. The program may contain instructions to perform certain operations upon the data that it reads. For example, it may compute an arc trajectory from data defining three points in space, either all on the arc, or the ends of the arc and the center of curvature, thereby significantly reducing the data that has to be stored when compared with that necessary to define the path on a point-by-point basis.

The program may contain subprograms when a cycle of events is to be repeated, usually with a modification in each case. For example, when a set of pins has to be placed in an array of holes, the manipulation of the pins from the feeder to the array is very similar in each case, only the final position is changed. The program for this operation can be written by using subprograms to define the movement of the pins; at each cycle of the subprogram, different information is obtained from a data set describing the array of holes. Similarly, if a sequence of operations is to be performed on an array of objects on a pallet, the process for only one object need be taught together with the data locating each of the objects. In the subsequent operation of the robot, the path executed for each object will be the taught path, referenced to the location data of that object. This facility reduces the effort required to program this sort of indexed, repetitive process.

Information on speed profiles, used to obtain the fastest movements without overshoot or oscillation, may be stored as tables or be computed as required. One method of using tables is to have one table for acceleration and another for deceleration. The former uses the distance from the last rest point to address the table and the latter uses the distance to the next rest point. The velocity used to control the motion is the lesser of the two values obtained. For large distances a constant velocity will be used between the profiled velocity sections.

7.3.1. *Teaching Methods*

The required path for a robot, with the actions of the end effector and pauses, etc., can be placed in the control system in several ways. These are by direct operation of the robot, by the use of a simulating robot structure, by the use of a control box, or by the development of the path in a separate computer, i.e., by computer aided design (CAD) of the operation.[30]

In the direct lead-through method, the end effector is held by the operator and the robot is forced to follow the required path. For this purpose the power to the drive units is removed and only the position sensors are active. In pneumatic and hydraulic systems all control valves are opened and in hydraulic systems it may be necessary to empty the cylinders to allow the robot to move freely. This technique is commonly used in painting and welding systems where the operator, skilled in the appropriate production technique, is able to produce the required motion only by performing the actions himself. The control system samples the joint position transducers at a preset time interval to form a list of positions. During playback the robot will repeat the path with the original speed variations. In a variant of this, the teaching is performed at one speed and the playback is at a different, usually faster, speed. The locations are stored in RAM, on magnetic tape or disk, or on paper tape.

Although removing the power from the drive units allows the robot to be manipulated, it may still be heavy to hold and difficult to move due to friction. A solution to this problem is to have a lightweight structure resembling the robot in shape and size but containing only the position transducers. This is moved through the required motions and the output of its transducers are recorded. Slight differences between the two sets of transducers are not significant in the applications where this technique is used, e.g., spray painting; however, if necessary, these could be reduced by calibration.

Most robots have a manual control box, with buttons or control knobs, which allows the operator to move the robot using the robot drive system. Often he will be able to choose which of the coordinate systems he wishes to use. It is rare that the operator can drive the robot smoothly along the required path by this method and therefore the time sampling system cannot be used. Instead the operator sets the robot in a series of positions and at each pushes a record button to cause the computer to store the data from the position sensors. The operation of the gripper or another end effector may also be controlled from the box and the state of these can be recorded. The box may allow control of the speed of operation and usually a low speed is used during recording.

Having programmed a robot in this way it is advisable to run a test replay at low speed to determine whether the operation is satisfactory. It is possible that in attempting to move from one programmed position to the next it may try to move through an impossible position, i.e., one in which it strikes another object or one that the limited motion of its joints will not allow it to achieve. If this occurs, positions may have to be modified or extra points added.

In simple systems the data will be stored directly, e.g., on tape, and no editing will be possible. In systems with computer control, facilities are provided which allow changes to be made and additional data to be added, such as pauses, waits for external responses, speed statements, and calculation of paths between programmed points. In systems with vision or other sensors, the program must contain directives to obtain position and other control information from the controllers of those sensors.

Many computer programs have been written which simulate the operation of specific robots, including maximum accelerations under given load conditions. This makes it possible to use CAD methods to produce the control data for the operation of a robot.[31–35] The control data can be generated off-line without interruption to the existing use of the robot system.[36] In the other methods, except for that using the simulating robot structure, the data stored is that derived from the position transducers of the robot. There will be nonlinearities and end-point errors in the transducers of the robot and in the off-line methods corrections must be applied to correct for these. One method is for the robot to be programmed to move to set positions and the actual errors in position in the world around the robot measured. These error values are incorporated in the simulation data for that particular robot (see Section 7.4.5).

7.4. Parameters of Robots

Various parameters must be considered when a robot is to be selected for an application. The data given in publications cannot be complete and care is required in interpreting that which is given. Robots that appear to be similar may perform differently in a given application. It may be advisable to use a robot simulation program to ascertain the performance to be expected from the robot when performing the specified task.[35,37–39]

7.4.1. *Load*

The load that can be carried by a robot depends upon the extension of its arm in a horizontal direction. The load will affect the dynamic characteristics of the system in that the maximum accelerations that can be achieved will change and the parameters of the servo-control loop may alter and affect overshoot and stability. The load causes a positional error due to flexibility in the structure, backlash in the drive, and the need for a small error signal from the position transducer to provoke the force from the drive to support the load. In some robots, self-adjusting systems, sometimes pneumatic, are provided to counterbalance the load carried by the robot; in others, the load-carrying and position-sensing members may be separated.[40]

It is important to realize that the load limit quoted for a robot often does not allow for the capabilities of the wrist joints. The quoted limit may be applicable to a load at the mounting flange of the wrist or to the normal configuration of the robot as supplied. The end effector has to be counted as part of the load. A large end effector or one which holds the load at a distance from the last joint may create large couples which exceed the capabilities of the robot.

7.4.2. *Work Space*

The work space is the volume within which a stipulated point at the end of the robot arm may be moved. It should not be assumed that at all points within this space all joints have their normal freedom to operate, nor that all paths within this space can be followed.

7.4.3. *Speed*

Speeds quoted for robots are usually the maximum, nonaccelerating velocities and consequently are little affected by the load. The maximum speed may depend upon the configuration of the robot and therefore will not be the same throughout the work space. Typical translation velocities are in the region of 1 m/sec and angular velocities 90°/sec but some of the new, smaller robots designed for assembly tasks are achieving speeds several times higher. In practice, the user is more interested in the time to move from one point of rest to another; hence, acceleration and deceleration times become important, particularly when the movement is fairly small. Some data sheets do give typical figures for accelerations.

7.4.4. *Rigidity*

The construction of the segments of a robot, the length of the bearings at each joint, flexibility and backlash in the drive system, and the feedback gain of the servo loop together determine the rigidity of the robot. The gain of the servo loop can be used to control the compliance of a joint. The rigidity, as measured by the force applied at the end effector necessary to produce unit displacement, will depend upon the direction of that force and upon the configuration of the robot. Naturally, it is usually worst at the maximum extension of the arm. The rigidity determines the positioning accuracy achieved when the arm is loaded. Also, it affects the oscillation and overshoot exhibited by the arm during a movement. In some robots, e.g., the SCARA, some lack of rigidity is introduced in some directions as a design policy (see Section 7.2.1).

7.4.5. *Repeatability and Accuracy*

The terms "repeatability" and "accuracy" are often confused, although their meanings are quite distinct. The ***repeatability*** of a robot refers to its ability to return to the same point when given the same control information in joint coordinates, irrespective of its path to that point and the speed used on that path. The figure given is normally the root mean square error of many movements from many directions, but the maximum error may be given. Spraying-robots often have a repeatability of a millimeter or more; for assembly robots, with low load capability and a small work space, it can be less than 10 μm. Due to lack of perfect rigidity the repeatability is load dependent; therefore, the figure quoted may not be valid if the load is changed. When the joint coordinates have been calculated from another coordinate system it is possible that the path by which the point is approached may affect the configuration assumed by the robot, i.e., the same final position of the end effector is achieved by the use of a different combination of joint coordinates. This may result in a significant difference in the rest positions.

Static friction can affect the repeatability when small movements are made. To overcome the friction, a minimum force is required and the servo system is unable to reduce the force to zero in the short time necessary to reach the specified point. As a result, overshoot occurs and friction now prevents the arm from returning to the point. Other causes of error are time dependent. In the short term, drifts may occur due to temperature changes in the position transducers and, in the longer term, wear will move the rest point.

The *accuracy* of the system refers to the relationship of the actual position reached and the required position expressed in some absolute form. This is important when the operation of the robot is programmed off-line or when position information is to be introduced from the world of some other system. This may be when the object to be handled is on a moving conveyor or the position of the object is being determined by a vision system. Another situation in which accuracy is important is when the robot is to be replaced by another of the same type, e.g., for servicing, and the same control program is to be used to avoid retraining. Accuracy errors are a combination of the repeatability errors and systematic errors introduced by nonlinearities and zero errors in the position transducers. Also, the accuracy is limited by the minimum step when the drive is from a stepping motor, the position sensor has a minimum resolution, or there is a digital-to-analog, or vice versa, conversion in the control loop.

The effect of that part of the error not due to the repeatability error itself can be reduced by measuring the error components and incorporating them into the control information. A common method used to achieve this is that of calibration.[41–44] Where the correction is to be between the robot and an absolute world around it, the robot is moved to accurately known reference points in that world and the transducer information recorded (see Section 9.2.2). Several systems are available for this purpose. The difference between the actual and the predicted values are stored and used to correct, by interpolation, all data for the control of the robot. Sometimes it is not necessary to relate to an absolute world, for example, when a robot is working with a vision system (see Section 9.2.2). In this case, the robot is driven to many positions in the common working space and the vision system observes the positions of the end effector of the robot. A differential error table can be produced and corrections applied to all programmed points by interpolation between the appropriate measured errors.

7.4.6. *Environment*

The environment in which the robot is to operate can restrict the choice of robot. Robots with electric motors using commutators should not be used in flammable atmospheres; even the use of brushless motors should be carefully considered due to the risk of an accidental break in the conductor path. Clean environments may be contaminated by the oil from hydraulic systems and from pneumatic systems in which the air is released into the environment. Maximum and minimum limits will exist for ambient temperature. This may be satisfied by ducting air into the robot. The robot can be protected from radiant heat by screens, cooled if necessary, mounted on or around the robot. Dust, particularly if abrasive, and paint spray can damage the robot. Piston shafts on hydraulic and pneumatic systems can be vulnerable. In most cases the addition of flexible covers will bc adcquatc but in extreme conditions positive air pressure within the covers can be used. Similar techniques can be used to provide protection against high humidity. Usually the control equipment can be placed in a less exacting environment.

7.4.7. *Safety*

There are three aspects of safety; the safety of the robot, the safety of the environment, and the safety of the humans in that environment.[45,46] Although a robot has been programmed to follow a specific path, failures in the control system may cause it to move to any point within its reach or to release loads while it is in motion, causing them to be thrown some distance. These failures may be caused by an electronic fault, by a fault in the storage system, by the transmission of incorrect data from another system, e.g., a vision system, by electrical interference from other factory equipment, or by a power failure. Fortunately such events are rare, but they are possible.

The risk of faults can be reduced by using good quality equipment in all parts of the installation and by ensuring that it is properly installed. Connectors should be securely fixed, cables protected from unnecessary movement, and all units effectively earthed to reduce pickup of electrical interference. Earthing is very important in the presence of high-current equipment such as welders. Programs for the robot and other control equipment, e.g., sensor systems, can be written so that data is checked and dangerous actions, such as releasing a gripper while the arm is moving rapidly, are prohibited. Most robots are designed so that in the event of a power failure the robot is stopped and does not go limp (see Section 7.1.2).

Eliminating the possibility of impact between the robot and its environment or humans is more difficult.[(47)] Surrounding the robot with safety fencing having interlock switches and using pressure-sensitive mats are standard procedures to protect operators, but because these can be bypassed, operators have been injured.[(48)] The determined human cannot be stopped. Mechanical screens may be inconvenient; screens of infrared beams are an alternative. Since few injuries have been caused by malfunction of the robot system, the dangers to operators of robots should be set against the short-and long-term risks involved when the robot's task is performed by the operator himself.[(49)] Typical problem areas include press operation, grinding, and dangerous environments, including those with high levels of noise or vibration.

Damage due to impact between the robot and other objects can be reduced by providing surfaces with crushable layers. Also, various proximity sensing devices, mounted on the robot arm, can be used to detect close approaches, but most of these have a short operating range. Typical proximity sensors are switches, and magnetic, ultrasonic, and optical devices (see Section 6.1.2). In principle, vision systems can be used to monitor the operation of the robot and its surroundings, for example, to detect the intrusion of a human, but the ensuing additional costs and low speed may limit their use.

7.5. Programming Robots at Different Levels

Some of the different ways in which it is possible to program robots have already been mentioned briefly. The emphasis has been on the direct specification of the points through which the tool or end effector of the robot is to travel. The robot is led through that path by hand during the teaching phase, and it follows the identical path in the operating phase. At this level of operation the robot has no knowledge of the space in which it is operating. Thus, the program is a stored record of the joint angles that correspond to the taught trajectory and the speed of motion between the specified points. When robots were first introduced for industrial use they were nearly always programmed in this "low-level" and "on-line" manner. Although this is a very direct method of programming that allows the user to get started very quickly, it creates a number of significant problems. These problems will be discussed in the next section. Other higher-level and off-line programming methods will be described in subsequent sections.

7.5.1. Problems in On-Line, Low-Level Programming

On-line, low-level programming entails the following problems:

1. The typical trajectory required to perform an industrial operation is likely to contain many points. This is not a problem if the robot has been taught in "master-slave" manner to follow a human teacher. If, however, each point has to be entered manually into the robot controller's memory, programming can take a long time and be very tedious.
2. A description of a trajectory in terms of individual points can be rather imprecise, since the trajectory between the specified points is usually not defined. Suppose, for example, that one wants the trajectory to follow a specific analytic form such as a circle. Unless one separately calculates a large number of points lying on the circle and teaches each of these to the robot, the trajectory between the specified points will deviate from the circle.

3. Alterations to a low-level program can often cause difficulties. Clearly, it is easy to alter a few points. However, any program in a low-level specification language is generally difficult to interpret by anyone except the person who created it. So if one has a long program, it may be difficult to identify exactly what change has to be made. Furthermore, a change that in conceptual, or high-level, terms is very simple may become complicated if it has to be directly specified at a low level. Suppose, for example, one wishes to shift a whole sequence of points by the same amount in a specific direction. At a high level this may involve changing a single parameter, while at the low level each point may have to be changed individually.
4. On-line programming, if not carefully controlled, can be potentially dangerous. For example, when one is trying to teach the robot some precise action such as inserting a pin into a hole, it is often necessary to look very closely at the relative positions of the pin in the robot's gripper and the hole. This may involve placing one's head very close to the robot. Although specific precautions such as limiting the movement, speed, and power of the robot during teaching operations can be taken, there is always a risk that some error or malfunction could cause the robot to hit the programmer.
5. Programs which contain specific actuator values or joint angles are in general not transferable from one robot to another. Even robots that are nominally identical will in practice be significantly different. In general, they will have slight variations in the lengths of their component parts and in their joint sensors. This means that two robots following the same sequence of joint angles will follow different paths in space.
6. A low-level program is not by definition expressed in terms of the task to be performed, the objects to be manipulated, or the Euclidean space in which the manipulator is operating. The task, however, is likely to have been conceived, and is therefore much easier to understand, in some high-level terms. This mismatch between the high-level specification and the low-level implementation is bound to be a source of problems and errors.
7. An increasing number of components that the robots will have to handle, will have been designed and made using a CAD/CAM

(computer aided design/computer aided manufacture) system. An appropriate description of the object will therefore exist in a computer database. It would clearly be convenient to be able to use this description to simplify the programming task. A low-level program, however, cannot really make use of such CAD/CAM data.

8. Effective use of robots is becoming increasingly dependent on the use of external sensors, which are themselves becoming more complex. Unless the sensor is very simple or is tied in very closely to the robot, it is difficult to interface low-level programs with sophisticated sensors such as vision systems.

7.6. Higher Levels of Robot Programming

The low level of programming discussed above may be called "joint level" because the basic unit of command to the robot is essentially of the form "drive joint 4 to 28.4°." Above joint level is "manipulator level." At this level, the basic unit of command can be considered to be of the form "move robot arm to PICKUPOINT," where PICKUPOINT is a location in a particular Euclidean space. It is clear that a manipulator level instruction will generally involve multiple joint level instructions. Similarly, an "object level" program can be regarded as the next step up and will involve multiple manipulator level commands. An archetypal object level command could take the form "Insert pin A into hole B." The highest level is task level at which subtasks become the basic unit of instruction. Such a subtask command could take the form "assemble motor housing" and would, in this hicrarchical structure of programming levels, involve a number of object level commands.

The higher the programming level, the more the language interpreter is required to do. The type of capability required at each of these higher levels will now be discussed.

7.6.1. Manipulator Level Programming

There are a number of manipulator level programming systems available. They are generally tied to specific robots or to specific manufacturers. Good examples are IBM's AML[50] and Unimation's VAL[51]

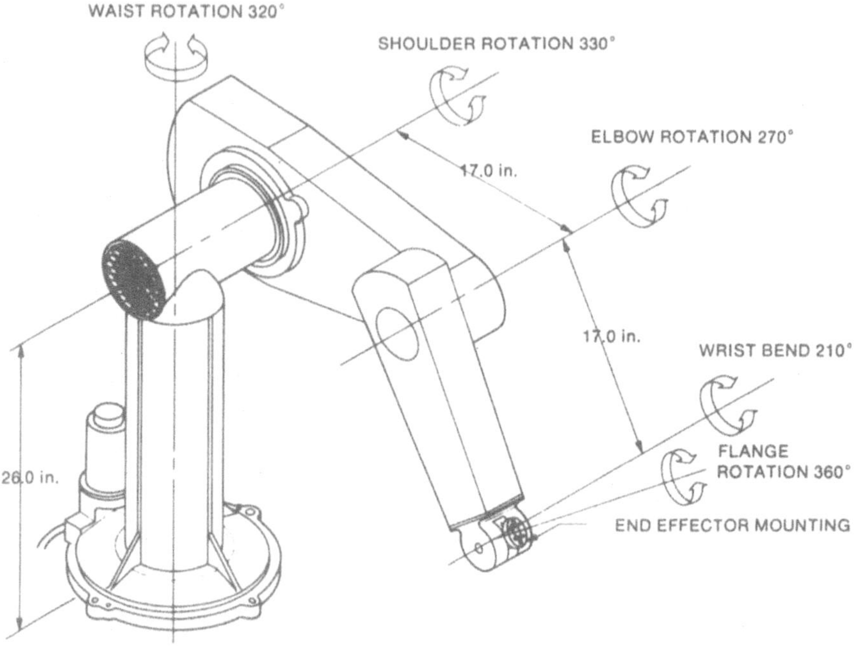

Figure 7.12. PUMA robot. (Courtesy of Unimation Inc.)

that was initially used to control their PUMA robot (Figure 7.12) but is now more widely used. These two languages are in practice quite dissimilar and each has its own strengths. However, for the purposes of this discussion, it will be sufficient to use VAL to illustrate the main features of a manipulator language.

7.6.1.1. Locations in Coordinate Frames. VAL allows locations to be defined in a number of three-dimensional coordinate frames of reference. Each location consists of six elements: three positions giving x, y, and z coordinates and three angles. With these six parameters one can completely define the position and orientation of a point in space.

The two most obvious coordinate frames are a "world frame" that is essentially fixed to the base of the robot, and a "tool frame" that is conceptually attached to the end effector of the robot. The z-axis of the tool frame may, for example, be defined as being parallel to the jaws of the gripper. The use of tool coordinates allows movements to be specified with respect to the tool. Consider a pin which has been picked up so that it has been aligned with the jaws of the gripper. Wherever the robot happens to be, it is possible to make it move along the axis of the pin by specifying a movement along the z-axis in the tool frame.

It is also useful to define other frames which may be attached to objects within the work area. Suppose one has a rectangular tray which contains objects arranged in a regular pattern. It is clearly easier to describe the positions of the objects in terms of axes which are aligned with the sides of the tray. In such coordinates, the position of the objects is independent of the tray in world space.

For a system to be able to handle these various frames it must contain a means for transforming between them, and for transforming locations in any frame to joint coordinates to drive the robot.

If locations are specified in a world or tool coordinate frame, it is much easier to relate sensor information, such as that obtained from a vision system, to robot movements. "Vision VAL" has specific facilities to allow the user to establish the relationship between the position of a stationary camera or vision coordinate frame and the base frame of the robot. A disk is placed at a number of locations in turn on a flat surface within the field-of-view of the camera. At each position, the coordinates of the disk in the robot frame are established by moving the robot to the disk; the coordinates in the vision frame are established by using the vision system to locate the center of the disk. Once a number of such dual positions have been found, the transformation between the two frames can easily be calculated.

A single camera can only generate a two-dimensional projection of the three-dimensional world. The principle described above can, however, be extended to relate robot coordinates to three-dimensional information obtained from more than one camera. A practical example of this is described in the case study in Section 11.7.

7.6.1.2. Specifying Robot Movement and Trajectories. VAL allows the user to instruct the robot to move between locations in a variety of ways. A simple MOVE command, specifying a destination location, will first cause VAL to calculate the joint angles required to bring the end effector to the destination. The motion control software within VAL will then cause all the joints to change smoothly to their required values. The trajectory of the end effector between start and destination is, however, undefined. If an alternative command called MOVES is used, the change of joint angles is constrained to make the end effector follow a straight line.

A further development of VAL, called VAL II, has more sophisticated trajectory control. It allows one to compute trajectories within the program. One can, for example, write a small procedure which will define a circular trajectory formed from any desired number of segments.

Although many other features are available in manipulator level languages, such as being able to define movements of specific joints, and to specify the behavior of the robot as it passes through defined locations, the essential feature of a manipulator language is that movements are specified between locations defined in three-dimensional coordinate frames.

7.6.1.3. Control and Calculating Features. Manipulator languages generally have a number of facilities associated with communicating with the outside world, and synchronizing robot movements with external events. They also have a number of standard features of general purpose, high-level programming languages. VAL has very limited arithmetic and flow control facilities compared to VAL II and AML.

7.6.2. *Object Level Programming*

Object level programming implies a language which can be used to specify the solutions to application problems, such as assembly, in terms of objects and their static and dynamic relationships. An object level system must also have a means to analyze these descriptions and decompose them to a manipulator level. One example of a system of this type is RAPT[52] which is being developed at Edinburgh University. We will use RAPT to explore the principles of an object level system.

A RAPT program can be considered in three parts. The first stage is to create body models for all the objects, including the manipulator, that are involved in the application. The next stage is to specify the object relationships in the static situations that can be identified as occurring during the course of the application. Finally, one must describe the actions required in order to move between these static situations.

7.6.2.1. Building Models of Objects. RAPT does not require the body model to be a complete description of the object. The model needs to contain information about those parts of the object which will interact with other objects during the execution of the task. The partial body models are built-up from geometric entities such as points, lines, and circles, that can then be combined to produce a description in terms of features. These features include faces, holes, edges, shafts, and vertices.

A partial body model of the cube in Figure 7.13 could consist of the four points, p0 to p3, and the three faces, f1 to f3, each defined by three of the points. Associated with each face is a vector normal to the face, which specifies which side of the face corresponds to the solid object. These three faces are sufficient to locate the cube in the world.

7.6.2.2. Creating Static Situation Descriptions. Static situation descriptions are defined in terms of feature relationships such as "against," "coplanar," "fits," "aligned," and "parax." For example, two planes may be "against" each other and two shafts may be "aligned" with one another. In addition, there are the attachment relationships "tied" and "untied," which specify whether two bodies should be treated as being joined together or separate. Each of the four static situations shown in Figure 7.14 could be described in terms of these relationships.

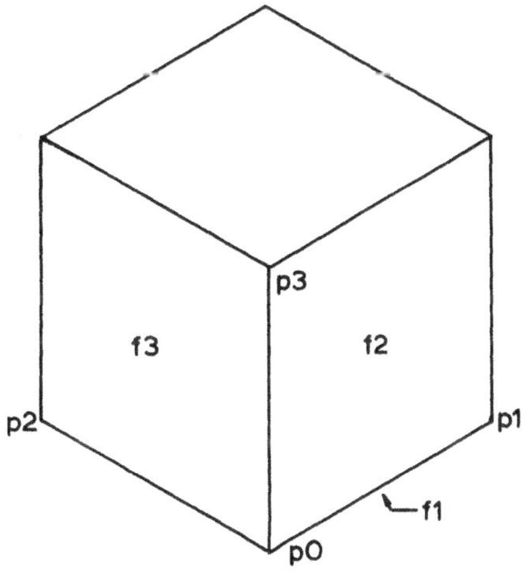

Figure 7.13. Partial body model of a cube is defined in terms of three faces.

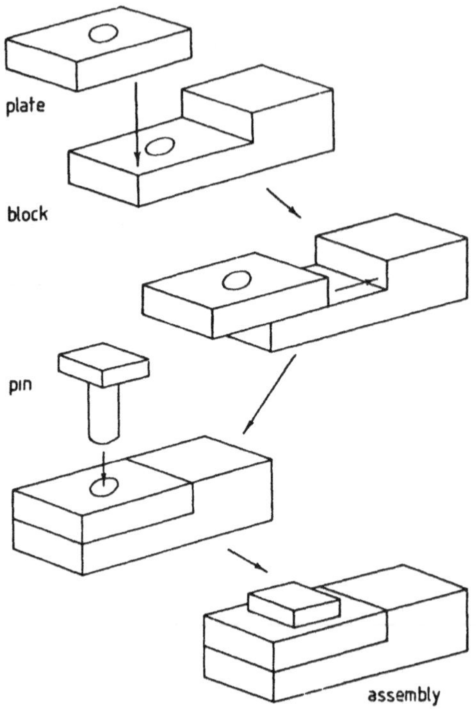

Figure 7.14. Situations within a RAPT assembly task. (Courtesy of Department of Artificial Intelligence, University of Edinburgh.)

7.6.2.3. Creating Action Descriptions. Having described each static situation, the body movements needed to get from one static situation to the next must be described. These movements are defined in terms of functions such as "move" and "turn."

7.6.2.4. Creating an Assembly Program. The combination of body models, situation descriptions, and action descriptions provide sufficient information to describe the complete assembly task. Since the robot manipulator is one of the bodies known to RAPT, the RAPT system can now compile all the descriptions to produce a robot control program to perform the assembly. A program for a specific robot such as the VAL for the PUMA can be produced by using an appropriate post processor.

7.6.3. Problems of Off-Line Object Level Programming

7.6.3.1. Infrastructure. It is simple to write a trivial robot program at the joint level. As discussed earlier, the problems of low-level, on-line programming arise when more complicated and realistic situations are encountered. It is only at these more complex levels that object level programming becomes useful, because it requires considerable overheads or "infrastructure" to write even the simplest program. One must, for example, have models for all the bodies involved. In an established system, models for standard bodies, including the robot, would exist in a library. As objects are increasingly designed by CAD systems, body models will be created as a byproduct of the design process.

7.6.3.2. Sensor Interaction. In the form that has been described above, RAPT is entirely deterministic. There is no provision for the use of sensors to provide positional data. It is, however, possible to have both a partially calculated robot program and also to allow certain locations to be specified by sensors at execution time. RAPT is being extended to allow the inclusion of sensory data.

7.6.3.3. Visualization. One of the prime advantages of on-line programming is that the robot trajectory can be immediately verified by direct observation of the robot itself. This is clearly not possible with off-line programming. In order to overcome this problem, off-line systems such as RAPT are often provided with graphic visualization systems in which the programmer can "see" the robot performing the programmed task on a CRT screen. With sophisticated "3D" solid-body graphics systems it is possible to see a very realistic representation of the assembly process.

The computer-graphic representation of the assembly process could also be used to design vision programs to locate objects. One can generate the view of an object that would be seen by a television camera so that off-line verification of vision programs is possible.

7.6.4. *Task Level Programming*

There have been a number of attempts to write systems which allow programming at the task level. Such systems involve an ability to perform a number of high-level tasks. A capability to understand natural languages is required to interpret the task description. The system must then be capable of generating a plan to execute the task efficiently. This will involve the generation of subtasks which lead to the ultimate goal. The generation of this plan involves geometric modeling, geometric reasoning, and knowledge of mechanics and dynamics. Without these capabilities, even simple problems such as object collision cannot be avoided.

7.7. Creating Complete Systems

Many programming systems assume that one is using equipment from a single manufacturer. It is generally more complicated to combine components from a variety of sources into a single system. However, most visually controlled robot systems have similar structures. Each one contains a task description program which specifies the task to be performed. The task description program invokes commands to the vision system to determine identity or location information, which is then converted to the reference frame of the manipulator to perform the required mechanical actions. There are generally various facilities for sequencing the operation of the various parts of the system, and for coping with foreseeable errors and failures.

Although different systems are structurally similar, the elements of which they are composed are not. Each system probably has a different vision processor, a different manipulator, and different numbers and types of computer. It is not easy to combine such systems in a coherent manner. ROBOS, developed at Philips Research Laboratories in the U.K.,[53] is an attempt to produce a programming structure to allow this.

ROBOS is conceived as a robot operating system by analogy with a computer operating system, in that it contains a number of facilities which are generally required in the development of sensor-controlled robot systems. ROBOS also provides a framework and discipline to guide the producer of the task-specific parts of the software, giving the whole activity greater coherence. It does not, however, contain computer operating system facilities such as handling files or running compilers.

At the core of ROBOS is a data structure which enables the user to model both the processing structure and the physical structure of the system being developed. The user can describe the processing structure in terms of "machines," which are virtually processors, and the "functions," which are to run on each machine. The physical structure can be described in terms of "devices," which are collections of sensors and effectors, and "frames of reference," in which positions and motions can be described.

For example, consider the two robot systems described in Section 11.7. Each system has TV cameras attached to it, but is able to share the same vision processor. From a processing point of view, the cameras are part of a single machine capable of executing vision functions. From a physical point of view, they are separate devices, each with its own frame of reference.

The structure of ROBOS is shown in Figure 7.15. At the center is the system supervisor and the data structure. The supervisor is connected to the task description program and also to the specialized machines, each of which is controlled by a software module called a machine controller. Machines are connected by standard ROBOS communication channels, which are implemented by shifting data in memory, or by a physical connection if the machine controller is installed in a separate computer.

When the task description program requires a function to be performed it calls the supervisor, which first examines the data structure to determine the "runnability" of the function. Two options are available when a function is not runnable. Control can be returned immediately to the task description program, or the supervisor can wait for a specified time to see if the function becomes runnable,

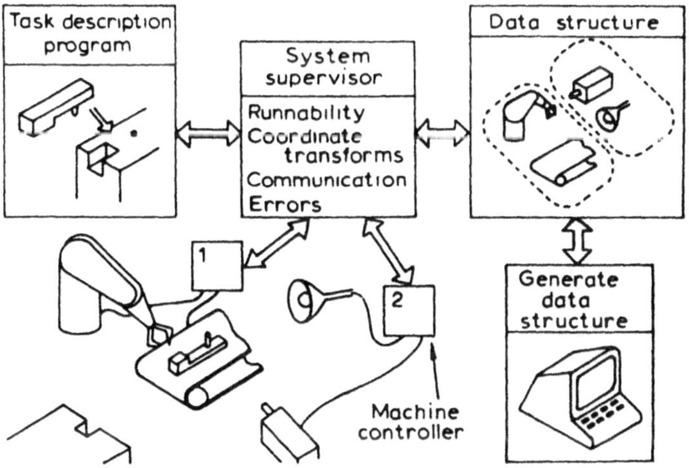

Figure 7.15. Structure of ROBOS.

failing which, it times out and creates an error condition. The supervisor performs all necessary transformations between frames of reference, as each machine controller expects to receive location coordinates in its own base frame rather than that specified in the function.

Machine controllers do not have fixed structures. Some controllers may be programmed quite independently. For example, a PUMA robot with a suitable VAL program is quite acceptable as a ROBOS machine controller. In general, a machine controller must be able to accept commands with locations expressed in its own base frame, return results, and indicate when the execution of a function has been completed and whether it was completed successfully. The manipulator controllers which we have written contain a specific "geometric model" for converting between the base frame and actuator values. This makes it simple to change from one robot geometry to another.

Once a function which involves motion of part of the system has been executed, the relationship between frames of reference may no longer be valid. ROBOS automatically updates these relationships whenever necessary.

The handling of errors or exceptions in flexible assembly systems often presents difficulties. For example, it is not easy to decide at which level error recovery procedures should operate. ROBOS distinguishes between fatal errors, that halt the system, and nonfatal errors, from which recovery is possible. Recovery strategy tends to be task dependent, and therefore ROBOS contains a mechanism to return control to a specified part of the task description program in the event of a nonfatal error.

References

1. F. Lhote, J-M. Kauffman, P. Andre, and J-P. Taillard, *Robot Technology,* Vol. 4, *Robot Components,* Kogan Page, London (1983).
2. H. Makino and N. Furuya, SCARA robot and its family, *Proc. 3th Intl. Conf. on Assembly Automation,* 433–444, Boeblingen (May 1982).
3. Anon., Gadfly—the answer to electronic component assembly, *Assembly Automation 3*(1), 20–22 (Feb. 1983).
4. *Members' Handbook,* British Robot Association, Kempston, Bedford MK42 7BT, U.K.
5. *The Specifications and Applications of Robots in Japan, 1984,* Japanese Industrial Robot Association, Tokyo (1984).

6. J. W. Clasper, Robotics information, *Database* (USA), 7(4), 39–42 (Dec. 1984).
7. A. M. Stanescu, G. V. Banu, V. I. Vlad, Th. Borangiu, and L. D. Serbanati, Architecture, control and software of a modular robot system with pneumatic stepping motors and vision, *Proc. 12th Intl. Symp. on Industrial Robots and 6th Intl. Conf. on Industrial Robot Technology,* 143–154, Paris (June 1982).
8. N. Barber, Coming to terms with brushless servo drives, *Electric Drives and Controls 1*(10), 25–29 (April/May 1985).
9. N. Wavre, Developments in electric motors for modern robotics, *Bulletin de l'Association Suisse des Electriciens* 75(12), 673–675 (June 1984).
10. M. Katayama, S. Nara, and K. Yamaguchi, Newly developed ac servomotor RA series, *Powerconversion Int.* (USA), *10*(5), 12–30 (May 1984).
11. P. P. Acarnley, Stepping motors: A guide to modern theory and practice, in: *IEE Control Engineering Series No. 19,* Peter Peregrinus, London (1984).
12. *Philips Data Handbook, Components and Materials, Book C17,* Stepping motors and associated electronics.
13. B. H. A. Goddijn, New hybrid stepper motor design, *Electronic Components and Applications 3*(1), 31–37 (Nov. 1980).
14. R. Cassinis, L. Schnickel, and M. Tomaini, An economical and powerful microcomputer based stepping motor driver, *Proc. 10th Intl. Symp. on Industrial Robots and 5th Intl. Conf. on Industrial Robot Technology,* 89–100, Milan (March 1980).
15. P. Lawrenson, Switched reluctance drives—a fast growing technology, *Electric Drives and Controls 1*(10), 18–23 (April/May 1985).
16. Harmonic Drive Gears, data sheets from Harmonic Drive Ltd, Billingshurst, West Sussex, U.K.
17. P. Grunewald, Car body painting with the Spine spray system, *Proc. 14th Intl. Symp. on Industrial Robots and 7th Intl. Conf. on Industrial Robot Technology,* 633–641, Gothenburg (Oct. 1984).
18. A. Rovetta, I. Franchetti, and P. Vicentini, On a general prehension multipurpose system, *Proc. 10th Intl. Symp. on Industrial Robots and 5th Intl. Conf. on Industrial Robot Technology,* 191–201, Milan (March 1980).
19. F. Y. Chen, Force analysis and design considerations of grippers, *The Industrial Robot 9,* 243–249 (Dec. 1982).
20. G. Bancon and B. Huber, Depression and dual grippers with their possible applications, *Proc. 12th Intl. Symp. on Industrial Robots and 6th Intl. Conf. on Industrial Robot Technology,* 321–329, Paris (June 1982).
21. R-C. Luo, Automatic quick-change gripper finger for assembly automation, *Proc. 5th Intl. Conf. on Assembly Automation,* 215–224, Paris (May 1984).
22. J. P. Bourrieres, P. Jeannier, and F. Lhote, Intrinsic compliance of position-controlled robots—applications in assembly, *Proc. 5th Intl. Conf. on Assembly Automation,* 133–142, Paris (May 1984).
23. M. S. Ohwovoriole, J. W. Hill, and B. Roth, On the theory of single and multiple insertions in industrial assemblies, *Proc. 10th Intl. Symp. on Industrial Robots and 5th Intl. Conf. on Industrial Robot Technology,* 545–558, Milan (March 1980).
24. T. L. De Fazio, Displacement-state monitoring for the remote centre compliance (RCC)—realisations and applications, *Proc. 10th Intl. Symp. on Industrial Robots and 5th Intl. Conf. on Industrial Robot Technology,* 559–569, Milan (March 1980).

25. F. Caillot and M. Kerlidou, Air stream compliance, *Proc. 5th Intl. Conf. on Assembly Automation,* 225–233, Paris (May 1984).
26. A. Fakri, A. Jutard, and G. Liegeois, Passive compliant wrist with two rotation centres for assembly robot (DCR-LAI device), *Proc. 5th Intl. Conf. on Assembly Automation,* 235–241, Paris (May 1984).
27. H. J. Warnecke, M. Schweizer, and D. Haaf, An adaptable programmable assembly system using compliance and visual feedback, *Proc. 10th Intl. Symp. on Industrial Robots and 5th Intl. Conf. on Industrial Robot Technology,* 481–490, Milan (March 1980).
28. T. L. DeFazio, D. S. Seltzer, and D. E. Whitney, The instrumented remote centre compliance, *The Industrial Robot 11,* 238–242 (Dec. 1984).
29. M. R. Cutkosky and P. K. Wright, Position sensing wrists for industrial manipulators, *Proc. 12th Intl. Symp. on Industrial Robots and 6th Intl. Conf. on Industrial Robot Technology,* 427–438, Paris (June 1982).
30. M. Parent, *Robot Technology, Vol. 5, Robotic Languages and Programming Methods,* Kogan Page, London (1983).
31. R. L. Tarvin, Considerations for off-line programming a heavy duty industrial robot, *Proc. 10th Intl. Symp. on Industrial Robots and 5th Intl. Conf. on Industrial Robot Technology,* 109–117, Milan (March 1980).
32. J. Anderson, The benefits of using CAD as a base for manufacturing, *Proc. 2nd European Conf. on Automated Manufacturing,* 67–72, Birmingham (May 1983).
33. J. J. Craig, Anatomy of an off-line programming system, *Robotics Today* 7(1), 45–47 (Feb. 1985).
34. R. N. Stauffer, Robot system simulation, *Robotics Today 6*(3), 81–90 (June 1984).
35. A. Liegeois, *Robot Technology,* Vol. 7, *Robot Performance Evaluation and Computer-Aided Design,* Kogan Page, London (1983).
36. S. J. Kretch, Advanced off-line programming for robots, *Proc. 2nd European Conf. on Automated Manufacturing,* 55–58, Birmingham (May 1983).
37. T. Fohanno, Assessment of the mechanical performance of industrial robots, *Proc. 12th Intl. Symp. on Industrial Robots and 6th Intl. Conf. on Industrial Robot Technology,* 349–358, Paris (June 1982).
38. L. Vecchio, S. Nicosia, F. Nicolo, and D. Lentini, Automatic generation of dynamical models of manipulators, *Proc. 10th Intl. Symp. on Industrial Robots and 5th Intl. Conf. on Industrial Robot Technology,* 293–301, Milan (March 1980).
39. J. H. Gilby and G. A. Parker, Laser tracking system to measure robot arm performance, *Sensor Review 2*(4), 180–184 (Oct. 1982).
40. A. Zalucky and D. E. Hardt, Active control of robot structure deflections, *Jnl. of Dynamic Systems, Measurement and Control 106*(1), 63–69 (Mar. 1984).
41. B. Scheffer, Geometric control and calibration method of an industrial robot, *Proc. 12th Intl. Symp. on Industrial Robots and 6th Intl. Conf. on Industrial Robot Technology,* 331–339, Paris (June 1982).
42. P. G. Ranky, Test method and software for robot qualification, *The Industrial Robot 11,* 111–115 (June 1984).
43. C. Morgan, The rationalisation of robot testing, *Proc. 10th Intl. Symp. on Industrial Robots and 5th Intl. Conf. on Industrial Robot Technology,* 399–406, Milan (March 1980).

44. Anon., Robot calibration and error detection, *National Engineering Laboratory Newsletter,* (13), 3 (Jan. 1985).
45. R. J. Barrett, Practical robot safety measures within a legal framework, *Proc. 6th British Robot Association Annual Conf.,* 33–39, Birmingham (May 1983).
46. R. R. Schreiber, Robot safety: a shared responsibility, *Robotics Today* 5(5), 61–65 (Oct. 1983).
47. M. Linger, How to design safety systems for human protection in robot applications, *Proc. 14th Intl. Symp. on Industrial Robots and 7th Intl. Conf. on Industrial Robot Technology,* 119–129, Gothenburg (Oct. 1984).
48. V. Weatherby and S. A. R. Pike, The safety implications of a new technology, *The Industrial Robot 10,* 185–188 (Sept. 1983).
49. Y. Hasegawa and N. Sugimoto, Industrial safety and robots, *Proc. 12th Intl. Symp. on Industrial Robots and 6th Intl. Conf. on Industrial Robot Technology,* 9–20, Paris (June 1982).
50. *IBM Robot System /1: AML Concepts and User's Guide,* 1st ed., IBM Corporation, Boca Raton, FL (Sept. 1981).
51. *User's Guide to VAL—A Robot Programming and Control System,* Version 12, Unimation Inc. (June 1980).
52. A. P. Ambler, R. J. Popplestone, and K. G. Kempf, An experiment in the off-line programming of robots, *Proc. 12th Intl. Symp. on Industrial Robots and 6th Intl. Conf. on Industrial Robot Technology,* 491–504, Paris (June 1982).
53. P. Saraga, B. M. Jones, and D. J. Burnett, ROBOS—Towards a robot operating system, *Philips Research Laboratories Annual Review 1983,* 85–87.

8

Vision: Human and Machine

8.1. Overview

The principal objective of this chapter is to present the principles of computer vision which are applicable in intelligent automation. However, vision by machine is essentially a replacement for human vision, hopefully with improvement. Thus, we start by explaining how the human eye–brain system works, so far as this is known. Machine vision is such a vast subject that we can mention only a few topics that are particularly useful in intelligent automation. Even for these, reference to selected textbooks[1–3] will be necessary for in-depth information. However, we do provide a general survey of machine vision, specifying the various subdivisions and indicating their relevance to automation, before concentrating on the selected practical topics.

8.2. Nature of Images

8.2.1. Gray Scale

An image is a continuous distribution of light energy over a two-dimensional surface; it is a form of signal as defined in Chapter 2. Before they can be processed by either man or machine, images must be converted into a sampled form. For machine processing, the samples are normally taken at uniform spatial intervals which are the same in the X- and Y- directions, i.e., are over a square grid (Figure 8.1). The samples are called *pixels* (picture elements or picture cells). In the human eye, however, the pixels form an expanding spiral centered on the fovea centralis. For machine processing, a hexagonal grid offers the advantage that the distances between each pixel and its six nearest neighbors are the same. This minimizes distortion as an image is

(A) (B)

Figure 8.1. Alternative pixel grids. (*A*) shows a square grid in which each pixel not on the edge has four neighbors at distance *d* (the pixel separation), indicated by thick arrows, and four neighbors at distance $1.41 \cdot d$, indicated by the thin arrows. For the hexagonal grid (*B*) each pixel has six nearest neighbors at distance *d*.

rotated with respect to the grid axes, but is difficult to implement practically and is seldom used in automation applications. The spacing between pixels must be close enough to satisfy the Nyquist criterion (i.e., the sample spacing d must be less than $1/2f$, where f is the highest spatial frequency at which appreciable energy is present) or information will be lost on sampling, manifest as aliasing error. The amplitude of a pixel is termed its "gray level" for a monochrome image.

The input to a viewing system is called the "scene." If not self-luminous, the light reflected from each point of the scene is the product of the light incident on that point, and its local reflectivity. The distribution of light intensity acquired by the viewing system is termed the "image." This is generally corrupted by noise and distortions introduced by the vision system, and is hence different from the scene. Simple distortions in geometry (e.g., a point which should be imaged at point x,y in image place actually appears at x',y') and in amplitude (e.g., a pixel at point x,y has amplitude a' instead of a) are easily corrected computationally, if necessary in real time. More complex corruptions, such as blurring caused by poor focus, crosstalk between pixels, and motion within the scene are much harder to correct.

Images are almost invariably sampled over a uniform grid and quantized before being processed, yielding an array of N numbers vertically by M horizontally, in which the I,Jth element is the amplitude of the I,Jth pixel. It is useful to handle this as a matrix, although it does not have the normal interpretation in mathematics in which the elements are the coefficients of equations whose linear simultaneous solution is desired.

8.2.2. Color

Images generally comprise energy at many wavelengths within an interval (the "passband" for the system); the energy at each frequency is the product of the energy in the illumination at that frequency, multiplied by the reflectivity at that frequency (Figure 8.2). The (subjective) color of an image is a measure of a human observer's ability to perceive the energy content for each pixel in the range 0.4–0.7 microns wavelength over which the human eye is sensitive.

The observed properties of human color vision can be explained (see section 8.4) by assuming three kinds of sensors, whose regions of response correspond roughly to the "red," "green," and "blue" regions of the spectrum. The passbands of the regions are indicated in Figure 8.3. Thus, three numbers are required to specify each pixel in a color image. If the stimuli to the three sensors are regarded as defining the orthogonal axes of a three-dimensional space so that each pixel is a vector in the space, the color perceived is observed to depend only on the direction of the vector; its length determines intensity. Thus, we can simplify the description by considering the octant to be filled with similar concentric surfaces, each of which may be repre-

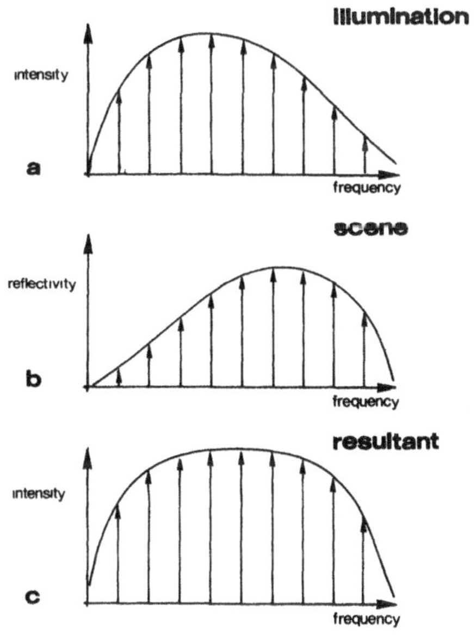

Figure 8.2. Color of an image as function of illumination and reflectance.

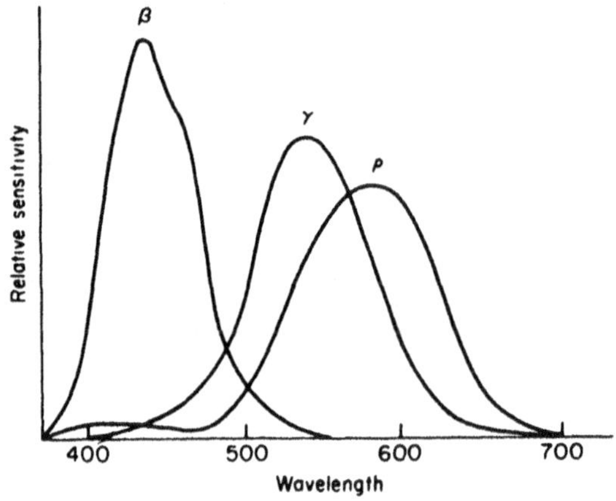

Figure 8.3. Color vision—presumed passbands for the three kinds of color sensor in the human retina. β indicates blue response; γ indicates response to green; ρ indicates response to red. The wavelength units are nanometers. From COLOUR—ITS MEASUREMENT, COMPUTATION AND APPLICATION by G. J. and D. G. Chamberlin. Reprinted by kind permission of John Wiley and Sons, Ltd.

sented by the diagram shown in Figure 8.4a. When developed onto a flat surface, this becomes the familiar "color triangle." Because any color may now be represented by a point on a two-dimensional surface, only two numbers are required to specify it. Coordinate transformations are used to obtain numbers which accord with subjective properties of the color, such as *hue* and *saturation* (Figure 8.4b), and *luminance* and *chrominance* (Figure 8.4a).

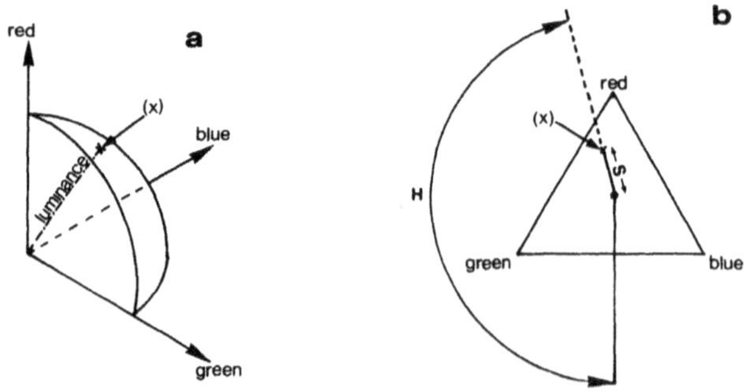

Figure 8.4. Color in a three-dimensional space. In (a), the stimulus represented by the vector (x) is seen to comprise proportions of energy exciting red, green, and blue sensors. The length of the vector is indicative of the total amount of energy in the signal (i.e., its intensity), and determines its *luminance*. The direction of the vector determines the *chrominance*, i.e., the perceived color. Chrominance can then be described on a surface as in (b), either using x- and y-axes as chromaticity coordinates, or the polar coordinate description shown, in which the radial distance S from the origin determines the *saturation* of the color, and the angle H determines its *hue*.

The equilateral triangle description is something of an oversimplification; a more accurate specification is given elsewhere.[4]

Recent research has shown that color perception is subjective in that the color perceived for any particular region depends on the color perceived in the other parts of a scene, rather than merely on the sizes of the local stimuli relative to the three types of color sensors. Thus, perceived color can remain invariant despite change in the distribution of energy with wavelength in the illumination.

For self-luminous sources, the *additive* description of color just given is appropriate. For the reflective surfaces more usual in automation problems, a *subtractive* description is more valuable. Printing inks (and almost all color films) generate color subtractively. This is achieved by starting with white light and removing appropriate proportions of the subtractive primary colors, yellow, cyan, and magenta. The latter fall midway between the additive primary colors in the color triangle.

8.3. Human Vision

Intelligent automation systems often use machine vision to replace the human eye–brain system. Thus, the properties of human vision are significant and will be discussed. Two aspects, physiological and psychological, are involved. The former regards the eye–brain system as akin to an electronic processing system, and is concerned with optical properties, signal paths, and processing methodologies, i.e., with low-level information. The latter aspect is concerned with what an operator perceives, i.e., with high-level information. We will discuss the physiological aspects first.

8.3.1. Physiological Aspects

The general organization of processing in the eye–brain system in humans is indicated in Figure 8.5.[5–7] The scene is sensed by the two eyes. A lens in each eye focuses an image of the visual field onto the retina. A network of cells of two types, called respectively "rods" and "cones," in the retina sense light intensity; they generate electrical signals comprising bursts of impulses whose frequency is less than 1000 Hz. The number of impulses in a burst is proportional to the

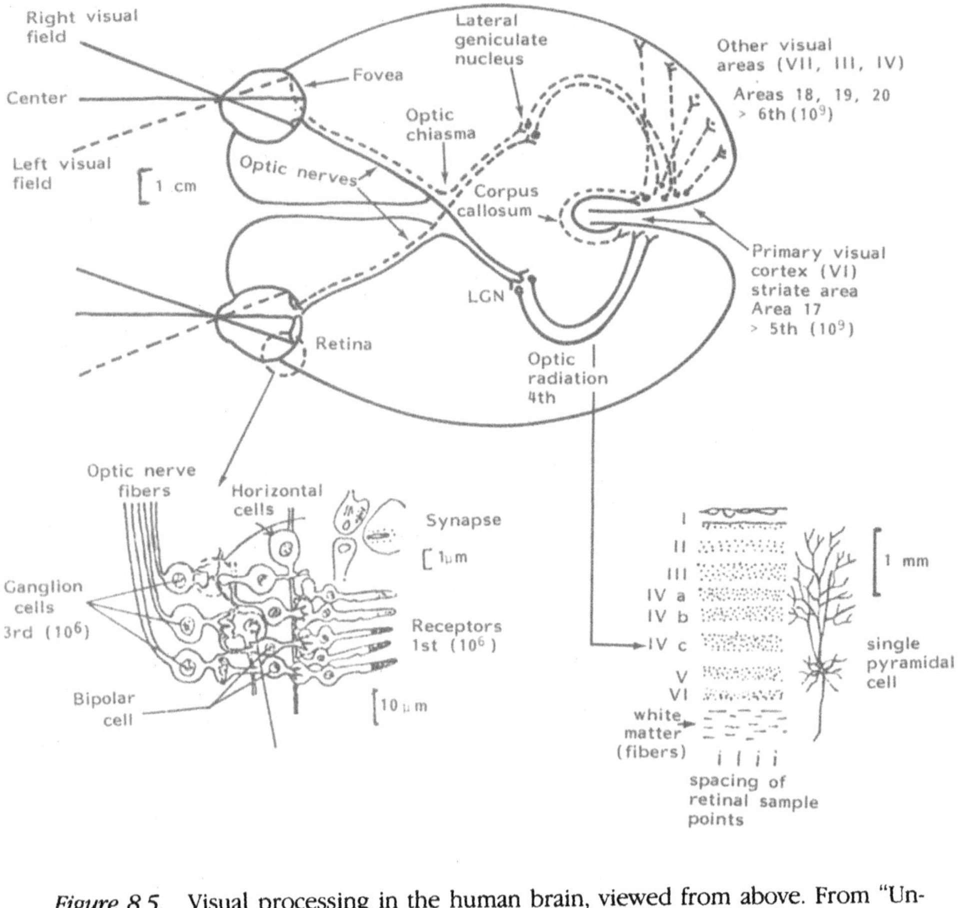

Figure 8.5. Visual processing in the human brain, viewed from above. From "Understanding Natural Vision" by H. B. Barlow in PHYSICAL AND BIOLOGICAL PROCESSING OF IMAGES, O. J. Braddick, and A. C. Sleigh, eds. Reprinted by kind permission of the author and publisher, Springer-Verlag.

time rate of change of light intensity. Rapid small motions of the eyeball (called "saccadic" motions) ensure that even a stationary scene produces a time-varying stimulus, and hence generates bursts.

These signals are processed further within the retina in so-called X-cells in which groups of excitatory sensors are surrounded by rings of inhibitory sensors; the latter tend to curb the activity of the former. The result is that the bursts of impulses generated by the X-cells depend also on the spatial rate of change of intensity of the retinal image. The effect is to filter the image with a band-pass filter whose response is circularly symmetrical (Figure 8.6). This response is a "difference of Gaussians" (DOG) form, which is very close to a doubly-differentiated Gaussian response. The consequence of this spatial filtering is to enhance edges (i.e., places where the rate of change of intensity with distance is high), without regard to the orientation of the edge.

The partially-processed signals leave the retina via a bundle of nerve fibers at the "blind spot." This bundle meets that from the other eye at the optic chiasm; half the signals from each eye are processed in each side of the brain to achieve stereoscopic vision. The visual signals are analyzed in the rear of the striate cortex in the cerebrum of the brain. Groups of cells present here are sensitive to edges of different orientation, both fixed and moving in particular directions. The outputs of these (and other cells, sensitive, for example, to general illumination level and color) are combined and processed further, for example, to analyze scenes and identify objects. This high-level processing is as yet far from understood.

It must be stressed that some information regarding the positioning of objects within a scene, and hence of depth, is obtained from single images using clues such as occlusion (some parts of the scene are partially hidden by others) and relative size. The processing required to extract and interpret these clues is extremely complex. In contrast, that required for depth perception using stereo vision operates even in the human visual system at the lowest level; it will work even with random textures.

For visual automation purposes, the most immediately interesting characteristics are those of the eye itself. Whereas man-made electronic cameras have only a single output channel and must transmit their data serially, the eye has hundreds of thousands of channels emitting

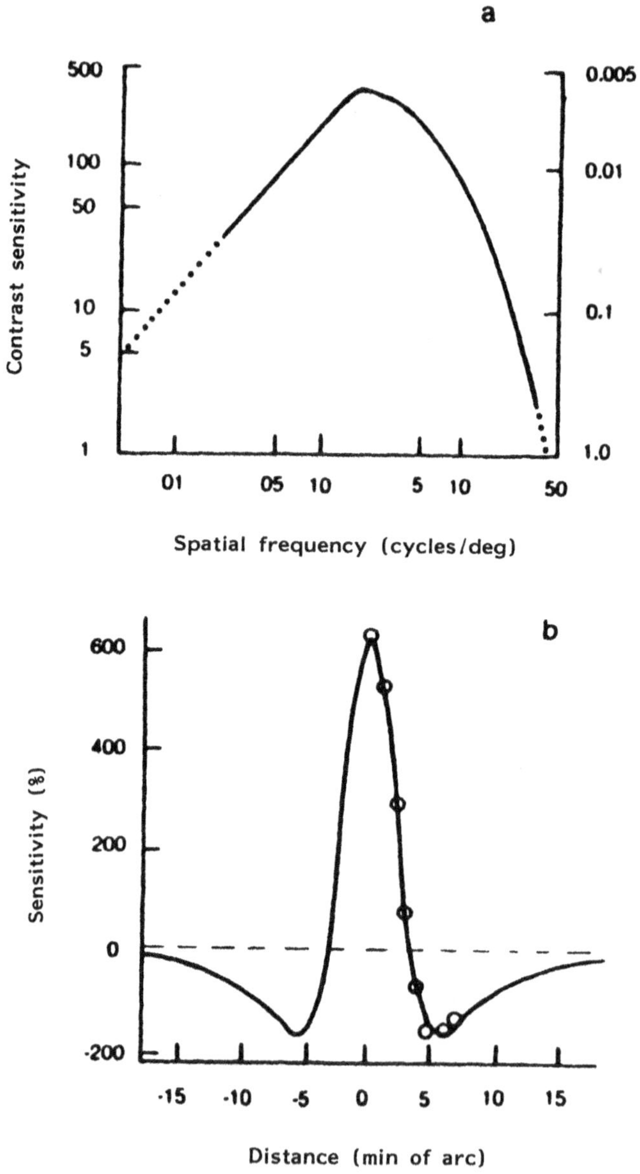

Figure 8.6. The modulation transfer function for the human eye is shown in (a) and the spatial response pattern (DOG filter) is shown in (b). Both functions are cylindrically symmetrical (the diagrams show sections only). The two functions are related by being Fourier transforms of one another. The negative sensitivities shown in (b) indicate inhibition rather than excitation. From "Visual Coding of Position and Motion" by S. Anstis in PHYSICAL AND BIOLOGICAL PROCESSING OF IMAGES, by O. J. Braddick and A. C. Sleigh, eds. Reprinted by kind permission of the author and publisher, Springer-Verlag.

in parallel though at a relatively low rate. The eye–brain system has a response time of about 1/30 sec.

The visual field has high resolution only around the *fovea centralis;* for a region of angular width about 2° in the fovea, the resolution is about 1 min of arc. The visual field extends outwards to more than 60° on either side, but the resolution degrades quickly with angle, being only about 6 minutes at 60°. The cones that sense color are present only in the high-resolution foveal region. Rods, that sense intensity only (but are very sensitive to motion) are present exclusively in the extrafoveal region. When examining a scene, the viewer moves his eyes so as to scan with his foveal region small regions of the scene containing the information needed to further his analysis.[8] Thus, the regions scanned depend on exactly what the observer is looking for, and a particular region may be scanned several times as the analysis proceeds. This arrangement nicely accommodates a weakness of the eye (compared to a man-made camera), in that aberrations due to the lens in the eye are very severe and a sharp image is obtainable only within a cone about 1° wide about the optic axis. In contrast, a man-made lens can have superb image quality over a field of 20° or more. The scanned search process is efficient and economical; information is not taken into the eye–brain system until and unless it is evidentally needed.

The eye focuses automatically to cover a field extending from 25 cm to infinity. Spatial resolution is essentially an angular quantity, thus it is maximum in distance terms at the near limit of the field; here it is about 70 microns for the foveal region used for critical work. Resolution may be specified in the space domain, as an ability to perceive that two dots which are close together are in fact distinct, or, alternatively, in the spatial frequency domain. This latter involves specifying its attenuation of pure sinusoidal patterns at various spatial frequencies. The variation of attenuation of the amplitudes of the patterns with frequency is termed the modulation transfer function (MTF). Generally, attenuation increases with spatial frequency. The MTF for the eye viewing a steady pattern is shown in Figure 8.6a. The eye is most sensitive to spatial frequencies of about 6 cycles/deg.

Related to resolution is *vernier acuity,* the ability of the eye to perceive that two lines which are almost collinear are in fact slightly offset. Vernier acuity at about 6 sec of arc is an order of magnitude better than resolution which is about 1 min of arc. These concepts

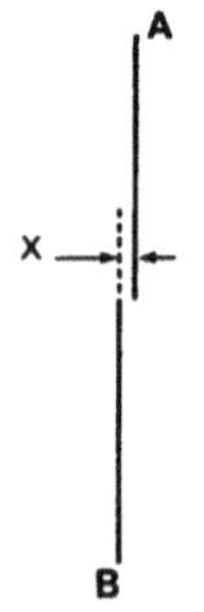

Figure 8.7. Vernier acuity defined as the smallest displacement x between the two lines that can be resolved.

are explained in Figure 8.7; Figure 8.8 shows how they are related to the spacing of the cones within the foveal region of the retina.

These characteristics limit human performance in operations such as detecting defects on moving sheet material. To examine defects 0.1 mm in diameter, the material must be almost at the closest possible distance of 25 cm. The visual field is then about 1 cm^2. If 1/30 sec is required to examine each field, only about 30 cm^2 of material may be inspected per second. Many industrial operations attempt to do much more.

The ability of the eye to detect change in brightness within a scene is logarithmic; an increase of 11% can just be detected. That is,

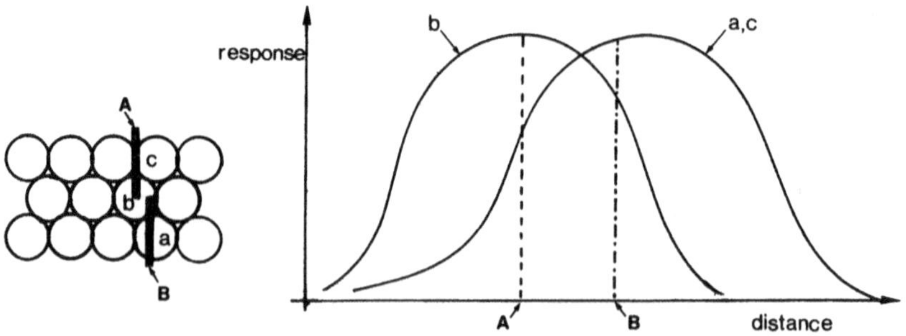

Figure 8.8. Visual acuity in terms of cell response. A line excites a different response according to its position within a sensor field. Thus, a vernier acuity of 6 sec of arc is obtained from cells whose fields are 20 arc-sec wide.

the smallest increase in brightness which can be detected is a fixed fraction of the original brightness. About 128 levels of gray can be distinguished in a monochrome image, but fewer (8 will suffice in certain circumstances)[9] often give adequate quality, and are more economical to provide.

8.3.2. Psychological Aspects

The psychological aspects describe what is consciously perceived by an observer. One expects that they will be consistent with the observed anatomical and physiological properties of the eye–brain system, and they are. Sharp edges and straight lines are perceived preferentially, and tend to dominate the viewer's perception, masking (inhibiting) perception of competing information. Not surprisingly, representations of objects as outline drawings are often very effective, as well as being economical. They have the drawback of not being so pleasing aesthetically as representations in continuous tone or color. High-pass filtering (unsharp masking) can be used to enhance the acceptability of continuous tone images. This accords with the high-pass (DOG) filtering observed in the retina, and with cells sensitive to edges found in the striate cortex. But this effect is sometimes a nuisance, since edges introduced, for example, as pixel boundaries in computer-processed images are intrusive and will mask (i.e., inhibit perception of) information which is more useful for interpreting a scene. It is often necessary to deemphasize such features, for example, by adding controlled amounts of random noise; this can enable gray-scale imagery to be plotted using patterns of binary dots.[10]

Contrast and size are important in determining how readily a feature may be perceived. This topic is discussed elsewhere by Rose.[11] Dark dots on a white background become more difficult to perceive visually as they get smaller and become of the order of the resolution of the eye. This may be offset by increasing the contrast between the dots and the background. A similar effect occurs for white dots on a dark ground, but the dots can go well below the resolution of the eye before they cannot be seen. At that point, a dot will be affecting only one sensor in the eye, but the magnitude of the signal for that sensor will still be significant when compared with those for surrounding sensors. This ratio is important in the detection of the dots. Stars which are well below the resolution of the eye can be seen at night, although birds, etc. having a much larger angular size are not seen in the daytime. This behavior is important in defect detection in inspection, and exactly that same behavior would be exhibited by a machine system. Whenever possible, defects should be made to appear as light objects on a dark ground.

The way in which shapes or objects are recognized is not understood, but it is plausibly argued that this involves separating them from their background ("segmentation"), then correlating them with stored prototypes, and assigning to the closest fit. Psychologically, recognition develops in two stages as learning proceeds. In the first stage, individual features are noted and compared systematically with possible matches until a fit is obtained. In the second stage, the object is perceived as a whole ("gestalt"), and there is no conscious process of feature-by-feature comparison. The correlation is apparently much faster when performed implicitly.

The actual analysis of a general scene probably involves tentative identification of particular objects, followed by semantic analysis to establish that the objects are connected by meaningful relationships. For example, an object tentatively identified as a table might reasonably be surrounded by items tentatively identified as chairs, and the presence of the two items thus related might be used to confirm the initial identification. Initial identifications will not always be correct, and the amount of information acceptable by the eye–brain system within a reasonable time is limited. The processes of recursion (going back to earlier stages to confirm or disprove an initial assignment), iteration (repetition of the various stages several times), and convergence (each successive iteration yields a semantic description which is more consistent than the previous one, and is therefore more acceptable) are almost certainly involved. The iterations are repeated until the degree of semantic consistency achieved reaches an acceptable level. The process is undoubtedly partially bottom-up (i.e., low-level clues are integrated to reveal more general truths), and partially top-down, i.e., high-level information such as knowledge of the general nature of a scene (is it inside a room?) is used to facilitate interpretation of the low-level clues. Undoubtedly, the vast majority of the information used to interpret a scene is contained already in the brain of the observer (for example, gravity acts downwards and chairs are usually found next to tables), rather than being extracted from the image of the scene. But the exact nature of the process is not truly understood, and science can at present offer only a plausible conjecture.

In the past, knowledge of the physiological and psychological properties of the human visual system has generally been of little use in devising instrumentation for computer vision systems applied to industrial automation. The flow of ideas has in fact been in the other direction; concepts developed by image processing engineers are being applied by researchers to enable them to understand the human visual system. However, important indications are provided as to why human vision is sometimes inadequate in industrial operations. The difficulty introduced by limited resolution has already been noted. Poor contrast and high speed combined with the small size of many defect indications render the eye–brain system inefficient for inspection. The fatigue resulting from performing boring tasks further inhibits precise inspection.

8.4. Image Processing

Four principal kinds of complex computer image processing operation have been developed. These are:

- Image restoration
- Image enhancement
- Image interpretation
- Image coding

All of the above are of use in intelligent automation to some extent; most, however, are too slow and too expensive to perform (a maximum entropy restoration may take 8 hours per image!) to be applicable to tasks in on-line automation. We discuss the methods in more detail below. The matrix description of an image introduced in Section 8.2 is used to make the presentation concise.

8.4.1. Restoration

When images are formed by sensors (such as electronic cameras), distortions are often introduced. A perfect sensor would produce a point image of a stationary point object, but practical sensors generally smear the image over a region of finite size, causing blurring. If the smearing relationship is known, the blurring may in principle be corrected completely by an inverse operation. The process is described mathematically (in bare outline), for a linear system in which the distortion is uniform over all of the image, as follows. We designate the corrupted image formed by the sensor by an $M \times M$ matrix $[B]$, and the original uncorrupted scene by another $M \times M$ matrix $[A]$. The objective of the restoration is to recover the scene $[A]$ from the image $[B]$. The process of image formation is represented by the matrix equation

$$[B] = [H].[A]$$

In the above, $[H]$ is the response, or point spread function, for the sensor; it is also an N-square matrix. If $[H]$ is nonsingular, then the inverse operation for correcting distortions becomes simply

$$[A] = [H]_{\cdot}^{-1}[B]$$

Such corrections may be implemented in practice if a unique inverse operation exists, and if the blurring fulfills constraints such as being relatively localized, so that the time for the computation is reasonably short. Most imaging systems also introduce random noise, which cannot be removed exactly even in principle, and frustrates inverse processing for correcting smearing. Techniques are however available[12] that enable a useful restoration to be achieved even for a noisy image with the blurring function not explicitly known, for example, maximum entropy estimation. However, techniques of this kind are very time consuming computationally, and thus are not really viable in automation applications, and certainly not on-line and in real time. If $[\dot{H}]$ is relatively simple and noise-free, and may be determined by measurement, then real-time compensation (a form of deconvolution; see Section 2.4.4) may be possible. For a reasonable value of N, say, 256, then 65,536 multiplications and 65,536 summations will be required for a noise-free, spatial-domain deconvolution. Some improvement is available using the fast Fourier transform and working in the frequency domain, but this would reduce the number of computations only to 4352 complex multiplications and 4096 complex additions. Thus, the computations required for image restoration are generally too complicated for the approach to be very useful in industrial automation; distortions, if present, should be dealt with in other ways.

8.4.2. Enhancement

Enhancement[13] involves doing things to an image to make it easier or more convenient to interpret visually. In mathematical terminology, we create a new image $[B']$ which is not necessarily closer to the original scene $[A]$ than $[B]$ was, but is considered more desirable. Examples of enhancement often used include edge enhancement using spatial differentiation, because (as we have already seen) human observers are more responsive to edges; modification of the histogram of gray levels (to even contrast over an image); and false color. The latter involves encoding each level of pixel amplitude for a monochrome image with a unique color. This enables the absolute amplitudes of various regions to be determined quantitatively by a simple visual examination. Spatial filtering (by selective rejection of energy at particular spatial frequencies) can be used to increase the contrast of interesting features, such as defects on a noisy surface, or blow holes in an X-rayed cast component.

Enhancement, however, usually implies the presence of a human operative in the processing chain, and systems for automation should avoid this kind of complication. Thus, enhancement is of limited applicability in on-line systems, though it is of some use in preprocessing, and may be very valuable during the research stages of an investigation.

8.4.3. Coding

The objective here is to remove redundant information from images, in order to improve the efficiency with which they may be processed, communicated, or stored.[15] Two distinct approaches have evolved:

1 *Reversible* encoding, (for example, the Huffmann code; see Section 2.2.2), that minimizes the effects of various pixel levels possessing unequal probabilities, and of pixel levels not being mutually independent.[14] All the information within an image is preserved by a reversible encoding; the original image may be completely reconstituted merely by inverting the processing.

2. *Irreversible* encoding (for example, transform coding), in which an orthogonal transform (such as the Fourier transform) is obtained of the image, and only the lower M of N original components are retained; the remainder are discarded. It is possible to eliminate a substantial fraction (sometimes 90%) of the code information without substantially degrading the semantic or pragmatic information that the image is required to convey.

Codes of both kinds are used widely in automation applications, though often in disguised form. Describing an image by features is a form of irreversible encoding.

8.4.4. *Interpretation*

Interpretation comprises processing to extract from the low-level (symbolic) information in which the image initially appears, the high-level (semantic or pragmatic) information which is really of use. This activity is fundamental in intelligent industrial automation, and (in simple forms) occupies much of this chapter. The complete interpretation of a general, real-world scene is well beyond the current state of the art. Analysis is profitable only when the scene is highly constrained (so that its content is largely known before analysis commences), or has been considerably simplified. Thus, most of our discussion considers simple images such as silhouettes. Processing often commences by converting a relatively complex (gray-scale) scene into silhouette form.

8.5. Silhouette Images

In a silhouette, or binary, image, each pixel can occupy only the two levels, 0 or 1. Silhouette images are obtained from gray-scale images by comparison with a threshold. The spatial quantization involved often introduces considerable noise at boundaries; images with smooth edges are made rough. This introduces error into any measurement which involves the boundary[(16)] (compare Figures 2.3a and 2.3b).

Automation commonly requires the following operations on silhouette images:

1. Measurement of particular dimensions.
2. Detection of defects, such as the absence of holes.
3. Determination of the orientation and location of the image.
4. Identification of the images (following training over all image classes which may be encountered).

The image is acquired and stored as a grid of samples. Section 2.2.2 indicated how coding techniques can be applied to improve efficiency in processing and representation.

8.5.1. Feature Measurements

Separation between points in a silhouette (and hence dimension) may be measured by determing the number of cells in the *X*- and *Y*-directions between the cells concerned, and applying Pythagoras' theorem (Figure 8.9). Simply counting the number of cells directly between the ends will not do, since the distance will then depend on the orientation of the line with respect to the grid axis. Distances measured along the diagonal of a square will appear shortened by 40%, since the diagonal length of a square is 40% longer than its side. Perimeter measurement is subject to error introduced by quantization effects; these may be corrected, but the process is costly. Measurement of area on the other hand involves merely counting cells at the "1" level, and

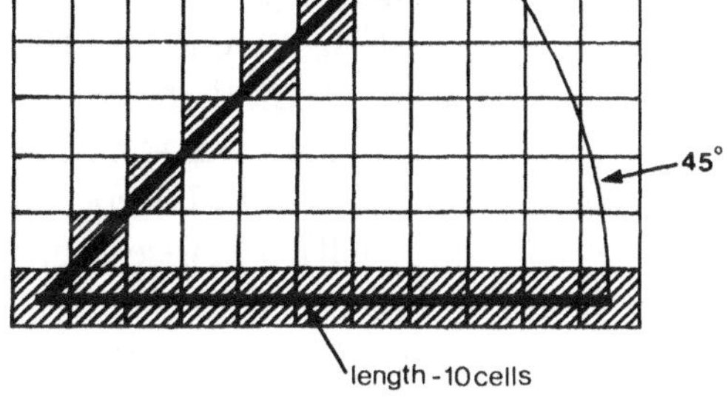

Figure 8.9. Apparent variation of line length (measured in pixels) with orientation.

is a fast, simple operation. It is a stable quantity since (unless the image is unusually thin) most of its cells will not be on the boundary, and thus will not be subject to quantization noise.

For classifying images, representation by a finite set of measures is necessary. Many alternatives have been proposed; to be useful, they must fulfil the following desiderata:

1. *Compactness*—the representation must contain far fewer bits than the original image, in fact, as few as possible while retaining the required ability to discriminate.
2. *Uniqueness*—the representation must possess the ability to represent each image class by a different set of values.
3. *Stability*—the representation must be unaffected by corruptions such as quantization noise.
4. *Invariance*—the representation must not be changed by rotation and translation, though it is often required that the representation give an indication of these quantities.
5. *Speed*—the time required for processing should be small enough to meet any requirement imposed by real-time operation.

Among the most important features for representing silhouettes are:

- Gross measures, such as area and perimeter
- Radii lengths from centroid to boundary
- Transforms of the shape (e.g., Walsh-Hadamard and Fourier)
- Transform of boundary
- Residue of convex hull
- Orthogonal projection histograms
- Moment invariants

The methods divide into two classes according to Pavlidis's principle.[(17)] The first class we term "asymptotic," the second, "nonasymptotic." Asymptotic methods have the property of being able to distinguish any two different shapes, however similar they may be, merely by increasing image (and hence feature) resolution. If an image 256×256 pixels will not suffice, then improving resolution to 512×512 may work, and so on. This is clearly very valuable in industrial applications. If the resolution parameters chosen for a system early in its development are later found to be insufficient to enable closely similar patterns to be discriminated, resolution improvement will guarantee success. Nonasymptotic methods (such as those relying on dimensionless measures such as perimeter squared over area) may well be unable to distinguish between particular classes of shape, whatever the resolution. If silhouette objects touch or overlap they are much more difficult to process.

The various methods are explained in further detail below.

1. Area may be computed for a silhouette merely by counting pixels whose amplitude is on the correct side of a threshold to be "inside" the shape. It is thus fast to compute, does not require that the silhouette be held in store, and is stable to errors such as boundary quantization noise. It is a very good feature.

   Perimeter is measured by counting cells at the silhouette boundary. Boundary cells are located by convolving the image with a mask at least 3 pixels^2 (2- × 2-pixel masks may omit certain configurations of points and add extra cells in other cases). But perimeter length is very much affected by quantization noise at the shape boundary, and is also a function of orientation. Algorithms[(18)] are available to correct these, but introduce undesirable complications.

   The ratio, perimeter^2/area, is a dimensionless measure which can distinguish, for example, circles (ratio = 12.6) from squares (ratio = 16) irrespective of their size. Figure 8.10 includes two shapes (B5 and B5*) which are clearly distinct but have the same ratio of perimeter^2/area; the existence of such ambiguities is a weakness of this approach. Although the compression obtained is excellent, the description obtained is not unique.
2. A shape specified by the lengths of radii from centroid to boundary, measured at specified angles anticlockwise from the longest radius, is shown in Figure 11.2. The position of the centroid is determined easily, accurately, and without a need to store the whole image, using simple equations below (Section 8.5.2). The coordinates of each point in the boundary, however, must be stored, and the angle (from the longest radius) of each, computed. Generally, a small number (e.g.,

20) of radii suffices to distinguish a large number of shapes, thus the approach provides good compression. It is clearly asymptotic in the sense of Pavlidis.

The method has weaknesses in that the longest radius may be ambiguous, and that if a radius approaches the boundary at an acute angle, then small errors in orientation can cause large errors in length estimation.

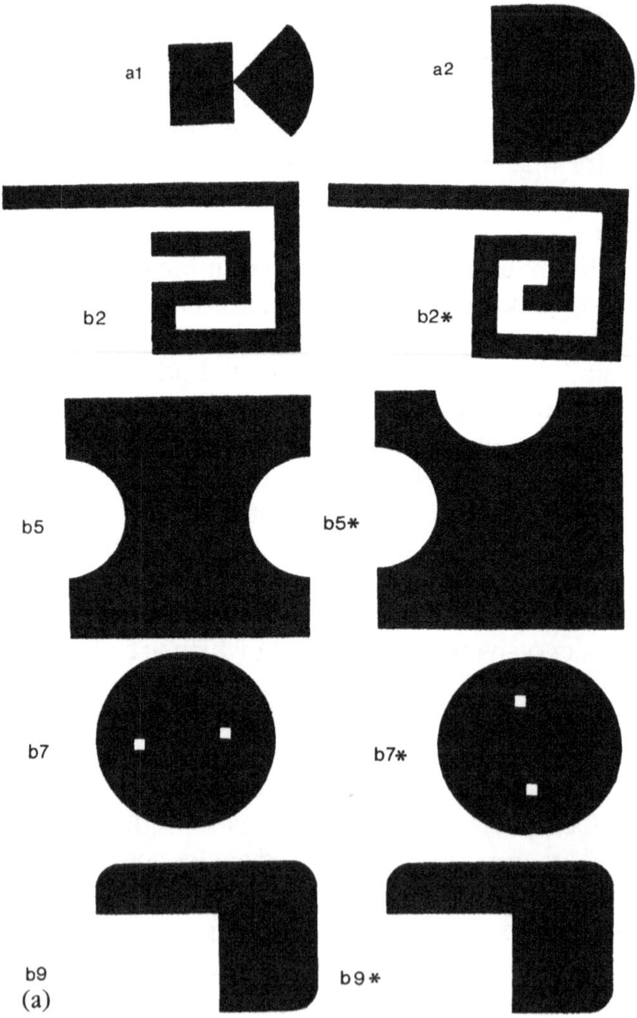

Figure 8.10. (a) and (b) are silhouette shapes from a set for benchmarking silhouette vision systems.

Figure 8.10. (*Continued*)

Alternatively, shapes may be specified by the angles at which the boundary lies at a specified distance from the centroid (Figure 8.11). This may be slightly more efficient computationally, since radius lengths may be handled as the sum of squared x- and y-distances from the centroid, whereas computation of angles is trigonometric.

3. In the transform of shape approach, the silhouette is characterized by a few low-order components of its two-dimensional transform. This is invariant to translation and location, and is also asymptotic, but the processing required is extensive if the

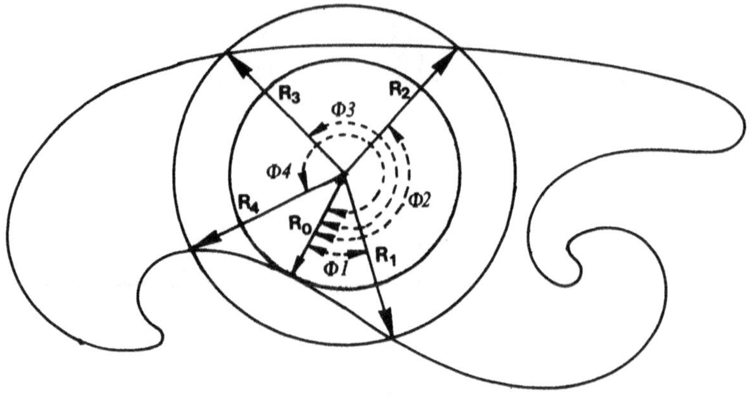

Figure 8.11. Shape described by radii from the centroid to the boundary, at points where the boundary lies at a specified distance from the centroid. The datum is provided by the shortest radius R_0 which in this case is unique. The shape is specified by the angles Φ1, Φ2, etc. of the equal radii R_1, R_2, etc. This should be compared with Figure 11.2.

Fourier transform is used, and the complete image must be held in store while it is being performed.

The processing required for the Walsh-Hadamard (W/H) transform is much simpler, and a commercial instrument has been made using this principle,[19] although the W/H transform is not translationally invariant.

4. In this so-called parametric approach, the distance around the boundary measured from a fixed point is specified as a function of the *x*- and *y*-coordinates of the point.[20]
5. The convex hull of a shape is a line touching its edge at every point, except that concavities are not followed. It is the figure which would be generated if an elastic band were to be stretched around the shape. An example is shown in Figure 8.12. The residue of the convex hull, i.e., the regions left between the convex hull and the shape itself, specified by their areas, is used to specify the shape.[21]

   Many simple shapes (e.g., circles, ellipses, squares) will have no residue for their convex hulls, and will thus be indistinguishable using this method. Other shapes which do have residues for their convex hulls (e.g., B4 and B4* in Figure 8.10) are also indistinguishable since their residues are identical.

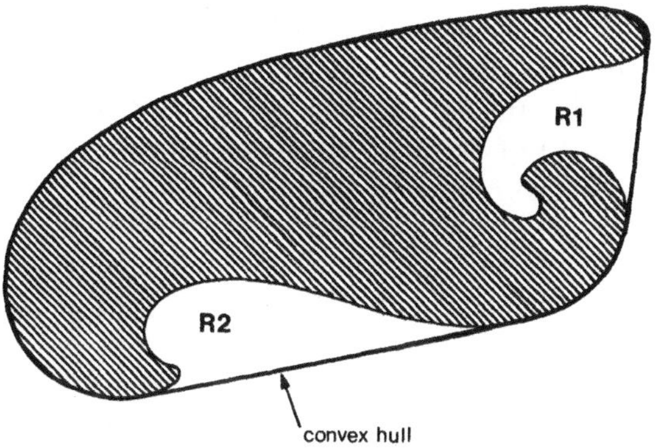

Figure 8.12. Convex hull of a silhouette shape. This is the thick line surrounding the shaded silhouette. The residues $R1$ and $R2$ would be used to characterize the shape.

Thus, the description provided is not unique, and confusion is possible. To compute the convex hull, the complete silhouette must be held in store. Information regarding the stability of the description with regard to the rotation and translation of the shape and quantization noise, and the relative time required to compute the measure, is hard to obtain.

6. Row and column histograms[22] are formed by summing the number of pixels in each row and column of the silhouette. This generates two histograms (Figure 2.7) which characterize the shape with economy. An unknown shape is recognized by finding stored shapes whose histograms provide a best fit with its own. The form of the histograms is, however, a function of the orientation of a shape, and their position is a function of location. Thus, the method may be used to find position and orientation (workable when only one kind of shape is being cxamincd), and is attractivc for its computational simplicity. However, taking row and column histograms throws away information (Section 2.2.2), and the method is nonasymptotic. Working with the Fourier transforms of the histograms rather than the histograms directly achieves invariance to translation.

8.5.2. *Datums*

Measurement of position and orientation, and of features, requires the provision of datums for reference. These must be unique, easy to compute (preferably without a need to hold the complete image in store), and robust to quantization errors.[23] Quantities fulfilling these requirements are hard to find. The *centroid* is a good datum point for specifying the location of a silhouette. It is the point having coordinates $x(O),y(O)$ so that the sum S of its squared distances from all other points within the silhouette is minimum. Let the silhouette contain N pixels; then, S is given by

$$S = \sum_{i=1}^{N}[(x(i) - x(O))^2 + (y(i - y(O))^2]$$

On differentiating this expression, we obtain for $x(O)$ and $y(O)$ the formulae

$$x(O) = 1/N\sum_{i=1}^{N}x(i) \quad \text{and} \quad y(O) = 1/N\sum_{i=1}^{N} y(i)$$

Thus, the coordinates of the centroid may be determined simply by summing and averaging the locations of the cells in the silhouette. Its position is clearly both unique and insensitive to errors caused by boundary quantization.

For referencing orientation, lines such as the longest (or shortest) radius vector from centroid to boundary may be used, but are often unsatisfactory since they are not unique. Moments (and moment invariants) are safer;[24] the m,nth moment $\mu_{m,n}$ of a continuous image function $f(x,y)$ about the centroid is defined as

$$\mu_{m,n} = \int\int f(x,y)\, x^m, y^n dx.dy$$

The zeroth order moment $\mu_{00} = \int\int f(n,y)\text{dx,dy}$ represents the total power in the image. The first order moments $M_{1,0}$ and $M_{0,1}$ locate the centroid of the image, whose coordinates $x(O)$ and $y(O)$ are given by

$$x(O) = \frac{M_{1,0}}{M_{0,0}} \quad \text{and} \quad y(O) = \frac{M_{0,1}}{M_{0,0}}$$

The second order central moments μ_{02}, μ_{20}, μ_{11} characterize the size and orientation of the image. Any image can be represented by a silhouette ellipse having definite size, orientation, and eccentricity whose centroid coincides with that of the original image. The angle σ between the semimajor axis of the ellipse and the x-axis of the coordinate system is (for example) given by

$$\sigma = 1/2 \arctan (2\mu_{11}/[\mu_{20} - \mu_{02}])$$

This gives a stable measure of image orientation which is insensitive to boundary errors. Higher order moments may be used to specify a shape in a similar way to Fourier components.

8.5.3. *Silhouette Processing*

The first stage in processing a silhouette is to distinguish it from the background. In many cases this is accomplished simply by thresholding; good contrast is obtained by well-arranged lighting. For more difficult problems, such as distinguishing track from substrate in a printed circuit board, more complex approaches are necessary, such as adaptive thresholds dependent on properties of the histogram of gray levels.[(25)] Two very general and very powerful approaches are available when this will not suffice; they increase complication and hence processing speed. The first detects boundaries by detecting discontinuity, usually by using an edge detection mask.[(26)] The second works by detecting similarity, for example, by using region growing.[27]

The most common discontinuity detector is the gradient detector. This detects the finite difference approximation to the spatial derivative of image intensity in the x- and y-directions. Gradient is a vector, so both magnitude and direction may be used together for tracing an edge in the presence of noise. The simplest and fastest gradient detector is the Roberts Cross operator, implemented by a 2- × 2-element mask, of the form

$$|b(i,j) - b(i + 1,j + 1)| + |b(i + 1,j) - b(i,j + 1)|$$

Though adequate for many applications, it does not provide a direction, and is the least effective mask. The Prewitt and Sobel operators are 3- × 3-element operators which do provide both magnitude and direction:

$$\begin{vmatrix} -1 & 0 & 1 \\ -1 & 0 & 1 \\ -1 & 0 & 1 \end{vmatrix} \quad \begin{vmatrix} 1 & 1 & 1 \\ 0 & 0 & 0 \\ -1 & -1 & -1 \end{vmatrix} \qquad \begin{vmatrix} -1 & 0 & 1 \\ -2 & 0 & 2 \\ -1 & 0 & 1 \end{vmatrix} \quad \begin{vmatrix} 1 & 2 & 1 \\ 0 & 0 & 0 \\ -1 & -2 & -1 \end{vmatrix}$$

(Prewitt) (Sobel)

Laplacian operators generate a finite difference approximation to the Laplacian function. Since this is a scalar, no direction is provided. Other operators are the parametric (Heukel) and template (Kirsch) forms.

The various operators can be evaluated on the following criteria (Peli and Malah)[28]:

- Fraction of detected edge points on the desired edge
- Fraction of edge points not on desired edge
- Signal-to-noise ratio (a/b)
- Measure of deviation of detected edge from ideal edge
- Type of contour, e.g., perfect or broken, single or double, deviation from ideal

A comparison of masks using continuity and thinness of the generated edge as goodness criteria is given in Reference 29.

8.6. Gray-Scale Images

The range of gray-scale image processing usable in automation is currently rather limited. Most images are gray scale when acquired, and are then processed to extract silhouettes. The following operations are commonly required:

1. *Segmentation,* i.e., subdivision of the image into regions representative of different objects and so on. These are then treated as silhouettes for subsequent processing. Segmentation is usually performed using thresholding or masking, as already described, but noise may cause the boundary to be indistinct.

The Hough transform[30] may be used to extract straight lines and segments of circles following noise segmentation, but its costliness in computation may preclude its on-line use.

2. *Detection, delineation, and identification of defects,* particularly on web material such as sheet metal, glass, and textiles. If the material is free of fixed patterns (like sheetmetal and glass), then defects can be detected by adaptive thresholding; the principle is explained in Section 2.6.1 and its application, in Section 11.5. The presence of a defect generates "trigger" pulses, which also give some indication of the extent and plan shape of the defect, and thus aid delineation (i.e., determination of the region occupied by the defect). Triggers can be assigned to the correct defect when more than one is present simultaneously within a field of view using a "clumping rectangle" approach.[31,32] Defects can be identified by extracting features such as area, perimeter, width, and length, and then using the standard methodology for pattern recognition developed in Chapter 3. For material having a fixed pattern, such as printed material or textiles, detection of defects which are not large compared with the scale of the pattern can be very difficult; no standard methodology exists. Some approaches worth trying are described elsewhere.[35]
3. *Identification of gray-scale patterns* having the same external dimensions, such as product labels. Attempts to identify by reading characters are generally unnecessarily complicated and expensive. Pixel-by-pixel comparison between pattern and perfect prototypes until a fit is found is unsatisfactory because of the need to register precisely in position and orientation, because of the vast quantity of information which must be handled, and because in most cases some variation in patterns is allowable which would cause mismatch.

   The approach recommended is to try all possible measures, with the simplest and quickest being investigated first. The simplest measure is probably the average gray level of the surface, i.e., the number $g(av)$ defined by

$$g(av) = \frac{1}{N}\sum_{i=1}^{N} g(i)$$

where N is the number of pixels over the surface, and $g(i)$ is the amplitude of the ith pixel. The second simplest measure is the amplitude of the kth bar in the gray-level histogram; the bar used is selected so that its height differs significantly between pattern classes but exhibits a minimum of within-class variation. It may be necessary to use several bars to obtain adequate discrimination. These two measures are invariant to change in the position or orientation of the pattern within the scene, provided it remains completely inside.

If these do not give adequate discrimination, then the pattern may be broken into blocks and the average amplitude and histogram bar heights used for each block individually. This approach is asymptotic in the sense of Pavlidis: it is guaranteed to work, provided the blocks are made small enough. But in an exceptional but unfortunate case, this could mean using blocks small enough to include only one pixel! Also, it is now necessary to compensate for variation in position and orientation, normally by finding the boundaries of the pattern; which can be well delineated.

4. *Determination of texture.* This operation is treated in the next section.

8.7. Texture

The texture of a gray-level image describes small-scale, repeated variations in pixel amplitude that convey information only when the variations change. The term "texture" derives from "textile." To measure texture, the properties of the pixels are measured within a small window which is moved over the surface. There are two kinds of texture: regular, in which the variations form a pattern which repeats cyclically over the surface (as in a plaid), and statistical, in which the variations never repeat exactly but maintain constant statistical properties. The first kind of texture is generated by a machine tool cutter moving steadily over a surface; the second is generated, for example, by shot blasting. The properties of the texture of the gray-scale image of a rough surface convey some information regarding the mechanical profile of the surface.[33] The occurrence of a defect will cause the texture to change abruptly, but the texture itself constitutes a noise signal that tends to hide the defect. The window must be just large enough to contain a complete cycle of the lowest spatial frequency in the texture in the x- and y-directions.

For a window 10 pixels square, with each pixel able to occupy 256 levels, 256^{100} distinguishable patterns are possible. Measures must thus be imposed on the windowed data to compress the information. For regular textures in particular, the powers of the components of the two-dimensional discrete Fourier transform (DFT) may be used; these will not change as the window is moved through a stationary texture pattern. Otherwise, statistical measures can be used.[(34)] The zeroth order statistic of the data within the window is its mean value. The first order statistic is computed as the histogram of pixel levels. The second order statistics are computed as the gray-level cooccurrence matrix (GLCM) for the data, which shows the number of times that a point having gray level i is a neighbor in direction d of a point having a gray level j. This is specified by the i,jth element of the matrix; note that separate matrices are required for the x-direction and the y-direction. For example,

$$\begin{vmatrix} 0 & 0 & 0 & 1 \\ 0 & 1 & 2 & 2 \\ 0 & 1 & 2 & 0 \\ 1 & 2 & 1 & 0 \end{vmatrix}$$

a. Pixel levels seen through window

$$\begin{vmatrix} & 0 & 1 & 2 \\ 0 & (0,0) & (0,1) & (0,2) \\ 1 & (1,0) & (1,1) & (1,2) \\ 2 & (2,0) & (2,1) & (2,2) \end{vmatrix}$$

b. General form of cooccurrence matrix

$$\begin{vmatrix} 6 & 2 & 2 \\ 2 & 2 & 3 \\ 2 & 3 & 2 \end{vmatrix}$$

c. Vertical cooccurrence matrix for data in (a)

$$\begin{vmatrix} 4 & 4 & 1 \\ 4 & 0 & 4 \\ 1 & 4 & 2 \end{vmatrix}$$

d. Horizontal cooccurrence matrix for data in (a)

The gray-level cooccurrence matrix is symmetric, which may be used to reduce computation or provide a check. But even the GLCM contains far too much information, and measures must be used to compress the information still further. Some possibilities are

$$\text{Entropy} \quad H = -\sum_i \sum_j p(i,j)\log\{p(i,j)\}$$

$$\text{Correlation } 0 = \frac{\sum_i \sum_j (i,j)\mathrm{p}(i,j) - \mu_n\mu_y}{\delta_n\delta_n}$$

$$\text{Contrast} \quad C = \sum_{K=0}^{n} K^2 \left\{ \sum_{|i-j|=k} p(i,j) \right\}$$

$p(i,j) = i,j^{\text{th}}$ element of GLCM

μ_x, δ_n = are the mean and S.D. of row sums of GLCM

Investigation of the utility of these measures for surface characterization, as well as suggestions of some other measures, appear elsewhere.[35]

8.8. Performance Specification and Evaluation

Many computing methods for silhouette shape identification have been produced, and a correspondingly large number of systems have been developed for the automated recognition of silhouettes. Some of these systems have been tailored to perform one specific task, for example, to find the orientation of a specific piecepart and inspect it for defects. Others are general purpose instruments intended to be applicable with equal facility to a wide variety of tasks, and are marketed commercially as such.

When faced with a problem requiring the processing of, for example, silhouette shapes, engineers are often in doubt as to whether existing equipment is suitable, or something new must be developed. Even when commercial instrumentation is available, it may be necessary to choose the best from a selection of alternatives or to establish whether any alternative fully meets the need. When equipment fails to perform at the level predicted by theory, the cause must be identified so that it may be corrected. It may be necessary to evaluate individual components of a system. Thus, there is a need for standardized and objective performance tests of various kinds.

There are two types of tests. The first examines the performance of particular components in a system, such as the optical sensor. The second evaluates the overall system. The development of standardized tests is as yet embryonic, possibly because developers of instrumentation systems or processing methodologies are reluctant to reveal weaknesses. Also, the best way of assessing the suitability of a system for a particular application is undoubtedly to examine samples of the material actually used in the application. This is, however, often impracticable, owing to the cost of preparing special software or sensing configurations. Much instrumentation is intended to be general purpose anyway, and is designed without any specific application in mind.

8.8.1. Specification of Optical Sensor Performance

An example of the first kind of measure is the modulation transfer function (MTF), originated by Schade and used to specify the properties of lenses and sensing chips. Instruments are available for measuring MTF directly. In his book on the subject, Schead explains how MTF affects image quality.(36) The boundary point scatter diagram (BPSD) goes a little further; it is used for assessing the performance of the whole of the sensing part of a silhouette vision system, and is particularly valuable for evaluating distortions in systems used to make fine measurements. It is obtained by scanning a disk which is as nearly as possible perfectly circular, with a bevelled edge to eliminate distortions arising from perspective effects. The centroid of the disk, and the edge pixels nearest to and furthest from the centroid, are determined. Ideally, all edge pixels should lie at the same distance from the centroid; in practice they do not, because of effects such as edge quantization noise, geometrical distortions in lens systems, and incompatibility between pixel spacing and belt speed when a linescan array is used to examine a two-dimensional object. To generate the BPSD, the position of each edge-point is plotted proportionally between concentric circles representing the closest and most distant pixels; the effect (Figure 8.12) is to exaggerate the distortions. The edge-points cluster about a closed curve which should ideally be a circle, but is generally distorted due to the above-mentioned imperfections. Typically (Figure 8.13), the cluster is circular in the center of the field, but degraded towards the edge. Part of the distortion

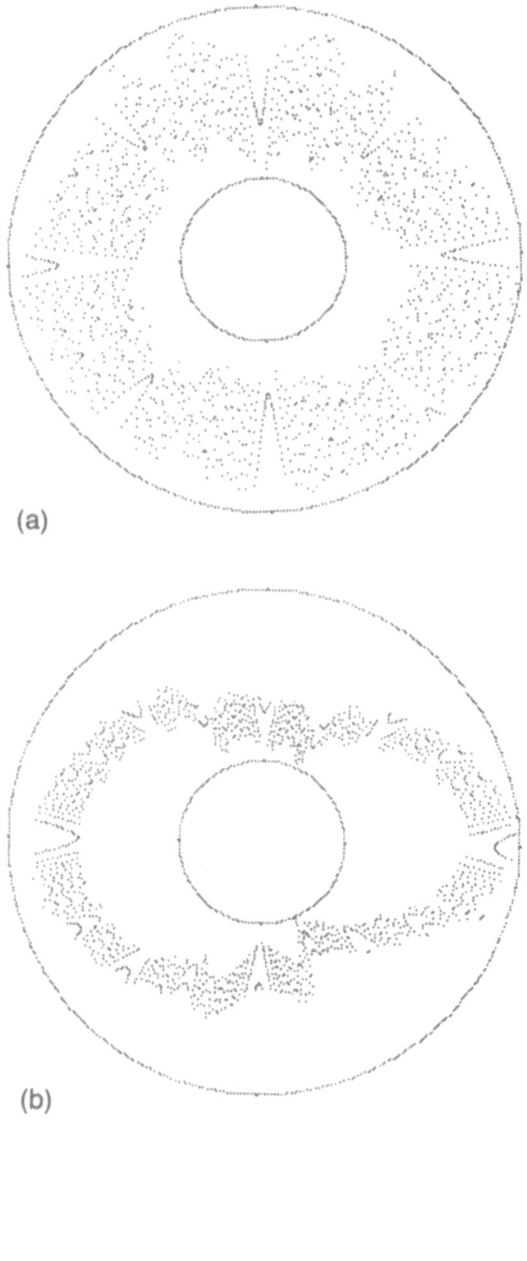

Figure 8.13. (a) and (b) are boundary point spot diagrams (see text) showing quantization error and distortion. (Courtesy of BUSM Co.)

appearing in the BPSD is due to the lens and will be revealed in the MTF; the remainder must be due to other parts of the system.

8.8.2. Specification of System Performance

The second type of test may be implemented using a carefully chosen standard set of silhouette shapes. In constructing such a set, the following must be taken into account

1. Instrumentation for processing silhouette images is generally required to perform a range of tasks. These include identifying shapes, determining orientation and location, and detecting and classifying any defects present. A good set of shapes for benchmark testing should cover all of these.
2. Some systems can handle only a few shapes, in extreme cases only one, and the system is designed specifically to handle these. Others may be required to handle many shapes, in some cases several thousand, whose form is unknown until they are actually presented to the machine. Systems in the latter category are generally self-training. They can learn to identify a previously unknown shape merely by scanning it.
3. Even with silhouettes, shapes from different task areas may differ considerably, so that methods or instruments optimal for one type would be quite unsuitable for another. For example, engineering components are generally quite different from parts of garments. The latter are much less likely to have boundaries which are straight lines, or to have internal holes. An instrument designed specifically for one kind of task might well perform poorly on another; benchmark testing must allow for this.
4. Many attributes contribute to the overall performance of a system. These include resolution, speed of identification, reliability (expressed by rejection and substitution rates), accuracy of orientation and location, ability to discriminate shapes which are closely similar, and tolerance of minor disturbances such as specks of dust. These attributes are strongly affected by resolution. Thus, specification of performance without including specification of resolution is meaningless. The varia-

tion of performance with resolution is a sound and useful measure.

5. The wide variety of methods available for processing silhouette shapes reflects the ingenuity and desire for originality of researchers, as well as the differing needs of the wide range of problems. Methods tend to have particular strengths and weaknesses, which the evaluation must reveal.

Two approaches have been used in developing sets of shapes for benchmark testing systems for handling silhouette images. That of Rosen and Gleason[37] uses shapes which are simple geometrical forms such as disks, squares, equilateral triangles, and rectangles. Some shapes contain internal holes. Because they are so simple, a potential user can easily draw his own. However, real silhouette objects are normally far from being simple shapes; they are often quite complex, and are rarely symmetric. Therefore, Rosen and Gleason's shapes may not reveal weaknesses that would appear when a system is applied to real shapes.

An alternative set[38] has been designed to overcome this weakness, at the cost of using specially devised shapes which a potential user must copy exactly if he is to apply a standardized and critical test. The number of shapes in the set is restricted to 40 to limit the cost, and to contain the time required to perform an evaluation. The shapes are of four types referred to as "A," "B," "C," and "D." Types A and B are carefully devised drawings; types C and D are silhouette photographs of real mechanical parts for which automatic assembly or inspection may be required.

The set is also divided into two categories on the basis of the required task. Shapes of types A and C are intended to be considered individually, and to test for precision in determining location and orientation. Shapes of types B and D are to be considered in pairs, and are intended to test the ability of the system to distinguish one member of a pair from the other.

The shapes in this set are illustrated in Figure 8.10. Shape A1 has been chosen so that the longest radius from centroid to boundary describes an arc subtending a substantial angle (about 90°) at the centroid. This may confuse systems which use the longest radius as datum for specifying orientation. Shape A2 has two identical longest radii, from the centroid (which lies at the center of the semicircle) to the remote corners of the rectangle. Naive processing might choose one or the other, leading to ambiguity.

Set B1 comprises a pair of patterns (B1 and B1*) which are right- and left-handed versions of the same shape.

Set B2 comprises two complex shapes which have a large ratio of boundary length to area. Though clearly distinct, the shapes are of similar construction, and have the same perimeter length and area.

Set B3 comprises two identical ellipse-like shapes. (Actually, they are a pair of semicircular segments joined by two straight lines.) Holes have been inserted so that the position of the centroid remains unchanged. The two holes in each shape are different. Thus, although the centroid is at the center of symmetry of the shape, and there are two identical longest radii diametrically opposed, the two radii are distinguishable. Also, the areas of the shapes (with hole area taken into account) are the same. Thus, the shapes should prove difficult to distinguish. The test could have been made even more difficult by designing the holes to have equal boundary length as well as area.

Set B4 comprises two shapes of equal area. The "bay" has the same area in both cases, though its shape is clearly different. Methods based on convex hull measurement are likely to find these shapes difficult to distinguish.

Set B5 comprises two shapes having the same ratio (perimeter^2/area) which are nevertheless clearly distinct. An infinity of shapes could in fact be devised having the same ratio of perimeter^2/area.

Set B6 contains two shapes which are identical except that they are scaled in size by 2% in linear dimension. In principle, any system with resolution 64 × 64 pixels or better should distinguish the pair, particularly if area is used as a feature. In practicc, a rather higher resolution will probably be required, because of quantization effects at the boundary.

Set B7 contains two shapes with a slight eccentricity so that a system may find it hard to identify the major axis. In fact, the shapes are semicircles joined by short straight lines, not ellipses. In shape B7, the holes lie roughly on the major axis; in shape B7* they are on the minor axis. The shapes have the same total area and length of perimeter. This pair might equally be included in type A, since the ill-conditioning generates doubt in fixing orientation.

Set B8 contains two almost identical complex shapes, one of which has a minor defect inserted. The area of the shapes is unchanged by the defect. The length of the boundary has been increased, but to an extent that is small compared to probable errors in boundary length introduced by quantization effects. The test here is to detect the presence of the defect.

Set B9 contains two identical shapes. However, B9* contains two "specks of dust," of area equal to one notional resolution cell.These are isolated corruptions which the system should ignore. Shapes B9 and B9* should be identified as being the same.

Sets C and D were produced by making silhouette photographs of small mechanical parts which are used in the electronics industry, including such objects as washers, springs, heat sinks, contacts, and component holders. These shapes may cause problems because the ratio of boundary length to area is high, and errors due to quantization into pixels occur at boundaries. Thus, feature values will be difficult to measure. Also, they tend to lack obvious features which may be used as datums for establishing orientation or position.

Most systems for recognizing silhouette patterns start by extracting feature measures from the image immediately after scanning. Some shapes have been included to test the analysis of both internal and external features; others investigate the resolution of systems. However, the overall performance of a system depends on much more than the features used, and the shapes test a system as a whole. The combination of tests plus a little thought should identify and locate reasons for particular shortcomings.

The following measurements, at least, would be needed to characterize a system or methodology for comparison purposes:

- Accuracy of orientation as a function of resolution
- Accuracy of location as function of resolution
- Processing time as a function of resolution
- Pair discrimination ability as a function of resolution
- Training time as a function of resolution, and of number of patterns already stored
- Time to process, as a function of resolution, and as a function of number of patterns stored.

For a system whose resolution cannot be varied, size of pattern could be altered instead. This permits the same assessment, but is much more costly.

References

1. D.H. Ballard and C.M. Brown, *Computer Vision*, Prentice-Hall, New Jersey (1982).
2. A. Rosenfeld and A.C. Kak, *Digital Picture Processing*, Academic Press, New York (1976).
3. R.C. Gonzalez and P. Wintz, *Digital Image Processing*, Addison-Wesley, Reading, MA (1977).
4. W.D. Wright, *The Measurement of Colour*, A. Hilger, London (1969).
5. O.J. Braddick and A.C. Sleigh (eds.), *Physical and Biological Processing of Images*, Springer-Verlag, Berlin (1983).
6. R.L. Gregory, *Eye and Brain*, World University Library, London (1974).
7. T.N. Cornsweet, *Visual Perception*, Academic Press, New York (1970).
8. P.A. Kolers, Reading Pictures—Some Cognitive Aspects of Perception, in: *Picture Bandwidth Compression*, T.S. Huang and O.J. Tretiak (eds.), Gordon and Breach, New York (1972).
9. D. Lines, Real-time histogram specification for ultrasound images, *IEE Conf. Pub. 214, Electronic Image Processing*, 11–15 (1982).
10. P.G. Roetling, Binary approximation of continuous tone images, *Proc. SPSE Toronto Conf*, 323–330 (1976).
11. A. Rose, *Vision:Human and Electronic*, Plenum Press, New York (1973).
12. B.R. Frieden, Image enhancement and restoration, in: *Picture Processing and Digital Filtering*, T. S. Huang (ed.), Springer-Verlag, Berlin (1976).
13. W.K. Pratt, *Digital Image Processing*, Chap. 12, Wiley, New York (1978).
14. Y. Yasuda, Y. Yamazaki, T. Kamae, and K. Kobayashi, Advances in FAX, *Proc. IEE 73* (4), 706–730 (April, 1985).
15. W.K. Pratt, *Digital Image Processing*, Chap. 23, Wiley, New York (1978).
16. D. Lavie and W.K. Taylor, Effect of border variations due to spatial quantisation on binary image template matching, *Electron. Lett. 18* (10),418–420 (May 1982).
17. T. Pavlidis, Algorithms for shape analysis of contours and waveforms, *IEEE Trans. Pattern Analysis and Machine Intelligence PAMI 1–2,* 301–312 (1980).
18. J-D. Dessimoz, Sampling and smoothing curves in digitised pictures, *Proc. lst EUSIPCO,* Lausanne (Sept. 1980).
19. J. Wilder, Application of a flexible pattern recognition system in industrial inspection, SPIE *182*, 94–101 (1979).
20. C.T. Zahn and R.Z. Roskies, Fourier descriptors for planar closed curves, *IEEE Trans. Comps. 21,* 269–281 (Mar. 1972).
21. B.G. Batchelor, Using concavity trees for shape description, *Computers and Digital Techniques 2* (4), 157–168 (1979).
22. K. Nakata, Y. Nakano, and Y. Uchikura, Recognition of Chinese characters, in: *Machine Perception of Patterns and Pictures, Inst. of Phys. Conf. Pub. 13,* 45–52 (1972).
23. L. Norton-Wayne, W.J. Hill, and L. Finkelstein, Image enhancement and pre-processing, *SPIE 130,* 29–35 (1977).
24. M.R. Teague, Image analysis via the general theory of moments, *J. Opt. Soc. Am. 70* (8), 920–930 (1980).
25. G.A.W. West, Ph.D. Thesis, The City University, London (1983).

26. G.S. Robinson, Detection and coding of edges using spatial masks, *SPIE 87,* 117–125 (1976).
27. S.W. Zucker, Region growing—Childhood and adolescence, *CGIP 5,* 382 (1976).
28. T. Peli and D. Malah, A study of edge detection algorithms, CGIP *20,* 1 (1982).
29. L. Kitchen and A. Rosenfeld, Edge evaluation using local edge coherence, *SPIE 281,* 284–298 (1981).
30. R.O. Duda and P.E. Hart, Use of the Hough Transformation to detect lines and curves in pictures, *Comm. of the ACM 15,* 11–15 (Jan. 1972).
31. I.G. Logan, and J.E.S. Macleod, An application of pattern recognition algorithms to the automatic inspection of steel strip surfaces, *2nd IJCPR Copenhagen,* 286–290 (1974).
32. L. Norton-Wayne, W.J. Hill, and R.A. Brook, Automated visual inspection of moving steel surfaces, *Brit. Jnl. of NDT 19* (5), 242–248 (1977).
33. K.J. Stout, C.D. Obray, and J. Jungles, Specification and control of surface finish—empiricism versus dogmatism, *Opt. Eng. 24* (3), 414–418 (1985).
34. R.M. Haralick, K. Shunugam, and I. Dinstein, Texture features for image classification, *IEEE Trans. SMC 3,* 610–621 (1973).
35. J. Weszka and A. Rosenfeld, An application of texture analysis to materials inspection, *Patt. Rec. 8,* 195–199 (1976).
36. O.H. Schade, *Image Quality*, RCA Corporation, Princeton (1975).
37. C.A. Rosen and G.L. Gleason, Evaluation of performance of machine vision systems, *Robotics International,* Paper MS 80–700 (1980).
38. L. Norton-Wayne and P. Saraga, A set of shapes for the benchmark testing of silhouette recognition systems, *Proc. 4th Intl. Conf. Robot Vison and Sensory Controls,* 65–74 (1984).

9

Robot Vision Systems and Applications

9.1. Tasks of a Robot Vision System

The conceptual structure of a robot vision system is shown in Figure 9.1. The object under consideration, together with the world around it, is scanned by the input unit of the vision system, and possibly examined by other types of sensors, and the outputs from these actions are passed to processors.[1] The principal task of the processors is to extract the information that is important from the usually much greater information that is redundant. The extracted information passes to the main control unit which holds a model of the world appropriate to the whole robotic system and has the program to execute the required process. It converts the information into the form required by the controller of the robot and of other mechanical systems. The main controller is aided in its task by a priori knowledge of the world, of the parameters of the sensing and actuating systems, and of the objects on which the system is to operate.

The design of systems such as these is complex and generally beyond the abilities of nonexperts. Fortunately, the development of expert systems and knowledge-based systems is beginning to simplify the task. These are effective in two ways: by assisting the design process, in that the designer is guided in the choice of techniques, and by incorporation into the software of the system[2,3] which then acquires greater self adaptability.

The processing units for the sensors may be aided in their tasks by receiving, from the main control, information stating when to take in data or from which area within the field of view to accept the data. For example, it may not be necessary to process the whole of a picture or to scan and process continuously. When discrimination is to be

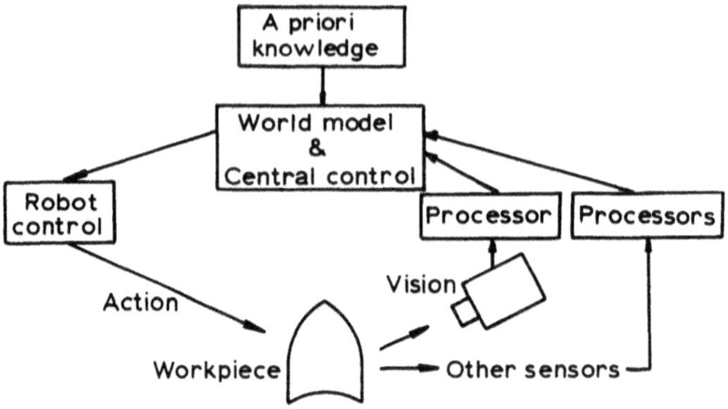

Figure 9.1. Conceptual structure of a robot system with vision and other sensory systems.

applied to a signal at a specific level, i.e., thresholded, this level may be provided from the a priori information.

An additional task for the sensor processors is to correct for the deficiencies of the sensors. This can be to correct for dimensional nonlinearities in the scanning, for variations in the spatial sensitivity of the sensor or the uniformity of the illumination, or for nonlinearities in the signal-to-measured-parameter characteristic of the sensor. Similarly, the robot control system may modify the coordinates which it receives, to match the known nonlinearities of the mechanical systems. In some systems, coordinate nonlinearities may not be corrected but are contained in the coordinate transformations that are used to move from the frame of a sensor to that of an actuator. In so doing it is possible that the application of the transforms will be made more complicated.

As a result of the sensory input and the set function of the whole system, the main control unit can initiate the operation of the mechanical system.[4–6] The control knows where the mechanical system is and what is its configuration; also it knows where it has to move to, but a robot system has considerable freedom in the movements that it may make. There are many paths by which the end effector may travel and many configurations of the robot segments during that movement. Two aspects need consideration during the movement. The move should be completed in the shortest time, but without excessive acceleration forces being applied to the object being handled, and the robot must not hit any object during the movement.[7] The main control system must handle this problem because it is the only part of the system that has, or can have, all of the necessary facts. It must have, in the a priori database, or obtain from the sensors, a picture of the world which includes objects that may be in the path of the robot. It may then plan a trajectory that avoids such obstacles.[8,9]

The data available at the start of the move may not be adequate to complete the move as required. The relationships between the various frames of reference may not be known with sufficient accuracy for the positional errors at the end of the move to be insignificant. More data is required. This can be obtained from the sensors. The vision system could make a high resolution scan over the limited area near the end of the end effector, or tactile sensors could be used. With the new information, the main control can give the necessary instructions to complete the operation accurately, and the robot can pick up, place, or perform an operation on the object.

As an aid to the safe movement of the robot it is possible to use the vision system to track its movements. In order not to slow down the robot significantly, the vision system must be fast. This can be helped if the task of the vision system is simplified, for example, by providing a high-contrast in the scene for the objects of interest. Other sensors, such as tactile or proximity sensors, on the arm or on fixed obstructions, may be used in a similar way.

9.2. Scanner Calibration

Due to imperfections in the scanning process, there will be differences between the actual positions of points in a scene and the apparent positions as determined by the scanner. As an example, the electron beam deflection in a vidicon does not produce straight line scans, the scans are not made with constant velocity, and the frame and line scans are not exactly orthogonal. It is possible that these effects are dependent upon time, temperature, etc. Thus the coordinate frame of the scanner is not perfect. Corrections must be made within the sensor system to remove these defects. Similar statements can be made about the other sensors and about the robot itself. Usually the errors, of linearity, etc., in the coordinate frame of the robot are relatively small; also, it is normal that the errors have already been measured by the supplier and written into the control program for the automatic correction of the joint coordinates.

Further, the coordinate frames of the sensors will not correspond to the coordinate frame (the world coordinates) of the complete robotic system. It is necessary to determine the values of these differences so that coordinate transformations can be applied to the data passed around the system, and the data is meaningful.

9.2.1. Theoretical Considerations

The transformations that need to be applied to a coordinate set to match it to a second set can involve several different processes. These are translation, rotation, scaling, skewing, and perspective. In Cartesian coordinates, skewing refers to changing the angles between lines, e.g., changing a rectangle into a parallelogram, and perspective refers to introducing a scale change, the magnitude of which is proportional to the distance from a point. For a camera, this point would be at the lens and the effect would be to correct for the perspective that is in the two-dimensional representation of the scene.[10]

These corrections require slightly different mathematical treatment in their application to the standard representations of the coordinate systems. A unified treatment can be used if they are converted to homogeneous coordinates. For simplicity, these will be discussed with respect to Cartesian coordinates. The coordinates x, y, z are replaced by wx, wy, wz, and w, where w is an extra variable, and a point in three-dimensional space is represented by a line in four-dimensional space. For simplicity, w can be given the value 1, so the coordinates become x, y, z, 1. It then becomes possible to use matrix multiplication to transform the coordinates. Writing the coordinates as a row matrix, or vector, the transform is

$$[wx' \; wy' \; wz' \; w] = [x \; y \; z \; 1] \, T$$

where T is a 4×4 transforming matrix. For rotation by an angle A about the z-axis, the matrix T is:

$$\begin{vmatrix} \cos A & -\sin A & 0 & 0 \\ \sin A & \cos A & 0 & 0 \\ 0 & 0 & 1 & 0 \\ 0 & 0 & 0 & 1 \end{vmatrix}$$

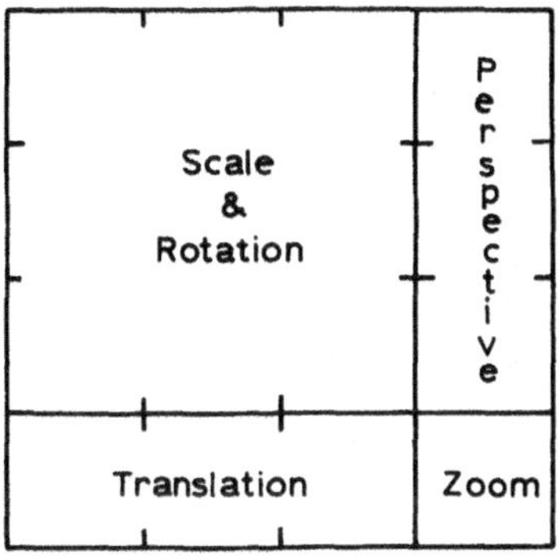

Figure 9.2. The processes effective within a 4 × 4 coordinate transformation matrix.

Matrix multiplication is performed by evaluating the equation:

$$R_{ij} = \sum_k P_{ik} T_{kj}$$

to obtain each term in the resultant matrix R. In the present application, in which R and P are row matrices, the expression simplifies to

$$R_{1j} = \sum_k P_{1k} T_{kj}$$

where j and k assume the values 1 to 4. Applying this to the above example produces the familiar expressions

$$\begin{aligned} w &= 1 \\ x' = wx' &= x \text{ cox } A + y \sin A \\ y' = wy' &= -x \sin A + y \cos A \\ z' = wz' &= z \end{aligned}$$

Similar matrices exist for the other transformations; by taking the product of these matrices, one general matrix can be produced which incorporates all the transforms. The coefficients in this matrix have specific effects as indicated in Figure 9.2. Rotation is produced by a combination of the effects of scaling and skewing. For an array of N points, the row matrices R and P can be replaced by $N \times 4$ matrices.

9.2.2. *Practical Methods*

Basically, the method is the obvious one of positioning a point in one frame and observing it by one or more sensors in their coordinate frames. This can be used to determine defects, e.g., nonlinearities, in the coordinate frame, and the transform to apply between the frames. For the trânsforms to be applicable by calculation it is necessary that corrections first be made for the nonlinearities, etc. It may be that some mathematical functions can be found which will allow the corrections to be calculated when the constants of the function have been determined by theory or by measurement, but more often it is necessary to store the correction values, the correction for each parameter depending, in general, upon the combination of values for several parameters.

To determine the nonlinearities, an object whose position can be sensed readily is placed in the field of view of the sensor. A high-contrast pattern, perhaps in the form of a Maltese cross, or a small light source are good objects for a camera. Its apparent position is measured by the sensor. The object is moved a known distance, as measured by standard metrological equipment, and the new position measured. This is repeated over the field of view of the sensor. Generally, the errors in the scan relate to the angular position of the object and therefore it is sufficient to move the object in a plane perpendicular to the optic axis of the scanner. When a robot that has been calibrated is available, it is convenient to use it to move the object. The robot can be preprogrammed to follow a pattern of positions, so that, with the vision equipment suitably programmed, the whole process can be performed automatically. The calculation of the nonlinearity factors will be simplified if the pattern of reference points are symmetrically arranged with respect to the principal axes of the vision system.

The nonlinearities of other sensors can be determined in similar ways, but if the device is range sensitive, as for a proximity detector or range finder, then variation of position in the third dimension will be required also.

Generally, higher accuracies are required, and are possible, from robots than from many sensory systems. One technique is to place accurately positioned markers in the working space of the robot. The robot is moved to touch these markers, and the coordinates within the control system for that position are compared with the known absolute positions. Thus, the errors for that position are determined. The robot can be programmed to perform the movements automatically. A low-compliance tactile sensor, e.g., a metal spike connected to an electrical circuit, can be used to sense the positions of plane metal surfaces which complete the circuit. Usually these surfaces will be set parallel to the principal axes of the robot's frame of reference. Other techniques are given in Section 7.4.5.

An alternative is to use short-range sensors, e.g., the optical sensor with the position-sensitive diode described in Section 6.2.3, or calibrated proximity sensors (see Section 6.1.2). Using these, the low-speed movement necessary to find the point of contact can be eliminated. Instead, the robot will move rapidly to a programmed position known to be just clear of the measurement surface, and the separation measured by the sensor.

The segments and joints of a robot are compliant, so loading will affect the positions taken by the robot, the degree of the effect depending upon its attitude. Consequently, the robot should be checked with various loadings. Also, the position reached may depend upon the direction from which the position was approached. In practice it is often possible to incorporate only a single average correction value for each position. The remaining variation serves only to indicate whether the robot is suitable for the task. Laser tracking systems, in which the robot carries a corner reflector, and a double galvanometer mirror scanner keeps the beam on the reflector, have been used for dynamic measurement of robot accuracy.

Similar techniques are used to determine the relationship between the coordinate frames of the robot, the sensors, and the world of the whole system, when the nonlinearity corrections have been applied. As before, the robot may move, in its coordinate frame, an object whose position is easily detected by the sensor system. There are, in general, 16 unknowns in the matrix which transforms from one frame to another. Therefore, an equal number of measurements should be sufficient to determine their values by the solution of a set of simultaneous equations. Since the making of the measurements will itself introduce some errors, it is usually better to perform additional measurements and calculate average values for the coefficients of the matrix. Standard procedures are available for the solution of the equations.

It is not always necessary to perform a complete linearity correction or calibration of the system. In some cases, a rough correction and calibration is adequate. The mechanical and sensor systems would then operate on a move-and-check basis. After each stage in the movement the mechanical system stops, and the difference between the required position and the actual position is measured by the sensor.

This is repeated for steps of reducing size until the accuracy in the estimation of the required last step is adequate.

9.3. Applications

Vision systems have been applied to the task of assisting robots to perform functions that are beyond the capabilities of the blind robot. A vision system provides the robot with the ability to determine where the parts are that it is using or operating upon, the condition of the parts, e.g., the wear on a tool or the quality of a component, and the quality of the final product. A vision system also provides guidance during the performance of the task; the robot can be guided around obstacles and helped to recover parts that it has dropped. In some cases, other sensory systems are also used and there are systems in which the support is given by nonvision sensors only. There are many systems under development in laboratories and several systems have been applied to tasks in industry. The examples described in this section were selected to illustrate points made elsewhere in this book.

Although the vision, or other, processing system may have a wide application, it is apparent that in the application of the system to a specific task, restrictions have been introduced, usually to get a better performance for a given price. These restrictions may apply to the presentation of the objects, the lighting of the scene (see Section 5.2.5), the positioning of the sensors, and the design of the object. An example of a small design change introduced to aid the sensing process is that of the three small holes added to the plastic frame of the deflection units whose handling is described in Section 11.7.2. Restrictions such as these and the use of simple processing techniques make it possible for a sensory system to be constructed which is not excessively expensive and can operate at a reasonable speed.

9.3.1. Objects on a Conveyor

A common problem is the removal of objects from a conveyor. It involves the location of the objects and the control of the robot so that it can track the moving object.[11] The latter part of the task is aided by a facility supplied in the software of many of the newer robots. It is possible to specify the position and orientation of an object, which is to be handled by the robot, in a reference frame which is moving at a known speed with respect to that of the robot. The robot controller will transform the coordinates to those of its own frame. One example of such a system has been described as a case study in Section 11.6.

Whether the system should locate every object on the belt depends upon the application. This will be so if no method can be provided to collect from the end of the belt those items that were not located by the vision system or could not be removed by the robot, perhaps because the flow rate was too high. In many component feeding systems the important aspect is that those components removed from the belt should be passed on in the correct orientation and position, and in sufficient number. Other components can stay on the belt and fall from the end, to be returned later.

As one direction of scanning is supplied by the motion of the belt it is possible to use a line-scanning device for the other direction. Linear array solid-state cameras and laser scanners are suitable devices. In principle, one scan is made for each conveyor movement equal to the size of a pixel. This implies synchronization of the scanner and the belt, which may be inconvenient, and often a slightly higher scan rate, e.g., +30%, is used and some scans are ignored so that the average interval is correct. Unless the speed of the belt can be guaranteed to be adequately constant, it is necessary to supply a device to measure the movement of the belt (see Section 6.1.3). So that the dimensions of the object, measured in pixels, do not change as the orientation of the object changes, the scanning pitch in the two directions must be the same. The use of area scanners, such as a vidicon, is not precluded, but usually the scan time for the area is much longer than the time for a movement equal to one pixel and blurring, due to light integration, can be avoided only by using flashed light to obtain a snapshot.

Two-dimensional pictures of the objects can be built-up as the scanning progresses and these can be processed. This is feasible if the number of pixels necessary to contain the object is small. In the example of Section 11.6, the scan across the belt had 1024 pixels and about 1,000,000 pixels would have had to be stored before a complete tube was in the picture. Consequently, it is common to follow the practice, as used in that system, of extracting information from the last scan, perhaps in conjunction with a few previous scans if the algorithm so requires, and forming a table of measures. The values in the table are updated as the scanning progresses. A table will exist for each object in view. A single object may cause the start of more than one table. For example, an E shape moving to the right would cause three separate starts. By monitoring the coordinates along the line scan for the extremes of each object, it is possible to detect their merging into a single object. The tables would then be combined. Of course the same result will occur for different objects which are in contact.

It is possible that when a series of items are being scanned there is insufficient time for the whole of the processing to be performed by a single computer. One solution would be to use a picture processing system in which very high speeds or parallel processing is used (see Section 10.3), but this may be expensive and involve hardware which is not standard in the factory, leading to servicing problems. An alternative solution would be to use several single-board computers with the task shared. Sharing of the processing of each item is possible, but often complex to organize. A simpler solution is to have a single vision input circuit and to switch its output to one of several single-board computers (Figure 9.3). The picture, or partially processed picture, is stored in that computer during the scanning of the item. At the end of the item, the video for the next item is switched to the next computer. The first computer then processes its data and its results are passed to the master computer before it is again required to receive new video data. Identical hardware with identical software is used in each of the processors and they are not interlinked. The only penalty is that the conveyor has to be slightly longer to match the delay in obtaining the processed information. This arrangement is often called concurrent processing. By having extra processors and a means to detect the failure of any of them, it is possible for the system to continue to operate with those remaining.

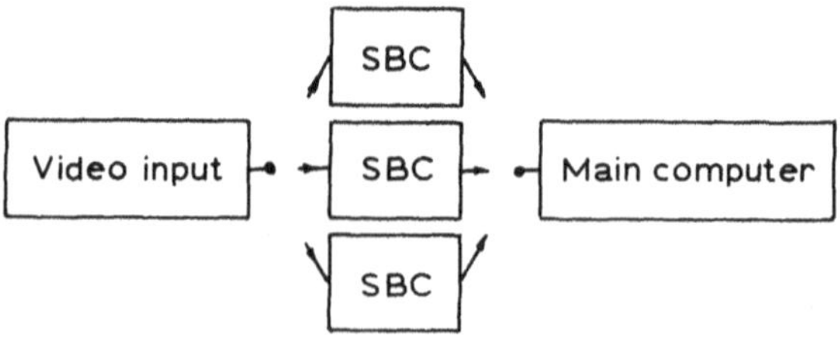

Figure 9.3. A vision system in which three single-board computers process separately the video from each of a cycle of three pictures.

For general applications, the typical data which might be extracted include the following:

1. The coordinates of the first point sensed. This can be used as a local datum which relates to the real world of the belt.
2. A pixel count, to give the area.
3. The coordinates of the extreme points, i.e., first, last, and those nearest the edges of the belt.
4. The sums of the x-coordinates and of the y-coordinates, from which the centroid of the shape can be calculated.
5. The length of the periphery and possibly the vector list of the periphery.
6. The sums of the squares of the x-coordinates and of the y-coordinates. From these, the second order moments, i.e., the moments of inertia about the centroid, X, Y, can be calculated. For simplicity, all these coordinates can be referred to the first point sensed. Note:

$$\sum_{1}^{n} (x - X)^2 = \sum_{1}^{n} x^2 - n X^2$$

7. Similar data about significant internal features such as holes.

These measurements may then be compared to stored reference data on the objects to determine their orientation and position on the belt.

Techniques such as these are used in the General Motors' Consight system.[12] The lighting system is described in Section 5.2.5, and in Figure 5.21. The camera is a 256-diode linear array, giving a binary output, and the processing is performed by a PDP 11/45, with the result being available about 200 msec after the object has passed from view. The system can operate with several objects across the belt, but will ignore objects that are in contact. The system can be trained by feeding sample components through the system. This produces reference data for the recognition and location of objects. During use, the measurements for each object are stored in tables and, after the passing of an object, compared with the reference data.

The system uses the techniques discussed in Section 9.2. to calibrate the camera and relate it to the robot. It is of interest here since it relates to the problem of the moving belt. If $[r]$ and $[v]$ represent the positions of the object in the coordinate frames of the robot and of the camera then

$$[r] = [v]\,[T] + s\,b\,[B]$$

where $[T]$ is the transform between the frames, b the distance moved by the belt, s a scale factor between this and measurements in the robot frame, and $[B]$ the direction vector of the belt with respect to the frame of the robot. $[T]$, s, and $[B]$ are the unknowns. The last two are found by placing an object on the belt and positioning the robot to grip the object. The robot is then taken clear, the belt moved, and the robot repositioned. To find $[T]$, an object travels on the belt past the camera and then the belt is stopped. The robot is positioned to grip the object. This is repeated for other positions and from the measured values, $[T]$ can be calculated.

Another system, developed at about the same time, is interesting in that a very simple algorithm is used, and the processing speed is improved by the use of a lookup table.(13) The processing requirement is small enough to allow a microprocessor (e.g., an 8080) to be used. Components travel on a standard translucent belt so that backlighting can be used (Figure 9.4). A binary output is obtained from the 128-diode linear array camera. The result is available 60 msec after the component has passed. Components in contact are ignored, as are different components simultaneously within the view of the camera. A modification could be made to overcome the latter restriction and give the location of one of them. This would be sufficient since the robot is not fast enough to remove both. This machine was intended for component feeding or magazine loading.

The processor determines the coordinates of the four extreme points of the object with respect to axes as shown in Figure 9.5. The position of the origin of the axes, in real-world belt coordinates, is also recorded. The microcomputer holds a reference table which contains, for 128 different orientations, the corresponding relative coordinates of a point, specified as the pickup point for the robot, and the 6 coordinates specifying the 4 extreme points. In operation, the microcomputer compares the six measured values with those in the table to find the combination giving the best, and adequate, match. The

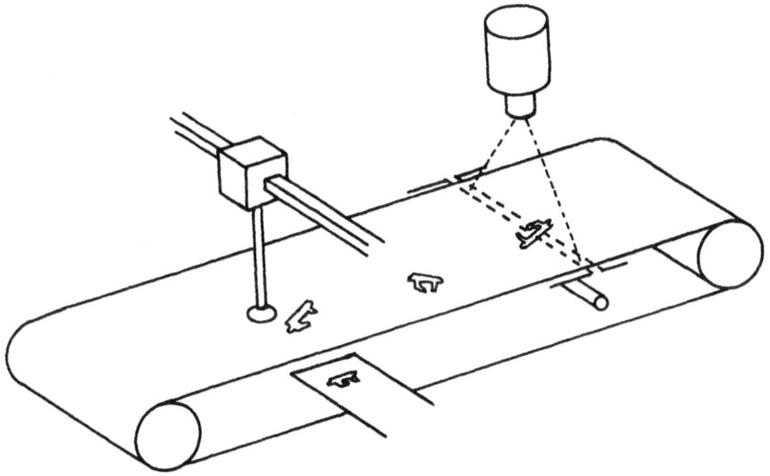

Figure 9.4. A machine for the orientation of components. Components located by the vision system are removed from the belt by the manipulator and placed in a jig.

corresponding values of orientation and pickup point coordinates are read, combined with the known origin of the axes, in belt coordinates, and the result fed to the robot controller. When the table is formed, if two lines possess similar values for the six coordinates, both are deleted. This results in an increased number of rejects, but this is better than getting a false result. The origin of the axes is not placed at the first point seen because the accuracy of estimating its position in the line scan direction is very low for a broad edge.

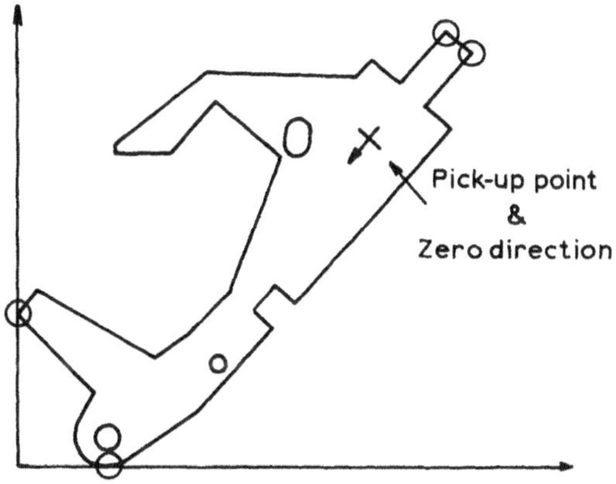

Figure 9.5. The four extreme points of a component, the first point seen being at the bottom of the figure, i.e., the component moves downwards past the line scanner.

9.3.2. *Three-Dimensional Objects*

In the preceding section, the objects might have been three-dimensional but a two-dimensional view was taken of them. It might be assumed that the object would be placed on a specified, stable, face, and other positions ignored. Alternatively, the system could accept several attitudes, effectively recognizing them as different objects.[14] There are many occasions when it is necessary to take a three-dimensional view to be able to extract all the required information. Possibly a view from a single position will suffice, although it is probable that complex algorithms will have to be invoked.[15] It is almost certain that gray-scale processing, or the processing of many binary pictures taken at different threshold levels, will have to be used. In the worst cases, stereo vision or incorporation of range finder techniques will be necessary. These complications add to the processing load and time.

Fortunately, in many applications the task can be simplified. The objects to be handled are known—the a priori facts—and it is known that they have a limited number of rest positions. This may help if the object is viewed from any direction but will be most effective in simplifying the problem if special directions, e.g., vertically down, are chosen. Views from several directions can be processed independently and the results brought together to define the object. The problem of partially hidden objects is more easily resolved in such a system. The picture may be treated in a two-dimensional fashion and lighting conditions can be arranged so that gray-scale processing can be avoided, e.g., by using contrasting backlighting. This approach was used in the system for handling deflection units, described in Section 11.7.

Surface shapes, e.g., the surface of a casting, have been measured by the use of a two-dimensional camera and structured light in the form of a thin light plane (see Sections 5.2.5 and 6.2.5). The camera sees a line of light on the surface, and from the position in the scan of each element of the line, the distance of that point from the camera can be calculated. In this application the robot plays a secondary role to that of the vision system. It is used to hold the camera and light unit in specified positions around the object. The dimensions are relative to the positions taken by the robot but any inherent errors can be removed by calibration against a standard casting which has been accurately measured by other methods.

9.3.3. Welding

There has been significant activity in the addition of sensors to robotic arc welding systems.[16–19] Robots without sensors are used for welding by manually, or otherwise, teaching them the path to be followed. Unfortunately, the accuracy obtained by this method is often inadequate, one reason being that the parts being joined change their shapes, and therefore their positions, during the welding process. The arc itself can act as a sensor in that the arc voltage/current relationship is a measure of the position of the metals being joined. Although it is useful in maintaining the arc at the correct distance, it can be used as a tracking system for only a few applications. The more widely applicable methods use contact probes,[20,21] vision, including stereo vision,[22] or inductive proximity sensors. The vision methods can be divided into the prescan and scan-weld systems. In welding applications, special consideration must be given to electrical noise generated by the high arc current and to ambient temperature. Also, protection must be provided to prevent metal and smoke from the arc collecting on the sensors. Glass windows and gas jet screens can be used for this purpose.

Inductive proximity sensors are sensitive to the distance between themselves and a conductive sheet and also to the thickness of that sheet (see Section 6.1.2). The edges of sheets will affect the current induced in the sheet and again, this can be detected. By using several sensors in a line at right angles to the seam, it is possible to track the seam and control the height of the sensors from the seam. The sensors are mounted on the welding torch so that they precede it along the track. When the system is following a curve, the robot must move the torch so that allowance is made for the distance that the sensor is ahead, and that the torch continues to follow the seam. Only a moderate amount of screening is necessary to exclude electrical noise.

The most common vision systems use structured light, in the form of a thin sheet, and a two-dimensional camera (see Section 5.2.5).[23–26] The light source is usually a laser diode, and the beam may be spread by a cylindrical lens or by an oscillating mirror (Figure 9.6). The image of the scene may be formed directly on a very small camera or ducted by a coherent fiber-optic bundle to a camera. Due to its size, ruggedness, and resistance to excessive light, the solid-state, two-dimensional array is favored for this task. In one implementation, using an oscil-

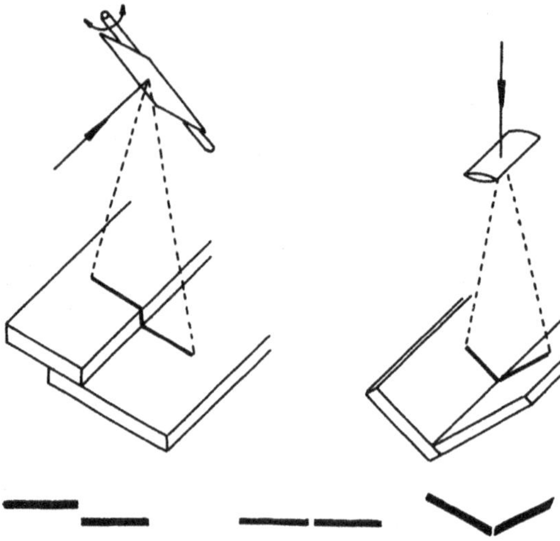

Figure 9.6. Structured lighting techniques in welding, using an oscillating mirror and a cylindrical lens.

lating mirror to produce the plane of light, the scene is imaged onto a linear array through a mirror mounted on the same oscillating shaft.[27,28]

The image to be processed usually consists of two well-defined lines to the sides of the seam, with a more complex structure from the seam position itself. From this, the path to be followed can be calculated. In multipass welding, the shape of the line at the seam will reveal the amount of weld that has been placed in earlier passes. As the system is also a range finder, the position of the image in the field of the camera gives a measure of the distance of the camera from the workpiece. The camera is mounted on the torch, and some ingenious systems have been devised to build the camera and lighting units into a housing surrounding the torch.[29] The range finder based on the position-sensitive diode (see Section 6.2.3) has been used for welding control. The laser and sensor are fixed to the welding head but they view the seam via an oscillating mirror. In this way, a distance profile is obtained across the seam.[30]

Two modes of operation are used. In both, the approximate path to be followed is loaded into the robot, by leading through or from a computer. In the prescan method, the robot follows the path, using the sensor but with the arc off. This allows the measurement to be made at the center of the arc position. The deviations from the approximate path are stored and used during a welding pass. In the second method, scanning and welding occur at the same time. Besides the problems from electrical interference, there is also the problem of seeing the line in the presence of the brilliant arc. The use of a laser, with its narrow optical bandwidth, helps in this respect in that a corresponding narrow-band filter may be placed in the path to the camera. In some systems the arc is not maintained continuously and attempts have been made to shield the camera during the on-time and view while the arc is off, but this is a difficult process to perform.

The most advanced of these systems actually use the arc itself.[31,32] Although this requires more complex processing, it does provide more information about the progress of the weld. The positions of the arc and the weld pool are visible to the camera. The shape and size of the pool can give information regarding the position of the underlying seam, but, to simplify the location of the seam ahead of the weld, the structured light system may also be incorporated. It is also possible to monitor the efficacy of the welding process.

9.3.4. Assembly

Although much research has been performed on the use of vision in the assembly process per se, its use in factories is, at present, limited, or at least not publicized. The status of tactile systems is similar, but their potential does appear to be greater. In many assemblies, it is difficult to view the mating of the parts due to the parts themselves or to other components near these parts. As is the casc for human operatives, it is more effective to monitor the process by touch. For this reason, output from force sensors in the wrist or gripper, including tactile sensors on the fingers, can give more useful information by which the operation of the robot can be guided.

Passive compliant wrists (see Section 7.2.1) and compliance in the robot itself, e.g., the SCARA robot (see Section 7.2.1), are being used to aid the insertion of parts, particularly rod-like objects, into assemblies. The compliance allows slight positional errors to be accommodated and avoids jamming during insertion. With force and tactile feedback, active compliance, i.e., by feedback through the control system, can be used with the same results, but the process is extendable to more complex assembly actions. The directions and magnitudes of the assembly forces can be controlled.

A close view of the mating process may be possible if a camera can be built into the gripper. This can be achieved, in effect, by imaging the scene onto a coherent fiber-optic bundle which leads to a remote camera, or, in actuality, by a very small camera.[33] A commercial system is availale with a 32 × 32 solid-state sensor array, the camera being 15 mm in diameter and 17 mm long. Although the resolution is low, the camera can be positioned very close to the scene and the scanning pitch in the scene can be adequately small. For either system, light can be brought to the scene by a fiber-optic bundle.

Vision most frequently enters the assembly system before and after the mating action.[34,35] It is used to determine the positions and orientations of the components while they are apart, and to inspect the result after assembly has been completed. An example of this is the placing of deflection units on the necks of television tubes, as described in Section 11.7. The position of the deflection unit is determined visually so that the robot may grasp it correctly, and the position of the neck found so that the robot can bring the parts together.

Another application is that of placing wheels onto a car.[36,37] The wheels are fed to a pickup position near the robot, where their position is defined by stops. They are rotated until the fixing holes are detected by simple sensors and then the nuts or bolts are placed in position. A vision system, with a television camera, scans the car as it passes on a conveyor, and determines the position and orientation of the hub. Using information on the conveyor motion, the robot brings the wheel, in the required orientation, to the hub, where motor driven wrenches place and tighten the bolts. A similar system places wheels onto the other side of the car.

The robot assembly of components is simplified if they can be fed to the robot in well-defined orientations and positions. In many cases this is possible, for example by leaving formed metal or plastic parts attached to a waste strip from which they are cut when required by the robot, or by carrying them on tape, e.g., as for resistors for component insertion. Another solution is to place components into magazines or onto trays having cells that retain the component orientation. Even so, there are situations were such solutions are not practical or economical. Some components are made on a batch basis and may have to be stored and transported. For these, magazines or trays could be expensive. Others may undergo deburring or plating treatments in which their orientation is lost. Consequently it is necessary to provide feeding systems for randomly orientated components. Many types of feeders exist, e.g., vibratory bowl feeders, which can feed such components. The tooling of these feeders, to effect the reorientation, can be expensive and difficult for some components. For these components, optical selection systems can be used, either in conjunction with a vibratory feeder or a conveyor belt.[38,39] Such systems are discussed below and in Sections 9.3.1. and 11.8.1.

One vision system, used to replace the tooling on a feeder, projects a narrow light beam onto the component on the track.[40,41] Scattered light is received by a detector and the shape of the output waveform is related to the shape of the component. By comparison with waveforms obtained from known orientations, the orientation of each component can be determined. It is necessary that the number of possible orientations on the feeder be limited, but most components would have only a limited number possible on a feeder track. The system also provides a limited inspection facility. For example, the approximate length of the thread on a bolt can be determined. This is an example of a very simple optical system which does not result in a picture in the normal sense, but generates a waveform which is, in effect, a signature.

Other systems exist in which a bowl feeder supplies a small conveyor on which components are carried past a linear array scanner at a constant speed and in a limited number of orientations.[42] Backlighting or a contrasting background is used to enable a binary video output to be taken. At specified distances along the component, the width is measured directly by the scan. These are compared with values derived from scans made during a learning phase. Other features may be measured and compared automatically in a similar way. This technique limits the processing that is required and highlights those features which are important in recognizing the components and determining their orientations. Also, this system allows a limited amount of inspection to take place.

These systems aid the robot in that it receives the components in a well-ordered fashion. It is also important that the components be of good quality. This may be satisfied by applying tighter specifications to the component manufacturing process, but frequently some inspection must be performed. To inspect components while they are randomly orientated is to do things the hard way. For example, the scanning resolution must be increased to overcome the distortion of the shape by the differing angles of scan for each component. It is better to orientate the components, using relatively low resolution, and then to inspect at the high resolution that is usually required. It is then possible to do direct shape matching with stored patterns or to measure crucial dimensions of the component. When it is known by experience that faults are probable at specific points on the component, then picture processing can be applied to these areas and other areas ignored or checked at much lower resolution, e.g., by taking 1 in N points from the scan.

Tactile and force sensors in the gripper can play an important part in assembly since excessive force may be required to put faulty parts into an assembly. This may be due to burrs on, or distortions of, the component or the parts already in the assembly. If the forces measured exceed present limits, the assembly can be stopped. The robot can try a second component to see if that is better, but if insertion is still not possible, then the base part of the assembly is probably at fault and should be put aside. Tactile and force sensors also provide a safeguard against failures in the component feeding, the gripping of the components, and stability of the base part. These failures will result in the wrong force values for each position of the robot arm.

The other way in which vision can aid the robot is in the inspection of the work done by the robot, after each assembly stage or after the entire assembly is complete. This can be performed while the robot is out of the way, collecting a tool or another component. Greater freedom in the arrangement of the cameras and the lighting is then possible. By mounting these on the arm of the robot, the robot itself can, in effect, inspect its work, setting the camera at the best angle or angles to view each part of the assembly. Alternatively, the robot can take the assembly to a camera and present it for inspection in several orientations. When a fault is found in the assembly it is possible, in principle, for the robot to remove the offending part and either replace it correctly or discard it and fit a new part. An example of an inspection system applied after an assembly stage is described in Section 11.8.3.

9.3.5. Bin Picking

For convenience in storage and transporting, some components are placed in bins or stacked on pallets. They may be stacked in a rough sort of order or be in complete confusion. They may be of such size and nature that it is not possible to empty the bin into some sorting system, the equivalent of a bowl feeder, or onto a conveyor or spread out on a flat surface. Instead, each must be removed separately. One solution to the problem of random components is equivalent to the fairground crane, i.e., go in with the gripper and see what comes out. This is a feasible solution in some cases, particularly if the components can be lifted with an electromagnet. The component, or components, from each attempt can be placed on a table or conveyor and scanned to determine their orientations and positions. When the component positions in the bin or on the pallet are known approximately, it may be possible to locate and grip them by using a compliant robot arm or wrist, which will accommodate the positional errors.

For the vision-based solution, fixed cameras and lighting or units mounted on the robot arm, or both, may be used. A serious problem arises in illuminating the scene and getting a good picture. An example of this occurred in the unpacking of deflection units discussed in Section 11.7.1. Many of the applications that have been investigated involve castings.(43,44) These usually have rough surfaces which scatter the light and therefore appear dark against the darkness which lies below. Other features which complicate the vision processing are that, although the shape of the parts can be known in detail by the system, they may be seen in any orientation and some parts may be partially overlapped by others.(45) Even if the system is unable to determine fully the situation of the part, it is essential that it can recognize surfaces by which the part may be gripped. The following are three more examples of bin picking problems.

Thick wheel-shaped castings are stacked in a bin, in an ordered fashion, for transportation.(46) Each layer of castings is flat and the layers form a set of roughly arranged columns. The surfaces of the rims and hubs are flat. In order for a robot to remove each wheel, its location is established by a vision system. Mounted above the bin is a camera, with a 128 × 128 solid-state array, surrounded by lights which can illuminate all the layers as they are uncovered. Good reflections are obtained from the hubs and rims. The picture is processed with a high-pass filter and thresholded to leave the positive elements. After noise removal, a process is used to locate circular shapes having radii between specified limits. These circles can be in contact. Basically, the algorithm is that the perpendicular bisectors of lines joining three points on the circumference of a circle meet at the center of the circle. The outcome of the process is an array of clusters of points which could be the centers of the wheels. Clearly, valid clusters must be separated by at least the diameters of the wheels, and those that are shown to be false are removed.

After further processing, the coordinates fed to the robot are accurate to a half-pixel width. Because the camera is fixed, the diameters of the images become smaller as the bin is emptied. Although the system is told how many layers are in the bin, the size of the images could be used as an indication of the stack height.

Another task with roughly stacked items is the removal of crankshafts from a pallet.[47] Each layer of shafts is laid orthogonally across the previous layer. Magnetic grippers could not be used because magnetization of the shafts would encourage the collection of particles. A gripper with compliance is used so that disturbing forces are not applied to the stack, and a positional error of less than 15 mm and 2° is required for the safe removal of the parts. The scene is scanned by a newvicon camera, with a 256 × 382-pixel resolution and a scanning pitch of 4 mm. Although parts of the shafts are shiny, variations are caused by rust. Lighting is provided by lamps arranged around the camera. The size of the image changes as the stack is reduced and, since the image quality is not good, the vision system is aided by receiving data on the stack height. This is found by an ultrasonic proximity sensor on the gripper.

In picture processing, the picture is thresholded and areas of linked pixels found. The convex hull is delineated and attempts made to fit lines of length and orientation corresponding to the known size and orientation of the shafts. The areas around the positions of these lines are explored, at specific points along the lines, to determine whether the shape is of the correct dimensions. In this way, the positions of the shafts are confirmed. Before removing each shaft, whose location is known, the robot uses the ultrasonic range finder, at specified points along the shaft, to determine its height to within 2 mm. The complete cycle time for the transfer of each shaft is 20 secs, of which about 2 secs is used by the vision system. This is an example of two sensors cooperating to locate the parts, with one sensor being assisted by the robot.

The third system is an example of a bin containing randomly positioned parts. The contents of the bin are scanned by a television camera. The pictures are processed, using the gray-scale values of the pixels, to find lines of high gray-scale gradient. In the next stage, an "edge propagation/collision front" algorithm is used. From each high gradient point the next point is found along the rising direction of the gradient. This creates a new curve just on the brighter side of the original curve and of a similar shape. This is repeated until this "propagating wavefront" meets another wavefront. These meeting points form the next version of the picture. As the points with high gradients are probably along the edge of the object, these meeting points are probably the center lines of the elements of the objects.

In general, the processed picture will have some areas which are confused and others which are simple. From a priori knowledge of the shapes of the component's segments it may be possible to recognize shapes in the simpler areas. With or without this aid, the robot can be instructed to grasp a component with the gripper aligned with a line of meeting points. Provided that there are not too many components lying on the selected component, it will be possible to lift it from the bin. At the worst it will fall back into the bin and the scanning process will have to be repeated. As the removal of a component will disturb the others, each removal attempt will have to be followed by a rescan.

The use of vision for the bin picking of complex components is in the early stages of development. The picture processing task is difficult and, in the general case, must be performed on gray-scale pictures. Many useful algorithms are available, but the processing times are unacceptably long when using the hardware that has been available. With improved computing systems, with parallel processing and special picture processing modules, that are now becoming available, it will be possible to perform the processing in an acceptable time.

References

1. J.A.G. Knight, Sensors for robots: The state of the art, *Proc. 2nd European Conf. on Automated Manufacturing,* 127–132, Birmingham (May 1983).
2. P.J. Gregory and C.J. Taylor, Knowledge-based models for computer vision, *Proc. 4th Intl. Conf. on Robot Vision and Sensory Controls,* 325–330, London (Oct. 1984).
3. J. Foster, P.M. Hage, and J. Hewit, Development of an expert vision system for automatic industrial inspection, *Proc. 4th Intl. Conf. on Robot Vision and Sensory Controls,* 303–311, London (Oct. 1984).

4. D. Nitzan, Use of sensors in robot systems, *Proc. Intl. Conf. on Advanced Robotics,* Japan Industrial Robot Association, 123–132, Tokyo (1983).
5. P. Sholl and C. Loughlin, A practical solution to real-time path-control of a robot, *Proc. 4th Intl. Conf. on Robot Vision and Sensory Controls,* 209–221, London (Oct. 1984).
6. J. Amat, A. Casals, and V. Llario, Location of work-pieces and guidance of industrial robots with a vision system, *Proc. 4th Intl. Conf. on Robot Vision and Sensory Controls,* 223–230, London (Oct. 1984).
7. E. Freund and H. Hoyer, A fast control method for collision avoidance of industrial robots, *Proc. Intl. Symp. on Automotive Technology and Automation,* 1787–1803, Milan (Sept. 1984).
8. H.B. Kuntze and W. Schill, Methods for collision avoidance in computer controlled industrial robots, *Proc. 12th Intl. Symp. on Industrial Robots and 6th Intl. Conf. on Industrial Robot Technology,* 519–530, Paris (June 1982).
9. U.L. Haass, H. B. Kuntze, and W. Schill, A surveillance system for obstacle recognition and collision avoidance control in robot environment, *Proc. 2nd Intl. Conf. on Robot Vision and Sensory Controls,* 357–366, Stuttgart (Nov. 1982).
10. S. Ganapathy, Decomposition of transformation matrices for robot vision, *Intl. Conf. on Robotics,* IEEE Computing Soc. Press, 130–139, Atlanta (Mar. 1984).
11. W. Patzelt, A robot position control algorithm for the grip onto an accelerated conveyor belt, *Proc. 12th Intl. Symp. on Industrial Robots and 6th Intl. Conf. on Industrial Robot Technology,* 391–399, Paris (June 1982).
12. S.W. Holland, L. Rossol, and M.R. Ward, Consight-I: A vision controlled robot system for transferring parts from belt-conveyors, in: *Computer Vision and Sensor-based Robots,* G.G. Dodd and L. Rossol (eds.), 81–100, Plenum Press, New York (1979).
13. A. Browne, The orientation of components for automatic assembly, *Assembly Automation 1*(1), 30–35 (Nov. 1980).
14. D. Andree and A. Wernersson, Linear vision for finding the orientation of parts: learning procedures, *Proc. 2nd Intl. Conf. on Robot Vision and Sensory Controls,* 147–158, Stuttgart (Nov. 1982).
15. P. Horaud and R.C. Bolles, 3DPO's Strategy for matching three-dimensional objects in range data, *Intl. Conf. on Robotics,* IEEE Computing Soc. Press, 78–85, Atlanta (Mar. 1984).
16. R.N. Stauffer, Update on non-contact seam tracking systems, *Robotics Today, 5*(4), 29–34 (Aug. 1983).
17. M.P. Howarth and M.F. Guyote, Eddy current and ultrasonic sensors for robot arc welding, *Sensor Review, 3*(2), 90–93 (April 1983).
18. J.F. Justice, Sensors for robotic arc welding, *Proc. Automation and Robotics for Welding,* American Welding Society, 203–209, Miami (1983).
19. E.L. Estochen, C.P. Neuman, and F.B. Prinz, Application of acoustic sensors to robotic seam tracking, *IEEE Trans. Industrial Electronics IE-31*(3), 219–224 (Aug. 1984).
20. S. Presern, M. Spegal, and I. Ozimek, Tactile sensing system with sensory feedback control for industrial arc welding robots, *Proc. 1st Intl. Conf. on Robot Vision and Sensory Controls,* 205–213, Stratford-upon-Avon (April 1981).

21. J.G. Bollinger, Using a tactile sensor to guide a robot welding machine, *Sensor Review 1*(3), 136–141 (July 1981).
22. D. LaCoe and L. Seibert, 3D vision guided welding robot system, *The Industrial Robot 11,* 18–20 (March 1984).
23. H. Toda and I. Masaki, Kawasaki vision system—model 79A, *Proc. 10th Intl. Symp. on Industrial Robots and 5th Intl. Conf. on Industrial Robot Technology,* 163–174, Milan (March 1980).
24. Z. Smati, D. Yapp, and C.J. Smith, Laser guidance system for robots, *Proc. 4th Intl. Conf. on Robot Vision and Sensory Controls,* 91–101, London (Oct. 1984).
25. J.E. Agapakis, K. Masubuchi, and N. Wittles, General visual sensing techinques for automated welding fabrication, *Proc. 4th Intl. Conf. on Robot Vision and Sensory Controls,* 103–114, London (Oct. 1984).
26. W.F. Clocksin, P.G. Davey, C.G. Morgan and A.R. Vidler, Progress in visual feedback for robot arc-welding of thin sheet steel, *Proc. 2nd Intl. Conf. on Robot Vision and Sensory Controls,* 189–200, Stuttgart (Nov. 1982).
27. W.J.P.A. Verbeck, Arc welding process control by preview sensor, *The Industrial Robot 11,* 86–88 (June 1984).
28. A. de Keijzer and R.J. de Groot, Laser-based arc welding sensor monitors weld preparation profile, *Sensor Review 4*(1), 8–10 (Jan. 1984).
29. Anon., Leading the way on robot weld guidance, *The Industrial Robot 10,* 104–107 (June 1983).
30. Z. Smati, C.J. Smith, and D. Yapp, An industrial robot using sensory feedback for an automatic multipass welding system, *Proc. 6th British Robot Association Annual Conf.,* 91–100, Birmingham (May 1983).
31. N.R. Corby, Machine vision algorithms for vision guided robotic welding, *Proc. 4th Intl. Conf. on Robot Vision and Sensory Controls,* 137–147, London (Oct. 1984).
32. R. Niepold and F. Bruemmer, Optical sensory system controls arc welding process, *Proc. 2nd Intl. Conf. on Robot Vision and Sensory Controls,* 201–212, Stuttgart (Nov. 1982).
33. P.M. Taylor, K.K.W. Selke, and G.E. Taylor, Closed loop control of an industrial robot using visual feedback from a sensory gripper, *Proc. 12th Intl. Symp on Industrial Robots and 6th Intl. Conf. on Industrial Robot Technology,* 79–86, Paris (June 1982).
34. G.R. Archer, Vision in programmable assembly systems—results, *Proc. 5th Intl. Conf. on Assembly Automation,* 65–74, Paris (May 1984).
35. A. Osorio, K. Ben Rhouma, L. Henninger, A. Meller, L. Peralta, J. Rivaillier, and D. Teil, Workpiece identification, grasping and manipulation in robotics, *Proc. 5th Intl. Conf. on Assembly Automation,* 143–151, Paris (May 1984).
36. G. Schupp, Automated passenger car wheel mounting system, *Proc. 5th Intl. Conf. on Assembly Automation,* 111–116, Paris (May 1984).
37. H. Worn and H.R. Tradt, Automatic assembly with industrial robots, *Proc. 3th Intl. Conf. on Assembly Automation,* 525–545, Boeblingen (May 1982).
38. M. Uno, A. Miyakawa, and T. Ohashi, Multiple parts assembly robot station with visual sensor, *Proc. 5th Intl. Conf. on Assembly Automation,* 55–64, Paris (May 1984).

39. J.W. Hill, Programmable bowl feeder design based on computer vision, *Assembly Automation 1*(1), 21–25 (Nov. 1980).
40. R.J. Dewhurst and K.G. Swift, A laser electro-optic device for the orientation of mass-produced components, *Optics and Laser Eng.* (GB), *4*(4), 203–215 (1983).
41. Anon., Robot assembly component supply, *Assembly Automation, 3*(3), 130 (Aug 1983).
42. A.J. Cronshaw, W.B. Heginbotham, and A. Pugh, A practical vision system for use with bowl feeders, *Proc. 1st Intl. Conf. Assembly Automation,* 265–274, Brighton (Mar. 1980).
43. O. Ledoux and M. Bogaert, Pragmatic approach to the bin picking problem, *Proc. 4th Intl. Conf. on Robot Vision and Sensory Controls,* 313–323, London (Oct. 1984).
44. R. Kelley, J. Birk, J. Dessimoz, H. Martins, and R. Tella, Acquiring connecting rod castings using a robot with vision and sensors, *Proc. 1st Intl. Conf. on Robot Vision and Sensory Controls,* 169–178, Stratford-upon-Avon (April 1981).
45. R. Ray and J. Wilder, Visual and tactile sensing for robotic acquisition of jumbled parts, *Optical Engineering 23*(5), 523–530 (Sept.–Oct. 1984).
46. G.J. Page, Vision driven stack picking in an FMS cell, *Proc. 4th Intl. Conf. on Robot Vision and Sensory Controls,* 1–12, London (Oct. 1984).
47. J-P. Hermann, Pattern recognition in the factory: An example, *Proc. 12th Intl. Symp. on Industrial robots and 6th Intl. Conf. on Industrial Robot Technology,* 271–280, Paris (June 1982).

10

Hardware for Intelligent Automation

10.1. Overview

In this chapter, we first survey currently available hardware. This can be divided into two classes: one is used off-line during the research and development stages of a project; the other is used on-line when the processing to be used has been determined. Description of standard microprocessors is omitted since this is available elsewhere, particularly from manufacturers' data sheets. Many systems for intelligent automation are based on bus systems that provide a range of mutually compatible modules, and we review the various possibilities which are available. Then different architectures applicable to intelligent automation are considered, followed by discussion of alternative and nonelectronic forms of processing. Finally, some suggestions are offered for configuring systems.

10.2. Standard Systems

10.2.1. State of the Art

Let us begin by discussing vision systems currently available commercially.[1,2] Rapid development means that specifications are continually changing, so detailed descriptions of instrumentation offered by particular manufacturers will not be attempted.

The simplest systems are intended for guiding robots; they incorporate small, very lightweight sensors ("eye-in-hand")[3] that can be mounted on a robot arm. This gives better depth of field, better resolution, and less parallax error than larger fixed sensors mounted further away. Resolution is typically 32 × 32 pixels; the solid-state sensors used are often Charge-Coupled Device (CCD) memory chips with the opaque cover replaced by a quartz window. Such systems normally process binary images only; they will detect holes, measure simple features such as area and perimeter, and compute functions of these such as perimeter^2/area, with operation at up to 50 frames per sec. The programmability and adaptability of these is very limited, but the cheapest costs less than £1000.

Linear array cameras, with simple processing using microprocessors, are available for easy tasks such as the measurement of edge position. High-contrast lighting is normally used so that the position of the edge of the image on the sensor array is easily determined. Typical applications include measurement of the diameter of wire and the width of sheet material such as steel.(4) Precise measurements are possible due to the large number of sensor elements which can be provided on a single chip; chips with lines of over 4000 elements are available. For measuring sheet width, two cameras would be used, each looking at one edge. Such systems can also be used to measure the area and check the shapes of objects moving with constant velocity. High resolution linescan cameras are being used for the acquisition of documents, such as maps and engineering drawings, for which high resolution and accuracy are required. A camera with a line of 4000 sensing sites has a resolution of 125 microns across a width of half a meter. Linescan cameras can be mounted on robot arms and held in preset positions with respect to a workpiece, and noncontact measurements then made of significant features.

At the next level, we find systems for comparing images to detect discrepancy(5); the intended application is automated visual inspection. A typical instrument processes images 512 pixels^2 × 4 bits (16 levels) deep. An alarm is set to respond when the number of pixels whose levels differ between the stored and sample images exceeds an adjustable threshold. The instrument cannot be programmed, and indeed needs no programming. It is "trained by showing;" this makes it easy to use. But it is not possible, for example, to discriminate changes in some pixels as being more significant than in others, so the positions of stored and sample objects must be carefully controlled during comparison, or misregistration will cause mismatch. The system costs about £5500.

At the next level, we find programmable systems adaptable to a wide range of tasks.(6) These commonly accept both solid-state (CCD) and vidicon 2D cameras; a capability to handle linescan cameras is scarce but growing. Often, the outputs from several cameras can be handled simultaneously (permitting, for example, stereo vision); complete images are held in framestore until the system has finished processing them. Image size is from 256 to 512 pixels square, with 64–256 levels per pixel. Ability to handle color is apparently absent. Most versions use 68000 processors (generally more than one), often with the UNIX operating system. A convolutional processor is generally available as an optional extra to speed processing perhaps tenfold. A few systems use the PDP11. Libraries of subroutines to implement operations likely to be useful are available, integrated using a high-level command language. Usually, a general purpose language such as Fortran is provided, to enable additional compatible subroutines to be developed by the user. Operations such as moments and run length encoding are available, and recognition in 100 msec with orientation to $\pm 1/2°$ typically claimed. Such systems must be configured and programmed by the user (or by a specialist systems house) to perform each specific task. The cost of the hardware per system is typically £30,000–£40,000. Even the most complex systems are intended for operation on-line, rather than merely for development in a laboratory environment. This poses a dilemma for the intending user, since these architectures are frequently too general to be optimal for specific tasks. However, the cost of developing a new processor and interface hardware may well justify use of a general system, particularly when the number of identical installations to be produced is small. The most advanced systems have been publicly demonstrated performing extremely complex tasks involving gray-scale vision, like inspecting the brakedrum assembly for a motor vehicle, in a commercially viable time.(7) An instrument of this kind is understood to be installed on the production line for the Volkswagen Golf, enabling 100% inspection of a critical assembly. This model-based system is software compatible with CAD systems which can be used to aid the vision.

Then there are special purpose systems with both hardware and software custom built for each specific application. A few standard building blocks such as CCD cameras and single-board microcomputers may be used, but the rest is specially designed. Such systems cost upwards of £50,000, at least for the first one off. Their viability arises from the nature of automation tasks, in which factors like lighting, presentation, reduction of information, and software are dominant in engineering an adequate system. The cost of developing the instrumentation as a system far outweighs the cost of the hardware.

10.2.2. *Off-Line Systems*

There is a significant difference between hardware for use *off-line,* when alternative processing sequences are being evaluated as part of research and development, and *on-line* hardware, used to perform specific tasks in real time.

When approached in the most efficient way, the provision of an intelligent system for automation involves two distinct phases. In the first, alternative processing schemes are evaluated until the best is found for performing the required task. This is normally performed using a standard development system;[8] these are offered by several manufacturers. They generally follow the configuration indicated in Figure 10.1; a vidicon camera is used as sensor, with a standard microcomputer, such as a PDP11, as central processor. The resolution is high, at least 256 × 256 pixels with 64 gray levels, and framestore, video display, and keyboard video display unit (VDU), for operator interaction, are provided. A rig for investigating alternative configurations of viewing and illumination may also be provided. The system almost always includes a package of mutually compatible software subroutines, which implement standard operations such as threshold-

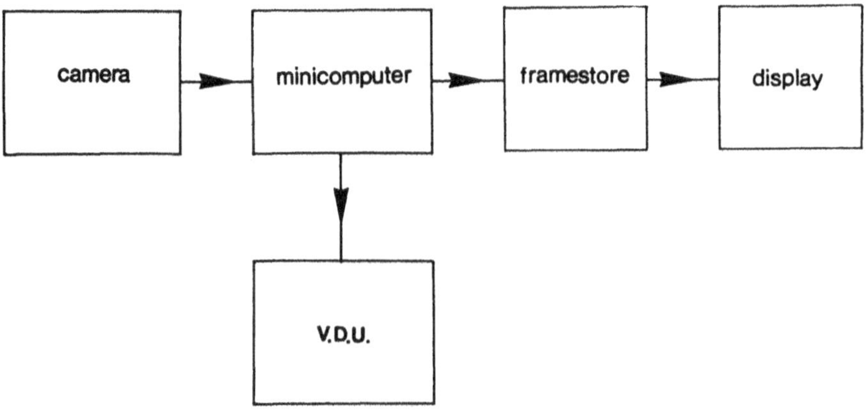

Figure 10.1. Off-line vision system.

ing, histogram modification, convolution with a mask, Fourier domain filtering (high- and low-pass), subtraction of frames, etc. Provision is also generally made for the user to add his own subroutines in a standard language such as Fortran. These programs generally run far too slowly (several seconds per frame) to be usable for most on-line applications.

Such systems are normally constructed to operate in an office environment. They allow an investigator to evaluate and compare the performances of a huge range of sequences of processing algorithms with a minimum of effort; the processing appropriate to a task may be determined typically after only a few days' work. Experimentation with alternative forms of viewing and lighting to highlight features of interest is catered to, but these systems, costing between £25,000 and £50,000, are very expensive, besides being too slow for operation in anything close to real time. Thus, their proper application is in a central research and development facility, in which the cost can be spread over many investigations. They do, however, have some direct application in off-line automation, for analyzing static images in medicine, metallurgy, and soil science, for example, to count particles which are within a particular size range.

A slightly different form of development system[9] is illustrated schematically in Figure 10.2 and pictorially in Figure 10.3. This was developed by one of the authors and his colleagues and has been used for several research investigations performed in academic environments. As far as is known, it is not available commercially, but may be worth copying for those with the necessary resources. Its key feature is the rig shown in Figure 10.4. The sensing device is a CCD linescan camera. High resolution is desirable and 1728 elements were available in the original, but useful results may be obtained with 256 elements, which is much cheaper to provide. The sample is placed on a table moved by a stepping motor in the direction perpendicular to the camera linescan, so that two-dimensional images may be obtained. The stepping motor is controlled by the computer, but indirectly, using a dedicated microprocessor, to avoid imposing a processing overhead; its minimum increment of advance is typically 25 μ.

The distance between the camera and the table may be varied, to set the scale (i.e., the distance between pixels). The advance of the table must be commensurate with the camera if the distance scales in the two directions are to be equal; since table advance can take place

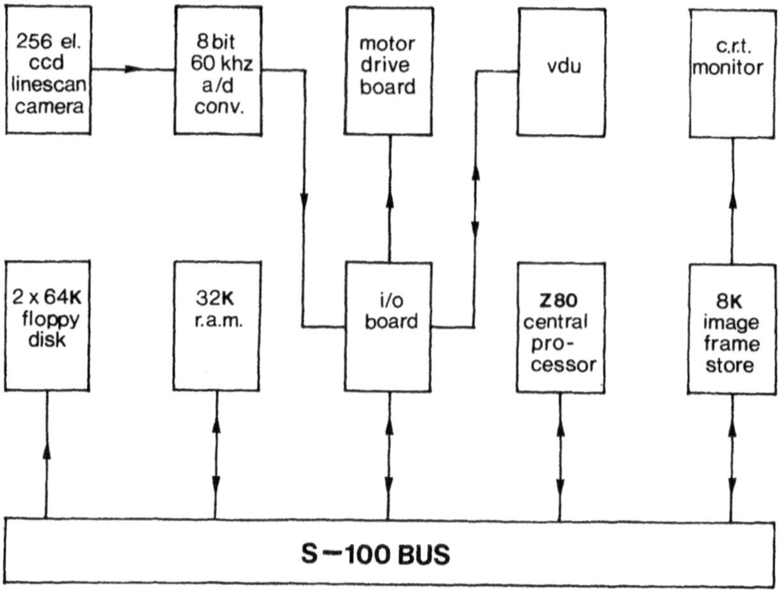

Figure 10.2. Low-cost system (later version) using an S-100 bus.

only in integer multiples of the minimum increment, only certain positions of the camera are suitable, and these can be calibrated (by scanning a circular disk) and marked. The angle of view of the camera may be varied by inclining the yoke. Illumination may be from below (to view transparent or silhouette objects), or from above, for objects which are opaque. This arrangement permits a wide range of alternative viewing conditions to be investigated very easily. Illumination is provided from fluorescent tubes, driven by high-frequency (ca. 20 Khz) ac, which give white light without ripple.

Use of a Tectronix 4006 storage display (1024 × 768 points) in the original system obviated the need for framestore, but the system could then display at high resolution only binary images. A later system, based on an S-100 bus, had framestore for a 240- × 256-pixel binary image or 120- × 128-pixel gray-scale image with 16 levels per pixel, selectable by throwing a switch. A useful feature of the original system was an industry standard (1/2 inch) magnetic tape, which permitted interchange of data and programs with virtually any computer system. Most current floppy disk systems are unsuitable for information in-

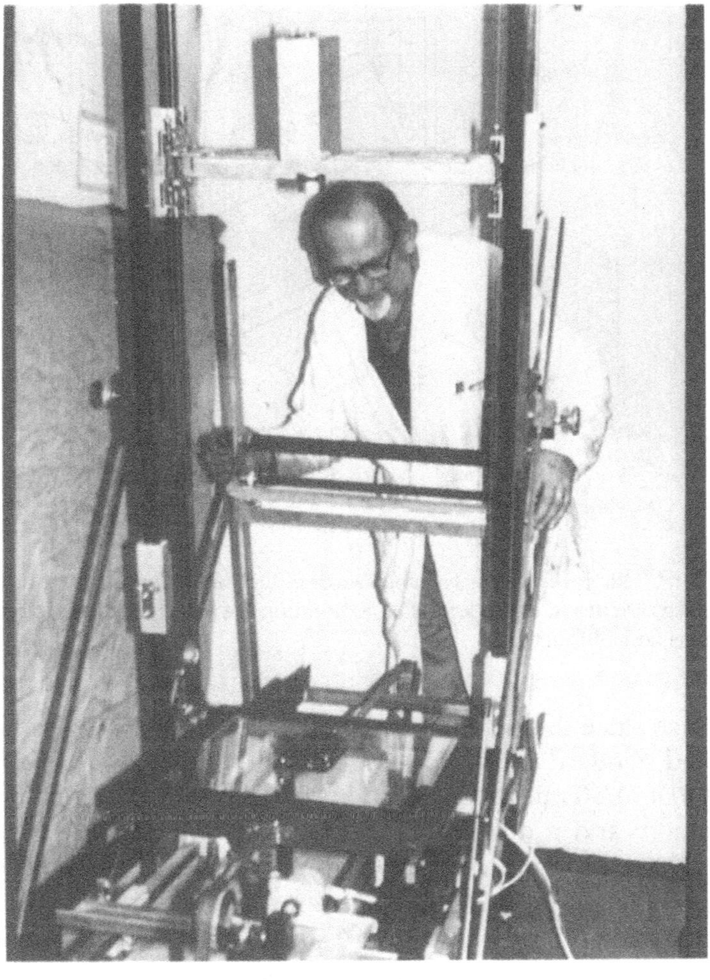

Figure 10.3. Rig for low-cost system.

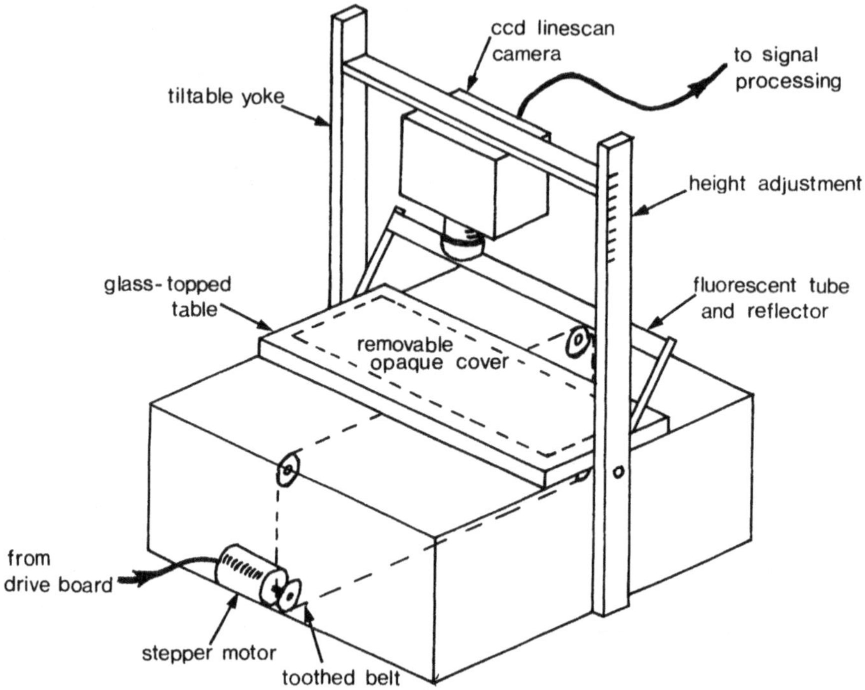

Figure 10.4. Rig for low-cost system. A second fluorescent lamp and reflector system (not shown) are mounted under the table pointing towards the camera for transparent specimens and silhouettes.

terchange since they use incompatible formats; there is no universally accepted standard.

A PDP11/10 micro was central processor on the original, with 24K of memory and a dual 8-inch floppy disk with the RT11 operating system. Most programs used Fortran, a few used assembler. With a 1-msec floating point multiplication time, this system was desperately slow, but admirable in all other respects. A high-speed preprocessor for extracting silhouette images boundaries at 5 Mhz was a useful addition. This system was used successfully for research into the automated inspection of steel strip, printed circuit boards, and glass bottles, and for the automated identification and positioning of silhouette shapes.

The latest version uses a Comart microcomputer with 64 Kbytes of random access memory (RAM), dual 5 1/4-inch floppy disks with capacity 400 Kbytes and a CP/M operating system. Programs are written in Pascal or the Z80 assembler. Standard S100 boards are used for framestore and input/output. This latest version is far better value for money than the original; it is also much faster. This may reflect 8 years of progress in hardware rather than astute selection. Its chief shortcoming is lack of resolution, and this is minor. The total value of "bought out" parts has been about £2500, since important components such as the camera (based on a Fairchild I-SCAN board) were constructed in-house rather than purchased. This version is demountable (by unplugging interconnections) and completely transportable. It may be placed in a van and transported on-site.

There may be a temptation to use "home" microcomputers for automation tasks, since their cost/performance ratio is often apparently excellent when compared with professional equipment. A 75% cost reduction may easily be obtained. Though performance in terms of memory capacity, processing speed, and so on is generally as claimed, home micros are mass-produced to achieve their minimum cost, and their reliability (never discussed in sales literature) is often so far below that of the professional article that they are best avoided.

10.2.3. On-Line Systems

For use on-line, the system architecture required is quite different, as indicated in Figure 10.5. In this case, the camera (or other sensing system) is chosen to suit the task, and vidicon-type TV cameras are rarely used. The output signal from the camera is passed to a preprocessor. This is generally a nonprogrammable device whose data throughput rate is much greater than a standard microprocessor. Its function is to drastically reduce the rate at which data is supplied to the main processor, by extracting and passing only essential infor-

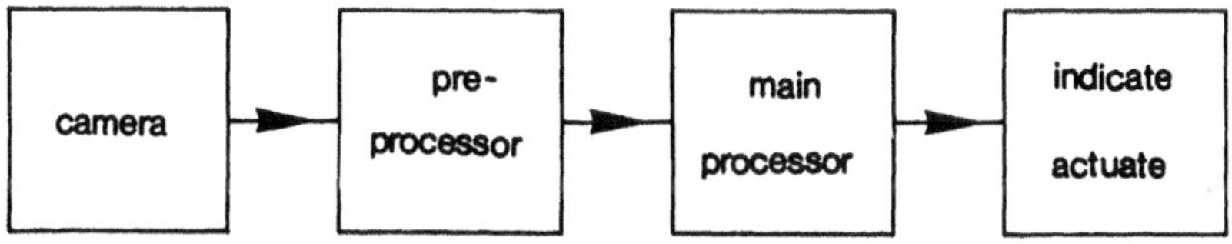

Figure 10.5. Architecture of a system for on-line vision.

mation. In a silhouette shape processing system, for example, the preprocessor would extract boundary points. The preprocessor would typically convolve some kind of mask with the K lines of image most recently gathered, locate interesting points, and pass the identification and location of these to the main computer, perhaps after a fixed number of scans had been processed. The main processor, which performs the bulk of the analysis, generally comprises one or two standard microprocessors (e.g., Z80 or Z8000; the 68000 is becoming very common), and uses a firmware program. Finally, some indication or actuation is provided, so that decisions made by the processing may either be used (for example) to mark or eliminate a defective product, or communicated to an outside operator. The system may maintain a record of part throughput or sample quality.

Note that an on-line system can generally perform one task only; once successfully installed, it is not intended to be reconfigured or even reprogrammed. Operation in hostile environments and at high speed is often required. Generally, cost must be kept down to a few thousand pounds per installation for instrumentation to be cost-effective. Thus, systems are tailored to specific applications in both hardware and software; they include only components that are absolutely necessary. Peripherals such as VDUs, keyboards, and backing store which are costly, unreliable, and permit tampering, are best avoided. Expensive components sometimes thought indispensible, such as framestores, can also often be eliminated by thoughtful design.

10.2.4. Bus Systems

When selecting a hardware configuration, use of a standard bus system[10,11] is worth considering. These enable instrumentation to be constructed largely from standard mutually compatible cards made (often competitively) by different manufacturers; this flexibility gives excellent value for money. Cards are generally available for operations such as A/D conversion, image storage, random access memory, peripheral driving, interface to networks such as Ethernet, and so on. Configurations using multiple central processors are possible, enabling a task to be shared. A bus is specified by its *structure* and its *architecture*. Bus structure comprises electrical and mechanical characteristics such as connector type and pin assignment, control protocols, and voltage levels. Bus architecture describes the various pathways available for communicating data and control signals. More sophisticated buses have separate pathways for the two kinds of signals, and may even have different versions of each kind of pathway. Pathways are usually parallel, but serial versions sometimes occur. A bus is, in fact, a special form of local network, which generally requires system components to be very close together, in the same console, and connected by fixed (not flexible) leads. If high-speed parallel pathways are not constructed very carefully, small variations in symbol speed may occur, causing skewing, i.e., pulses transmitted at the same instant arrive at slightly different times. One advantage of advanced multi-pathway architectures is that they can provide both high data rate, parallel pathways operating up to more than 40 MHz and lower-rate serial (e.g., 2 MHz) pathways. The latter provide standardized links to system components which are more distant on flexible leads. Simple buses having a single pathway may suffer from congestion caused by message contention.

Some examples (the listing is by no means exhaustive, since new buses are continually appearing) of standard bus systems and their chief characteristics are listed alphabetically below:

FUTUREBUS (IEEE P896). This is a bold attempt to provide (under the auspices of the IEEE) a universal bus system which can accept any foreseeable microprocessor, and uses the Eurocard standards. The maximum wordlength to be accepted is 32 bits, and the system is to be asynchronous and optimized for multi-processor operation. Like VMEBUS, but unlike most others, it is to have a separate serial bus. It has the unique and interesting characteristic of needing only one supply voltage (5V). Limitations set by TTL interfacing are to be avoided. But, alas, up to the time of writing this, a specification has yet to be agreed on.

MULTIBUS (main suppliers are Intel and National Semiconductor Corporation). This is an up-to-date, 16-bit system based on the 8088 series processor. It commonly incorporates the UNIX operating system and "C" language (note that these are not unique to MULTIBUS). A good range of cards for special purpose applications such as image framestore is available.

MULTIBUS II. This is an update of MULTIBUS to 32-bits and the 8086 processor. System components are mounted on Eurocards. Five kinds of data pathways are available.

NUBUS. This is based on the Texas Instruments 99000 series.

S100. This is an early system which is still widely used, particularly in microcomputers. A very wide range of compatible cards is available, at minimal cost. The CP/M operating system is commonly used with S100. The original bus was for 8-bit processors such as the 8080 and Z80, and could address only 64 Kbytes of memory. The latest (IEEE P696) version of S100 will support 16-bit microprocessors and can address 16 Mbytes of store.

STDBUS. This is a standard (IEEE P961) bus of maximum simplicity which will support any 8-bit microprocessor and is widely used in low-cost applications such as industrial control. It is highly modular, using small cards, with each card performing one function (e.g., RAM or I/O) only.

QBUS. This 16-bit bus is used on DEC LSI11 machines. It is well established but cannot support multiple central processors except in a master/slave configuration. The maximum address width is 18 bits; 16 bits is standard.

UNIBUS. This 16-bit bus is used on the larger DEC PDP11 machines. The widest possible range of compatible cards is available, including (for example) an array processor on two cards which speeds processing 20–50 times. Compatibles tend to be well engineered but expensive.

VERSABUS (Motorola is main supplier). This is a 16-bit system based on the 68000 microprocessor family.

VMEBUS. This is a recent (version "B," specified October, 1981) modification of VERSABUS (hence, 68000-based), to suit Eurocard standards, that is growing in popularity rapidly. It readily supports multiprocessing, and can accept 8-, 16-, and 32-bit processors, mounted on the same backplane. Maximum address width is 32 bits; 24 is standard. The VMS (serial) and VMX (32-bit parallel) pathways are part of the system, and provide versatility similar to MULTIBUS II. The family includes an interface for a 10-MHz Ethernet. The operating system used is Versados.

Simple 8-bit systems such as S-100 should not be despised if low cost is important and the task is simple. Several S-100 based systems for vision for automation have been produced (see Section 10.2.2 and ref 10.12). Many existing commercial systems are based on the DEC buses QBUS and UNIBUS; emergent systems seem to favor MULTIBUS. But VME is beginning to mount a strong challenge; for example, acquisition, storage, and display of a 384- × 512-pixel image 8 bits deep is available on a single board priced at £1500, with compatibility to standard CCD array cameras. Bus systems using the DIN pin connector standard for Eurocards are preferred to those using edge connector cards for greater reliability.

Digital processors for analog signals are available which are compatible with standard bus systems and offer processing efficiency combined with accuracy, stability, and programmability. A typical example is the SKY320 which is a 16-bit parallel system compatible with QBUS and is designed to work with PDP11 processors using the RT11 operating system. It is contained on a single card and incorporates a hardware 16- by 16-bit multiplier to achieve a high processing speed.

The languages most commonly used for programming bus systems for intelligent automation are Pascal, C, compiled Basic and, of course, assembler. Forth is used where high-speed response is important. Fortran, which is commonly used for general purpose off-line signal processing, is almost never used for on-line applications. Interpreted languages of the kind found on personal computers are of little use since they have run times typically one thousand times as long as the compiled version of the same language. Programs written in a compiled, high-level language may take ten times as long to run as the same program written in assembler, but are easier and quicker to write and debug. The time taken to produce a program is found to be proportional to the number of statements used, whatever the level of the language. High-level languages are cost-effective since they use far fewer statements than assembler for a given task.

10.3. Novel Architectures

10.3.1. Classification of Architectures

Let us now consider various processing architectures which may be applied to intelligent automation; many have been proposed and investigated in considerable depth by academics, but only a few are available commercially.

The standard architecture used currently in virtually all computers (including microprocessors) is the *Von Neumann* form, whose structure is indicated in Figure 10.6. Data and program instructions are held in a memory block which is addressed serially. A program counter-addresses each memory location in sequence; its contents are transferred to the single central processor (arithmetic and logical unit (ALU)) which interprets the instruction and executes it. The program itself must instruct the central processor not to try and execute the contents of locations being used to store data.

Processing images and other forms of signals, however, generally involves performing the same operation repeatedly on many samples of data in a segment of signal (regarded as a vector), and the positional relationship between the samples is significant. Images are two-dimensional, and two-dimensional relationships are obscured when serial storage is used. The obvious architectural development is to include more than one processor, ideally one for each pixel or data sample in a vector. It is then possible to process all elements of the signal simultaneously; each processor needs its own local memory, but the amount per processor may be small. The simplest such ar-

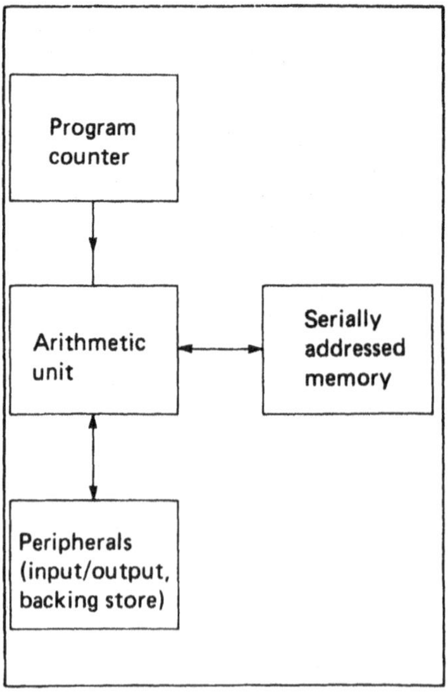

Figure 10.6. Von Neumann architecture.

chitecture is the single instruction, multiple data (SIMD) form, in which many processors perform the same operation simultaneously. Much more complex is the multiple instruction, multiple data (MIMD) form, in which the processors may perform different operations at each instant. Processing of images and similar vector data can be speeded up considerably simply by using the SIMD approach.

Data arrays used in intelligent processing are generally very large (images less than 256 pixels^2 are rarely useful), and configurations having one processor for each data point are very expensive. Parallel processing may be made more economical by using a processor array that is smaller than the data array, and is moved in stages through the data to cover it in a series of "blocks" by windowing.

10.3.2. Simple Developments

Alternative architectures are available for speeding processing which, though not as powerful as the massively parallel approaches just discussed, are very cost-effective. In a ***pipelined*** processor, each data sample proceeds through the same set of processes in a fixed sequence (Figure 10.7); each stage acts on the data output from the preceding stage. A separate and parallel path is provided for each component of a (one-dimensional) vector signal; normally, the processing provided in each path is the same, and processing for components and stages is synchronous. This approach is valuable for processing which can be regarded as matrix operations involving a single data vector, such as fast Fourier transforms. The *Systolic*[13] form of processor is an extension of the pipeline concept to more than one

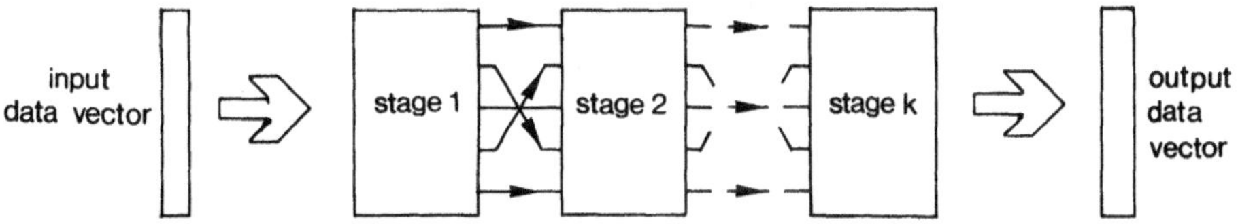

Figure 10.7. Pipelined processor (generalized). It comprises a sequence of stages each holding an N-sample vector of data. In synchronism with each clock pulse, the data in each stage proceeds (with processing according to the task in-hand) one stage to the right. A new vector of raw data is drawn into the left-hand side, and a new vector of processed data appears at the right. The processing between stages is often controllable by software.

dimension. Systolic processors can handle several streams of data vectors, and give improved efficiency for operations such as matrix inversion and multiplication. A *wavefront* array is a variant of the systolic form having asynchronous communication of data between adjacent processors. Several pipelined (one-dimensional array) processors are available commercially as cards compatible with standard bus systems and provided with software so they can be programmed easily in high-level languages such as Fortran; they are well worth incorporating if, for example, two-dimensional gray scale images are being processed. The more advanced forms are still in development.

Convolutional processors are normally arrays whose dimensions may range from 3 × 3 elements (2 × 2 is rare) to 32 × 32 elements. They are generally square for two-dimensional data, with the sides having an odd number of elements. The array is moved through the data, being positioned symmetrically over each element in turn; this element is replaced by a new value obtained by multiplying each "original" data element within the window by the corresponding array element (Figure 10.8). Notice the distinction from the systolic approach; in the convolutional processor, the mask elements operate

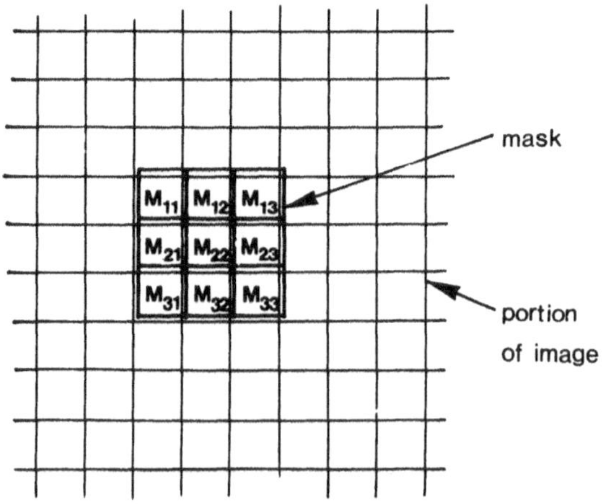

Figure 10.8. Action of a mask processor. The image pixels are designated by their row and column indices i and j, so that the (i,j)th pixel is $b(i,j)$. The processor output when centered over image pixel (i,j) is thus:

$$\sum_{l=1}^{3} \sum_{k=1}^{3} m_{l,k} \cdot b_{(i-2+l),(j-2+k)}$$

only on the original data, and not on one another's outputs. It is explained in Ref. 14 that convolutional processing and Fourier domain filtering are equivalent, subject only to approximation due to the finite size of the convolutional mask.

A convolutional processor would be used to implement the masks suggested in Section 8.5. It might, for example, replace each element by the mean value of the data in a 3-element square window centered on itself. A new two-dimensional data array is created, having the same dimensions as the original, or possibly slightly smaller due to edge effects. Processing of this form is valuable in on-line systems for use in preprocessors, to perform repetitive operations such as boundary determination. Special purpose hardware would then be used rather than a software program; a few bus-compatible convolutional procesesors are available commercially.

It is not necessary to hold a complete image in store to process it convolutionally; instead, the data can be moved conceptually through the convolver, instead of vice-versa. The architecture is shown in Figure 10.9. The processor includes a number of "first in–first out" stores,

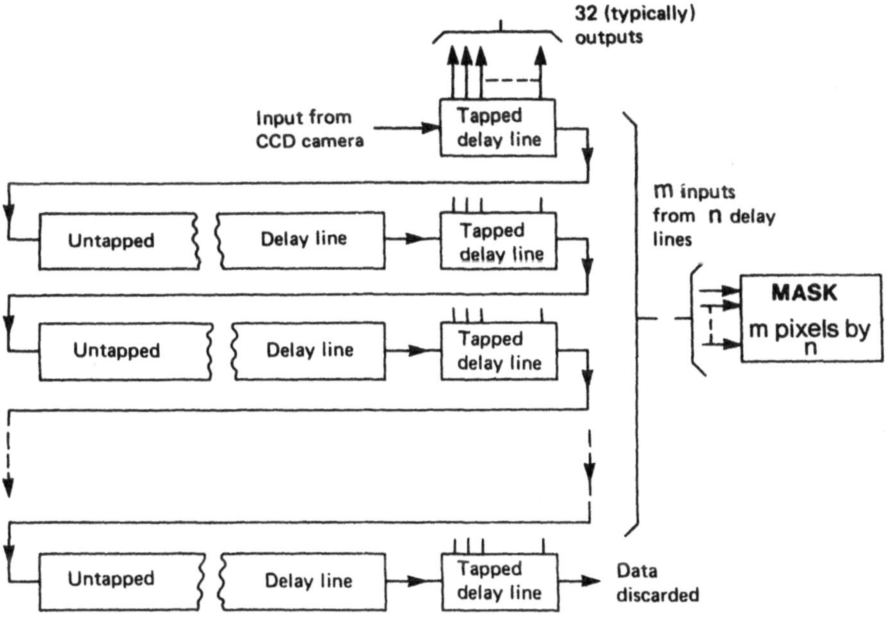

Figure 10.9. Arrangement for applying a mask processor to a continuous image.

which might be implemented using CCD delay lines or digital shift registers. The signal is transferred serially from the camera to the first line of store, then from one line of storage to the next as each new scan is gathered. The "oldest" line of stored data is discarded at each new scan. Portions of each delay line are tapped, and the convolution operates on the tapped signals. This approach is particularly useful for processing the signal from material moving as a continuous strip.

We shall next describe some architectures that, though not radically new, are commercially available and ready for application. Other architectures that, though both powerful and available, are not suited to on-line automation applications will not be covered. An example is the DAP array processor of ICL, which is a 64 × 64 array of central processors intended to be programmed in a high-level language as part of a mainframe system.

Many other special purpose architectures are being developed for processing vector data at high speed; these are potentially applicable to automation tasks in industry, but are believed not to be available commercially as yet. Notable systems include: DIP-1[(15)] developed at the Technical University in Delft, Holland, which is a dynamically programmable pipelined processor, the Cytocomputer (another pipelined approach),[(16)] being developed at the Environmental Research Institute in Michigan, USA, and the PICAP machine,[(17)] developed at Linkopping in Sweden. The Genesis system developed by Machine Vision International[(18)] is a pipelined configuration of processors which uses Serra's morphological approach to analyze images for applications such as inspection.

10.3.3. CLIP

The CLIP series of SIMD processors[(19)] have been under development by M.J.B. Duff and his team at University College, London since 1969. The first of the series (CLIP4) to be exploited commercially comprises a rectangular array of special purpose processors, each having a data pathway to its eight nearest neighbors. The resolution of the original CLIP4 was 96 × 96 points, and an image of these dimensions can be stored with each pixel at its appropriate spatial position. Thus, spatial relationships are preserved and may be exploited to speed processing. For economy and improved reliability, eight processors are contained in a single, special purpose PMOS chip. The

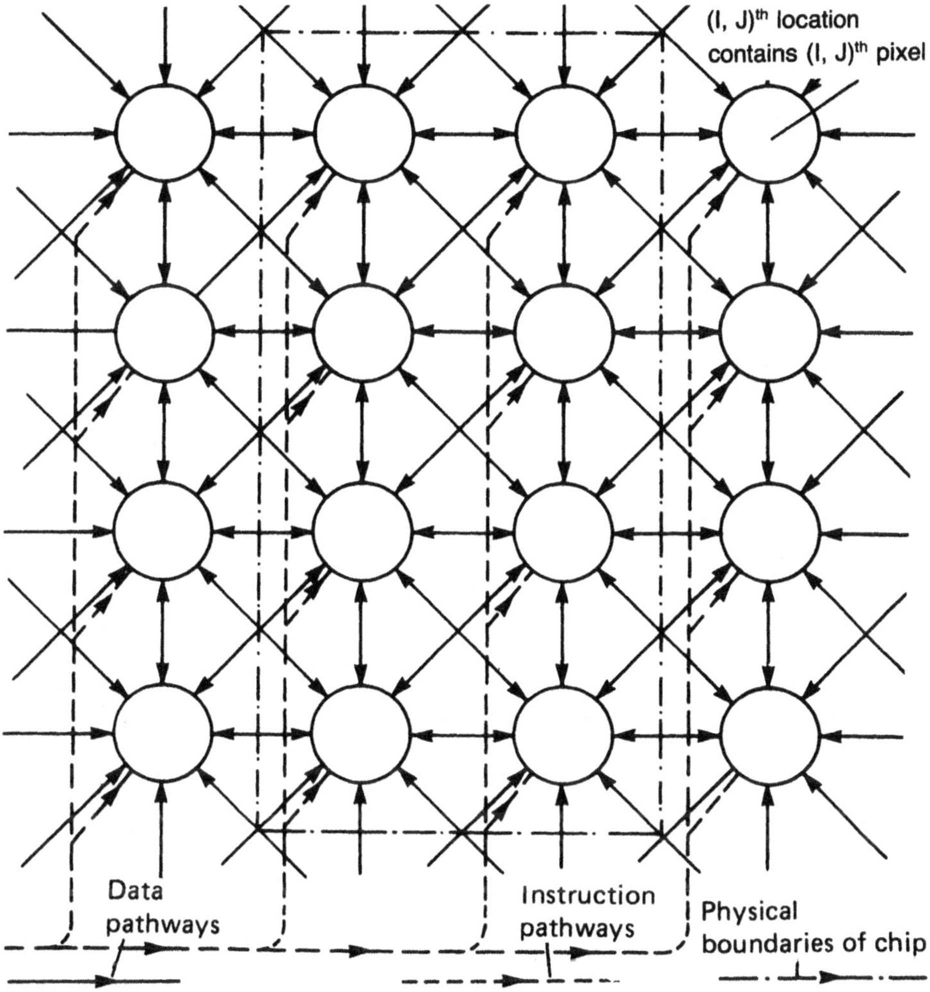

Figure 10.10. Operation of CLIP4.

structure of CLIP4 is illustrated in Figure 10.10. CLIP4 performs the same operation simultaneously on all processors. Individual operations are generally simple (e.g., subtract each element from that to its immediate left) but the parallelism gives extremely fast operation; some processes are sped up over a million times. CLIP4 is, however, loaded and unloaded serially. A single layer of CLIP4 can process only binary images (1 bit/pixel) but gray-scale images may be handled by using one layer for each bit, i.e., 128 level images would require seven bit planes. Processing would then take about seven times as long, since the planes have to be processed in sequence.

At least three CLIP4 systems with resolutions of 96 × 96 pixels and six layers, to give 64 gray levels, are being operated in research and development environments. A commercial version of CLIP4 is now available[20] in a range of sizes based on a building block 32 × 4 pixels (one card) with a maximum size of 128 pixels^2. It is driven by an Intel host computer and is thus able to use MULTIBUS components. Most important, it can use the vast library of software developed for CLIP4. The chief limitation to CLIP4 other than its cost is that the maximum resolution of 128 × 128 pixels currently available is too small for many applications. The outstanding exception is for a robotic eye, when images 32 pixels^2 are often adequate. One way of overcoming this is to configure CLIP as a windowed processor which deals with a large image by examining only those portions found by a preprocessor to contain something interesting. An arrangement for doing this is suggested in Section 11.3.

Meanwhile, the University College Group is developing CLIP7;[21] this uses a windowed array 512 × 4 pixels to achieve the same processing speed for an image 512 pixels^2, as does CLIP4 for an image 96 pixels^2. A prototype version is already operational.

The GRID processor being developed by GEC in the UK[22] is a SIMD system, somewhat similar to CLIP, that uses VLSI fabrication techniques to produce a chip with 64 processors in an 8 × 8 array.

10.3.4. *WISARD*

An alternative approach to processing data in parallel, and particularly for distinguishing two-dimensional signals from one another, is provided by the WISARD processor developed by Igor Aleksander and his team at Brunel University. This is radically different from all others considered so far; it is logical rather than numerical, and performs no arithmetic in the usual sense. Though totally unsuitable for some tasks, it is admirable for others. For distinguishing patterns whose location and alignment can be controlled but that exhibit statistical variation, it is excellent. Its classical demonstration is as a recognizer of faces, in which it can tolerate change of expression. Its operation can be virtually instantaneous, giving decisions in times of the order of microseconds. Most important, and unlike all other processors, it does not need to be told by an operator which quantities or features it must use for a given task. Instead, it contains built-in intuition to infer its own implicit decision criteria. In this respect it is very similar to the human brain.

The operation of WISARD is indicated in Figure 10.11. The signals to be recognized are stored within a two-dimensional memory block. Like CLIP, WISARD uses one layer of memory per bit plane (i.e., each bit plane handles a binary image). One such layer, called a discriminator, is required for each class to be stored. Connections are taken from randomly distributed groups of *N* pixels (called *N*-tuples) within a layer. Typically, $N = 8$. The binary outputs of the *N*-tuples are used

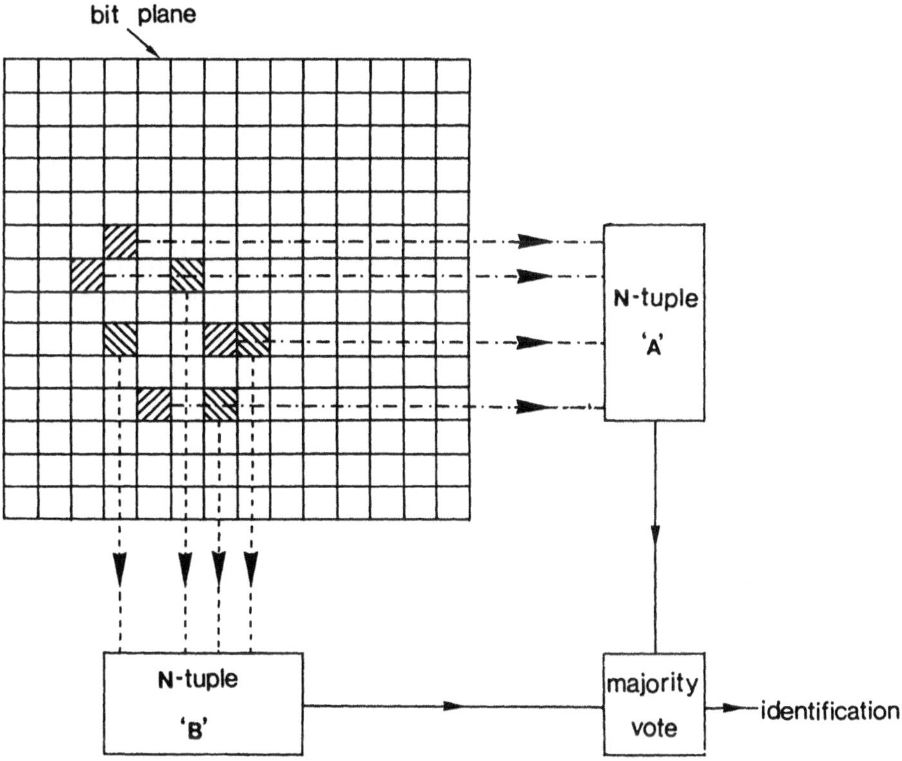

Figure 10.11. Operation of WISARD.

as address lines for RAM; if (for example), $N = 8$, then 256 distinguishable outputs are possible, and a binary digit of store must be provided for each of these.

When an image is stored, each pixel will be a "1" or a "0," defining an address N bits long. Before training, the contents of the complete store are set to zero. During training, one image of each class is viewed in succession, and a "1" is placed in the location addressed by the binary word held in the N-tuple, for the appropriate discriminator. During testing, i.e., identification of unknown shapes, the number of locations containing a "1" is counted for each discriminator. The unknown pattern is assigned to the class for which the number of "1"s is greatest, provided that number is sufficiently large. During testing, the states of all N-tuples will not be the same as during training, because of noise, distortions, and "within class" variations. The sum obtained from the discriminator trained to the correct match is normally sufficiently greater than those from the other discriminators to give correct classification nevertheless. WISARD can normally tolerate corruption of at least 20% of N-tuple outputs.

Thus, WISARD assumes implicitly that each class of image is defined by a single prototype which is subject to corruptions. When, however, a pattern is translated or rotated it becomes a new pattern so far as WISARD is concerned, and must be relearned as such. This can enable WISARD to measure the position and orientation of a shape, but the shape must still be learned as a separate pattern at each possible position and orientation, and the repeated training required is tedious and consumes a great amount of memory.

WISARD uses standard random access memory in vast quantity. But since RAM is already cheap and is becoming still cheaper, and special purpose chips are not required (they are required for CLIP), cost-effective implementation is possible. A commercial version[(23)] is available from Computer Recognition Systems with a maximum array size of 512 pixels^2 and 8 bits deep. This uses a programmable (rather than hard-wired) N-tuple configuration, and is thus rather slower than the prototype. It takes about 40 milliseconds to classify an image and can handle at least 16 classes. It is based on a VMEBUS microcomputer system. It is probably best suited to the inspection of small assemblies which can always be presented in the same position and orientation, to detect damaged or missing components. Only one discriminator is then required; if "go–no-go" limits are to be applied, then a discriminator can be trained on each limit. Its advantages in this application lie in ease of training (a conventional system would require a lengthy investigation to determine features to be used, etc.), very high speed, and adaptability; retraining by showing is the only modification required when the product being inspected is changed, plus some adjustment of viewing and illumination to highlight features of interest.

The chief shortcoming of novelties such as CLIP and WISARD is that they are rather expensive. It is often possible to achieve adequate throughput more economically by thoughtful engineering with conventional processors. Perhaps the economies of quantity production and the general trend of decreasing semiconductor prices will change this.

10.4. Novel Processing

The hardware for intelligent automation need not be digital or even electronic. Digital processing offers programmability, stability, easy data storage, and freedom from distortion, and is hence almost always used in intelligent automation, but is not fast enough for some tasks. Electronic systems require considerable skill in the diagnosis of faults, are prone to electromagnetic interference, and are expensive for simple tasks. Mechanical and fluidic processing were thus tried in early days but are now obsolete. The handling and processing of signals optically is a novel development which will steadily find increasing application.

10.4.1. Analog Processing

Analog processing can be very fast, and the signals generated by many transducers are inherently analog. Thus, there is scope for using analog processing, particularly at the front end of a system, to implement preprocessing to reduce the data rate to a level with which a conventional digital channel can cope. To achieve really high processing speeds (in excess of 20 Mbits/sec) it is necessary to exploit techniques developed for use in communications and radar. Transversal filters based on CCD delay lines[(24)] may be used, for example, to increase the contrast of noisy messages. The operation of CCD delay lines is explained in Section 4.3.4; these use a sampled analog signal which is easily digitized for subsequent processing digitally, and will clock at rates up to about 30 MHz (silicon) and 800 MHz (gallium arsenide).

Surface acoustic wave (SAW)[(25)] devices are also attractive; they offer a greater time bandwidth product than CCD devices and do not require a sampled signal. SAW devices are based on blocks of piezoelectric material such as quartz. An emf applied to interdigitated electrodes deposited on the surface of the block is used to initiate a transversal surface wave. Further interdigitated electrodes in the path of propagation generate electrical signals whose amplitudes are determined by the sizes of the electrodes, with a time delay set by the distance from the source divided by the velocity of propagation. Summing the weighted outputs with appropriate delays implements a transversal filter. Bandwidths of 10^8 Hz are achievable, but the center frequency is in the range 10^7–10^9 Hz, thus mixing may be needed to raise the signal frequency to this level and return to dc baseband afterwards.

10.4.2. *Optical Processing*

Information may be communicated, processed, and stored optically, though (except for optical communication) the technology is still embryonic. Programmable optical processing is not available. Communication using fiber optics is invaluable in a factory environment because of its immunity to electromagnetic interference.

Optical processors offer very high processing rates, since they comprise in effect thousands or even millions of channels operating in parallel. Incoherent optical processors use the intensity of light to encode signal amplitude. An example of an incoherent optical processor is the instrument using Walsh/Hadamard transforms to identify silhouette shapes described elsewhere.[26] Coherent optical processors[(27)] in which the signal is encoded as a complex quantity by the amplitude and phase of monochromatic light offer an alternative methodology capable of more sophisticated processing operations.

Coherent light may be used in precision measurement for determining the profile of a surface using an interferometer. But the most important use of coherent optical processing in automation is to exploit the property that a lens generates in its image plane a distribution of light whose variation in phase and amplitude is the

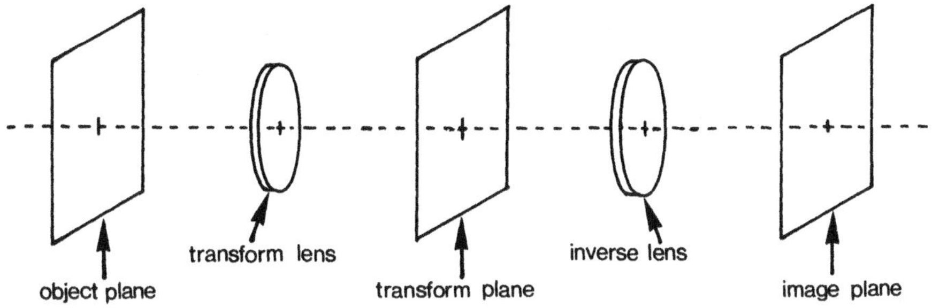

Figure 10.12. Coherent optical processor. Masks used (for example) to identify by finding a best correlation fit are placed in the transform plane.

two-dimensional Fourier transform of the complex distribution of light in the conjugate object plane. Operations which in the space (or time) domains are very complicated, such as convolution and correlation, reduce in the Fourier plane (see Section 2.4.5) to simple multiplication. Thus, an unknown pattern may be recognized by correlating it with masks representing possible matches, and assigning to the one which gives greatest transmission. A second lens is normally provided to perform an inverse transformation back to the spatial domain (Figure 10.12).

Coherent optical processors have been used (in a laboratory environment) for many real-time signal processing applications, such as spectrum analysis, with spectacular success. But optical paths correct to one-quarter wavelength (perhaps 0.15 μ) over the complete field of view are required, and dust and vibration can cause unacceptable disturbance. Thus, they are currently too expensive and too fragile for use on-line in industry. However, some of the principles are proving useful. The far-field pattern generated by a laser scanner is the Fourier transform of the distribution of phase shift (due to surface roughness), and absorption (assuming the spot has uniform intensity) over the region illuminated at any time by the spot. Thus, a scratch parallel to the x-axis appears in the far field as a line parallel to the y-axis; this has been used to enhance defect contrast and yield some recognition in a Swedish system for inspecting steel strips.[28] When illuminated by a coherent beam such as that generated by a laser, a periodic structure such as a textile diffracts the light as a grating would; the diffraction pattern is sensitive to changes in the structure such as defects, and may be used in automated inspection.

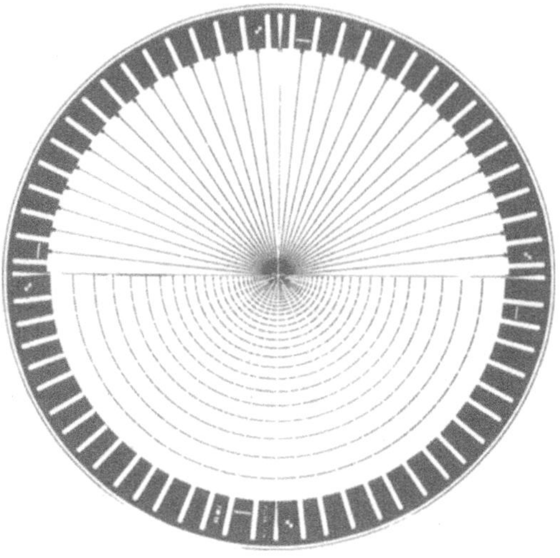

Figure 10.13. Solid-state sensor for radial and tangential distribution of light. From "Automated Visual Inspection by Optical Computing" by H. L. Kasdan and D. C. Mead in the PROCEEDINGS OF THE THIRD INTERNATIONAL CONFERENCE ON AUTOMATED INSPECTION AND PRODUCT CONTROL, Nottingham, April 1978. Reprinted by kind permission of IFS (Pubs.) Ltd.

Another kind of coherent optical system has been used on-line for inspecting the tips of needles.(29) This uses a special light sensing chip (Figure 10.13) which has two sets of sensors, one radial, the other circumferential. The pattern of light scattered when a laser beam strikes a needle point is changed if the point is defective, and this change is easily sensed using the special chip.

10.5. System Configuration

The configuration of a system to perform a required task on-line is as follows. The most difficult problems in intelligent automation systems are generally mechanical, concerned with actuation and so on; information processing is generally less troublesome and several alternative approaches are normally available. It is often best to select the mechanical configuration first, then to choose a processing methodology to be compatible with this. At the front end, we have to choose the sensor configuration. If an image is being sensed, this involves selection of a sensor type (e.g., solid-state, laser scanner, vidicon tube), a lens system, and an illuminator configuration. Detailed descriptions of the possibilities available together with information regarding their operation are provided in Chapters 4 and 5. The methodology selected for processing (and validated during the off-line stage of the investigation) should be designed to accomplish the following:

1. Do as much as possible at the sensor end using well configured optics, etc. to reduce the quantity and sophistication required of the electronic processing. It is, for example, better to get good contrast using good lighting rather than computationally, using a complicated masking operation.
2. Eliminate as much unnecessary data as possible, as early in the processing as possible, by using a preprocessor. This may well be analog, with digitization immediately following. Techniques such as filtering, used to increase the contrast to message signals with respect to accompanying noise, are often best implemented using analog processing.
3. Operate hierarchically, with the least expensive (albeit less powerful) processing applied first. For example, if defects are to be identified on strip material, then the first stage in the hierarchy is to establish that a defect of some kind is actually present. Only data known to arise from a defect is allowed to proceed to the more expensive processing needed for identification.
4. Avoid unnecessary hardware such as framestore, displays, etc., and not hold information in store that can be handled sequentially.

The electronic processing should use a standard bus system (VMEBUS, QBUS/UNIBUS, or MULTIBUS), with inexpensive possibilities such as the 8-bit S100 bus or STDBUS for simple systems. Multiprocessor operation should be used to spread the load if a single processor cannot cope.

References

1. G. Van Boven, P. Wambacq, A. Oosterlinck, and H. Van den Berghe, Industrial image computer systems: a comparative study, *Proc. SPIE,* vol. 435, 229–242 (1983).
2. W.B. Heginbotham, *An Assessment of Proprietary Vision Systems in Relation to Practical Production Engineering Applications,* Proc. 2nd European Conf. on Automated Manufacturing, Birminham, U.K. (1983), 21–36.
3. C. Loughlin and E. Hudson, Eye in hand robot vision, *Proc. 7th RoViSec,* 264–270 (1982).
4. M.S. Petty, Image analysis for product inspection at low cost and high speed, *Proc. 7th Conf. on Automated Inspection and Product Control,* 149–160 (1985).
5. L.A. Goshorn, Vision systems eye real-time speeds through multiprocessor architectures, *Electronics* (Dec. 15, 1983).

6. P.J. Gregory and C.J. Taylor, Knowledge based models for computer vision, *Proc. RoViSec 4,* 325–330 (1984).
7. B.G. Batchelor, Interactive image analysis as a prototyping tool for industrial inspection, *IEE Proc. Comps. and Digital Techniques 2*(2), (1979).
8. L. Norton-Wayne, V. Popovici, and W.J. Hill, Acquisition System for Surface Inspection Data, Report No. DSS/LNW/VP/WJH 105, Dept. of Systems Science, The City University, London (1976).
9. *Microprocessors and Microsystems 6*(9), Special Issue on Backplane Buses (Nov. 1982).
10. L. Teschler, Single-board computers pack more punch, *Machine Design*, 94–99 (April 26, 1984).
11. W.K. Taylor and G. Ero, Real time teaching and recognition system for robot vision, *The Industrial Robot,* 99–106 (June 1980).
12. S.Y. Kung and J. Annevelink, VLSI design for massively parallel signal processors, *Microprocessors and Microsystems* 7(10), (Dec. 1983).
13. F.A. Gerritsen and L.G. Aardema, Design and use of DIP-1: A fast, flexible, and dynamically microprogrammable pipelined image processor, *Pattern Recognition 14*(1–6), 319–330 (1981).
14. R.M. Lougheed and D.L. McCubbrey, The Cytocomputer: A practical pipelined image processor, *Proc. 7th Annual Symposium on Computer Architecture* (IEEE 80CH1494-4C), 271–277 (1980).
15. B. Kruse, The PICAP picture processing laboratory, *Proc. 3rd IJCPR,* 875–881 (1976).
16. S.R. Sternberg, Parallel processing in machine vision, *Robotica 2,* 33–40 (1984).
17. M.J.B. Duff, Parallel processors for digital image processing, in: *Advances in Digital Image Processing,* P. Stuki (ed.), 265–276, Plenum, New York (1979).
18. L. Norton-Wayne, CLIP offers superfast image processing, *Sensor Review 4*(3), 133–135 (July 1984).
19. T.J. Fountain, The development of the CLIP7 image processing system, *Pattern Recognition Letters 1*(5,6), 331–339 (July 1983).
20. A.G. Corry, D.K. Arvind, G.L.S. Connolly, R.R. Korya, and I.N. Parker, Image processing with VLSI, *Microprocessors and Microsystems,* 7(10), 482–486 (1983).
21. I. Aleksander, W.V. Thomas, and P.A. Bowden, WISARD: A radical step forward in image recognition, *Sensor Review 4*(3), 120–124 (1984).
22. J.D.E. Beynon and D.R. Lamb, *Charge Coupled Devices and their Applications,* McGraw-Hill, New York (1980).
23. M.B.N. Butler, S.A.W. Devices for Image Processing, Proc. IEE Intl. Seminar, Case Studies in Advanced Signal Processing, *IEE Conf. Pub. 180,* 21–33 (1979).
24. *Optical Data Processing—Applications,* D. Casasent (ed.), Springer-Verlag, Berlin (1978).
25. U. Sjolin, Optical Inspection of Metal Surfaces, Laser Focus, 67–70 (July 1972).
26. H.L. Kasdan and D.C. Mead, Automatic visual inspection by optical computing, *Proc. 3rd Intl. Conf. on Automated Inspection and Product Control,* 73–80, Nottingham, U.K. (1978).

11

Case Studies

11.1. Introduction

In this chapter we describe some projects conducted to produce instrumentation for intelligent automation. These examples, taken from the authors' own experiences, illustrate the principles expounded in the previous chapters, and show how they are applied to practical problems. Sections 11.2–11.5 present various projects in increasing order of difficulty.

11.2. Location and Identification of Silhouette Shapes

The first project we will discuss was undertaken in order to produce an instrument to identify and position the leather shapes from which shoe uppers are assembled. The investigation, funded by the British United Shoe Machinery Company (BUSM) (whose Research Department had no previous experience in machine vision or pattern recognition), is the subject of patents specified in Section 3.5.8.

When this project was started in earnest (1978), general purpose systems for recognizing silhouette shapes were just beginning to appear, and it was necessary to establish that no existing system could be adapted to do the job before committing resources to a new research investigation. This consideration initiated the activity in performance benchmarking which has been described previously. BUSM also had to select an outside organization to perform the initial work that had the required capability and was likely to give value for money. Twelve were considered (and visited); two were eventually chosen to receive contracts. The most reputable (and best qualified) organizations proved to be unacceptably expensive, and both contracts were given to universities that, although not able to deliver fully engineered instrumentation (even in prototype form) suitable for immediate trials on-line, could develop and evaluate methodology. The contractors were in fact providing know-how (some of which they would develop in the course of the investigation), rather than actual instrumentation.

The project was divided into phases; no subsequent phase was initiated until the previous one had been successfully completed. The initial phase was a preliminary investigation to determine methodology to be used and to establish that the project was feasible. Simulation and mathematical analysis were used in selecting the processing strategy; this took 6 months. Results were embodied in a report. The next phase (of 1-year duration) comprised implementation of hardware on a general purpose rig; its outcome was a demonstration that showed that the method worked, albeit very slowly. A third phase (again, of 1-year duration) involved the production of a hardware preprocessor that sped up processing sufficiently to enable the handling of a data sample large enough to yield statistically meaningful results; these results were presented and analyzed in yet another report. The first three phases were performed at the City University; subsequent work continued in the research department of BUSM.

Shoe upper components are stamped from leather and other flexible materials; they are of highly irregular shapes (examples are shown in Figure 11.1), and internal holes may be ignored. They range in size from about 1 inch^2 to perhaps 10×12 inches, and all items within this range must be accommodated by a single instrument without operator adjustment. Left- and right-hand parts must be distinguished, as must similar components scaled in size by 2% for successive half-sizes of shoes. The location of a datum point in the shape must be measured to 0.005 inch in both x- and y-directions, and orientation measured to $\pm 1°$. The instrument must store at least 5,000 shapes at any one time, and must be self-trainable to recognize new shapes by an operative without specialized computer expertize, and without requiring the rescanning of shapes already stored. No more than 1 shape in 1000 may be rejected, and no more than 1 in 10,000 may be substituted. The throughput required is four shapes per second.

Despite the apparently severe specifications for performance, this task was not particularly difficult; the main challenge was to solve it cost-effectively. The problem was simplified because shoe components can be regarded as being isolated silhouettes, and are easy to sense when illuminated from behind. Also, the pattern recognition required

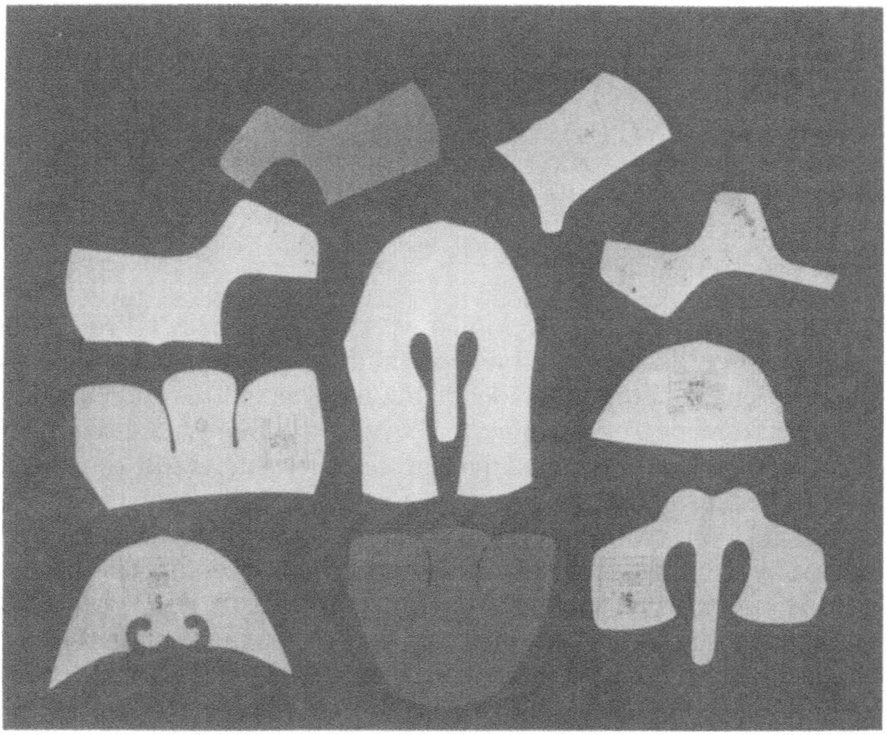

Figure 11.1. Examples of shoe upper components.

is essentially deterministic; all patterns from a particular class are identical and there is no within-class variation. Thus, the properties of each class can be learned from examination of only one sample.

The first task was to select a recognition methodology. Important here was a requirement to distinguish all possible shapes without constraint, as long as scaled similar shapes differed in size by at least 2%. Thus, some approaches can be eliminated immediately, such as measurement of parameters like perimeter^2/area. Pairs of shapes which are distinct yet have the same value for this ratio are quite likely to occur (e.g., shapes B5 and B5X in the benchmark set in Figure. 8.17), and mirror image shapes must be distinguished. It was clear that only a method that was asymptotic in the sense of Pavlidis could be used, since such methods guarantee to distinguish shapes (however close they may be) merely by using increased resolution. This eliminates

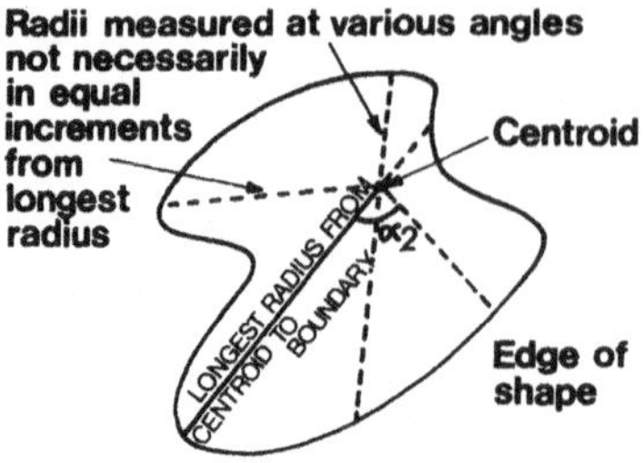

Figure 11.2. Silhouette shape with feature radii superimposed.

an unnecessary risk in developing what will eventually be an on-line system. Also, the approach must permit measurement of location and orientation, in addition to identifying the shape. There is, however, no requirement to determine that shapes are free of defects; this is not inspection. It is assumed that any defective shapes will be rejected as not being sufficiently close to a stored class.

The most promising method appeared to be to describe each shape by the lengths of radii measured from the centroid of the area to the boundary (Figure 11.2). The centroid is a good datum for location since it is unique, quick and easy to compute, and insensitive to quantization errors at the boundary (See Section 8.5.2). The longest radius is used as datum for orientation and for specifying the other radii. Area is also used as a feature, since it is stable and easy to compute. Mathematical analysis was introduced at this stage, to demonstrate that although the longest radius is not unique, cases in which ambiguity proves a nuisance are likely to be acceptably rare.

Analysis using computer simulation was then used to confirm the stability of these (and other) possible feature measures. Ellipses of varying eccentricity and cardioids were synthesized, and properties such as area measured by counting squares and compared with the result obtained from calculation. Figure 11.3 shows the error in measuring the perimeter and the area of an ellipse quantized onto a square grid, as the major axis is rotated through 360°. It is evident that area is a very stable measure and therefore a good potential feature; perimeter is unstable. Although several methods have been published for correcting errors in perimeter measurement due to quantization effects, these complicate and slow the processing, and are best avoided.

Having established the utility of a sufficient number of measures as features, it was necessary to provide a method for identifying patterns using sets of these features. Since the problem is deterministic,

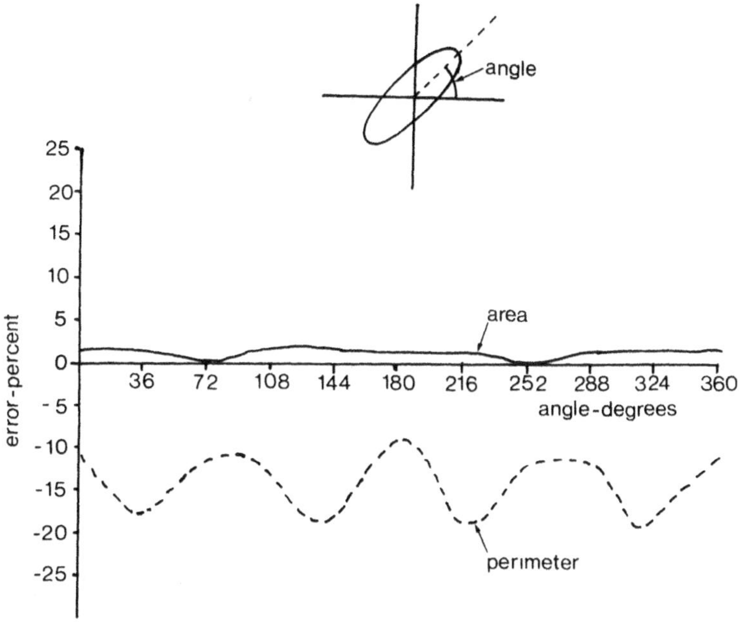

Figure 11.3. Error in measuring perimeter and area of an ellipse versus orientation.

and each class may be represented by a single point in feature space (i.e., by a single set of numbers representing feature measurements), the "nearest-neighbor" approach was a possibility. Suppose 20 features are used for each shape and 5000 patterns are stored. If so, then 100,000 squares, 5000 sums, and 5000 comparisons must be performed to identify each unknown shape. Because of errors in measuring features, a perfect match will rarely occur; a distance threshold must thus also be computed, and patterns falling outside this rejected. Also, selecting the feature set from the candidates is horrendous using this option.

Thus, the novel approach based on coding theory described in Section 3.5.8 was developed. On an average, only 2500 comparisons between feature vectors are required to identify each unknown shape, and comparison of features involves only a "same or different" check between data bytes that is simple and quick at assembler level. Most important, parameters can be adjusted and traded off against one another to optimize performance.

At this point, it became possible to consider the hardware configuration. As sensor, a high-resolution (2048-sample) CCD linescan camera was chosen, with the sample placed on a transparent moving surface and illuminated from behind using a fluorescent striplight.

For efficiency, it was necessary to ensure that only an essential minimum of data be stored and processed. The properties of a silhouette shape are specified completely by its boundaries, which can be stored very efficiently (as explained in Section 2.2.2). Thus, preprocessing was used, comprising extraction of the boundary points with each point being specified by its x and y coordinates. This information is necessary (and sufficient, once the centroid is located) to compute the features.

To validate this configuration, the general purpose vision rig described in Section 10.2.2 and illustrated in Figures 10.3 and 10.4 was programmed to perform the classification, with processing implemented by Fortran subroutines. Some images of shapes with radii superimposed obtained with this are shown in Figure 11.4. This confirmed that the approach proposed would actually work, but since it took 20 minutes to process each shape, only 7 shapes were considered.

To provide a demonstration convincing enough to justify the next phase, i.e., development of special purpose hardware by BUSM's Research Department, it was necessary to process and store at least 100 shapes, including some expected to prove difficult. The processing time was limited by the slow speed of the interface to the computer (which could accept digitized data at only 140,000 bytes/second), whereas the CCD array camera could scan (according to the manufacturer) 10,000,000 samples/sec. Thus, a high-speed preprocessor was constructed. This convolved a binary mask 3 pixels^2 with the data supplied by the camera after digitization at any clock speed up to 10 MHz, extracted and stored boundary points encountered during a scan at this speed, and fed them to the computer. This reduced the time to scan and process each shape to only 5 minutes, and made it possible to handle a database of 100 shapes. It proved to be impossible to operate the CCD camera above 5 MHz, since odd/even noise became excessive above this speed and no simple method was available for correction.

To establish that the number of features required to identify shapes was not likely to become excessive as the number of patterns stored approached the design objective of 5000, the number of features required versus number of patterns stored was measured; Figure 11.5 shows the result. This experiment of course *proves* nothing, but the

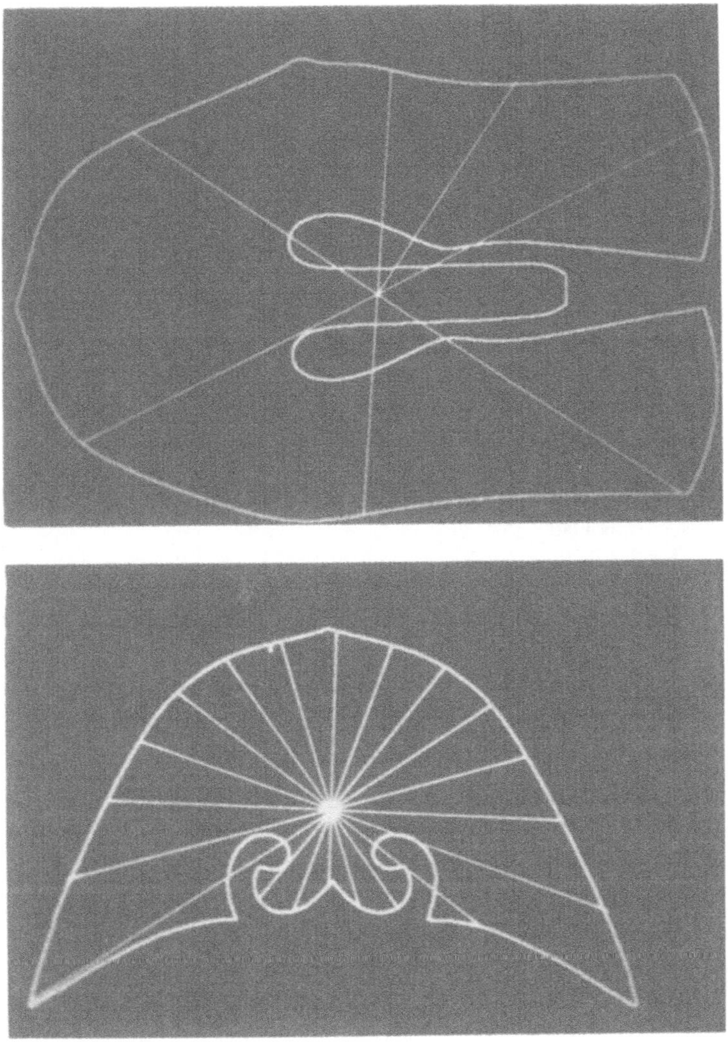

Figure 11.4. Shapes with radii superimposed as measured by rig.

very slow increase in number of features observed creates confidence that no problem will arise.

Another evaluation now possible was estimation (by extrapolation) of the rejection and substitution rates. Given a target rejection rate of 1 in 1,000, even a trial with 1,000 shapes (from the 100 classes stored), will not give a quantitative assurance that performance ap-

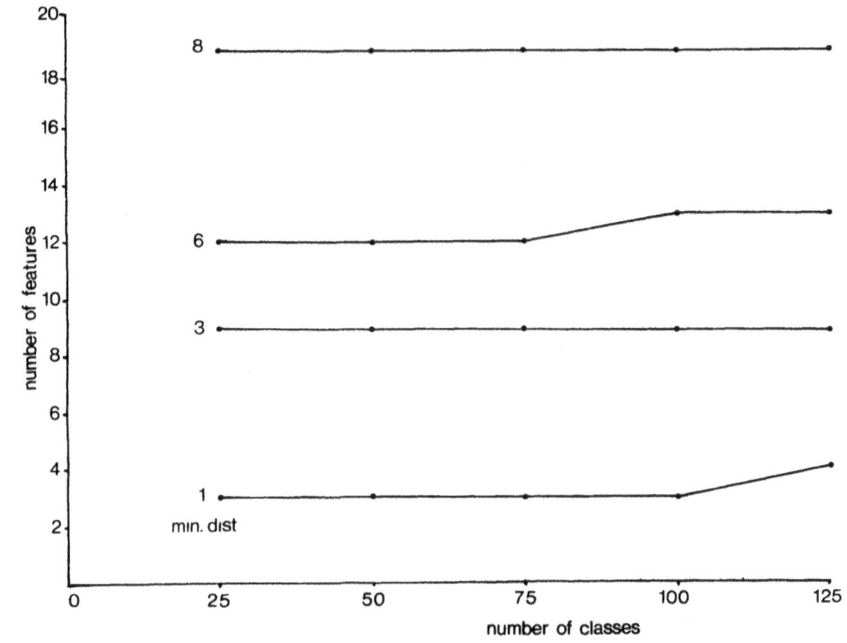

Figure 11.5. Number of features required versus number of pattern classes, for coding classifier used for identifying silhouette shapes.

proaches adequacy. Therefore, an indirect approach was used. The system will reject if more than a specified number of features mismatch; provided that errors in measuring the various features are statistically independent, the probability of this happening can be estimated from measurements which are practicable. Figure 11.6 shows the results of measurements of the frequency with which different numbers of features are out of tolerance; the frequencies that would be observed under the assumption of complete independence (from a binomial distribution) are superimposed, and agreement is seen to be good. The sum of the heights of bars above the threshold representing the number of features mismatched for rejection, divided by the sum of heights of all bars, gives the probability of rejection.

At this point it was felt that performance was good enough to justify continuation, and the investigation was transferred to BUSM's own plant. Here a new rig was constructed, designed specifically for this investigation and constructed soundly. It included a 2048-element

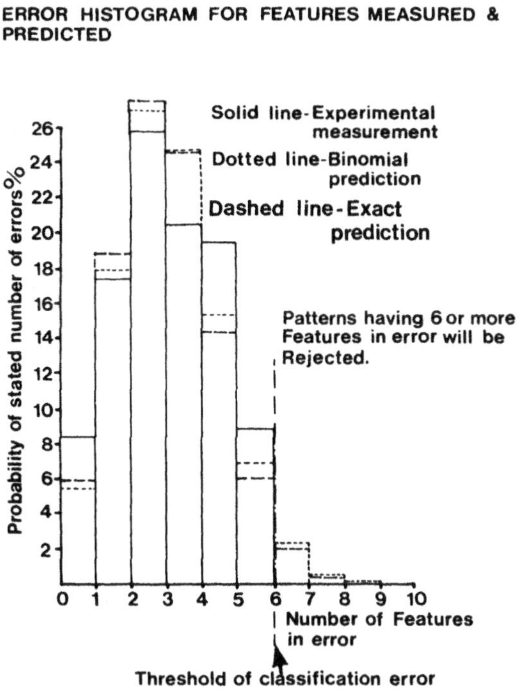

Figure 11.6. Distribution of errors in measuring features.

commercially-supplied CCD linescan camera, a redesigned preprocessor, and a 16-bit (Z8000) microprocessor to perform computations. This special hardware was interfaced to a Perkin-Elmer computer with a disk backing store, to hold the several thousand patterns for the duration, and to permit some of the investigation to proceed using high-level language software for flexibility.

The new processor was able to process a shape in 2 sec; use of the disk store in the P-E computer prevents the target of four shapes per second being achieved. Four and a half thousand shapes were successfully stored. The low rejection rate and almost undetectable substitution rates were confirmed. The feature matching was modified so that only prototypes with roughly the same area were considered in making a match; this considerably limits the number of comparisons which must be made. The longest radius was found to be inadequate as a datum (it is often nonunique), and other measures were added to give stability. With the 2,000-point resolution available, most camera lenses tried had too much distortion; a suitable lens was found only after a long search. The system was now ready for application.

11.3. Tack Inspection

Next, let us discuss a project to produce an instrument to inspect shoe tacks[1,2]. These are producted at 250,000 per minute by BUSM, by squeezing wire in a die. Their unit value is insignificant. Defects introduced during manufacture such as feathered ends (illustrated in Figure 11.7), bending, incorrect length, etc. will cause batches (which may amount to several tons) of material to be rejected by customers,

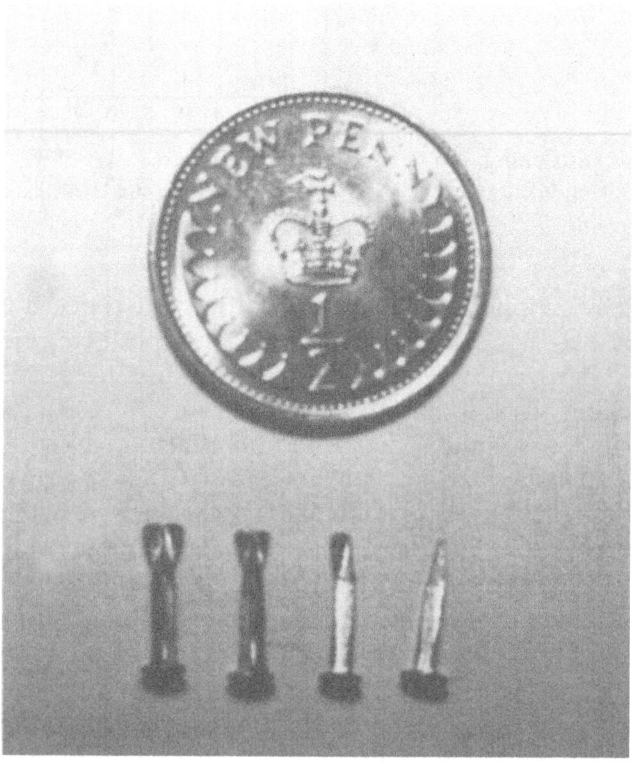

Figure 11.7. Tack with a good point (extreme right), with a single feather (second from right), with two small feathers (second from left), and with two large feathers (extreme left).

since they may cause the machinery that automatically inserts the tacks to jam. The visual inspection by human operatives currently used is expensive and may be applied only to a small sample of total product. It would be preferable to spend research money to prevent the production of defective material in the first place, but the causes are so obscure that improvement of inspection by introducing automation is worthwhile as an interim measure.

Since feathers are the most troublesome defect, initial efforts were focused on instrumentation intended solely to detect these. Several methods for sensing the feathers were considered, such as weight, magnetic reluctance, and profile as determined by machine vision. Measurement of weight was soon rejected, since the variation of weight resulting from the 15% variation in overall length allowed as a manufacturing tolerance is much greater than the weight increase caused by the feathers. Further, it would be difficult to measure the weights of individual tacks presented in a continuous stream. In fact, presentation of tacks to the sensor is much more difficult than actually sensing the feathers. Study of the profile of the tack with and without feathers (Figure 11.8) showed that even when the feathers appear at their most unfavorable orientation, the change in signal produced should be much larger than any noise. Accordingly, the simple optical sensor shown in Figure 11.9 was devised, in which the tacks are dropped between a slit of light and a photodiode. The decrease in the light

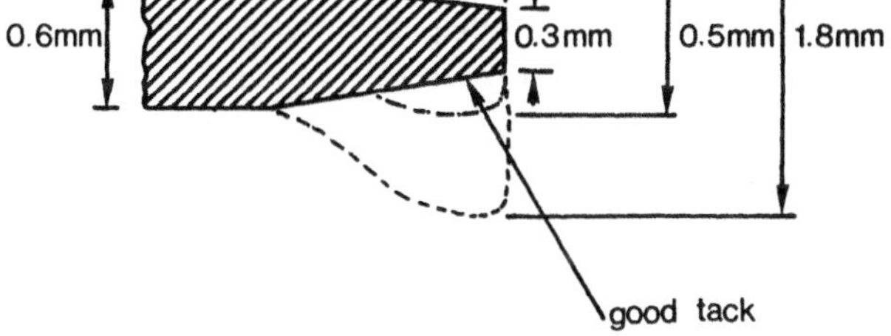

Figure 11.8. Profile of a tack point.

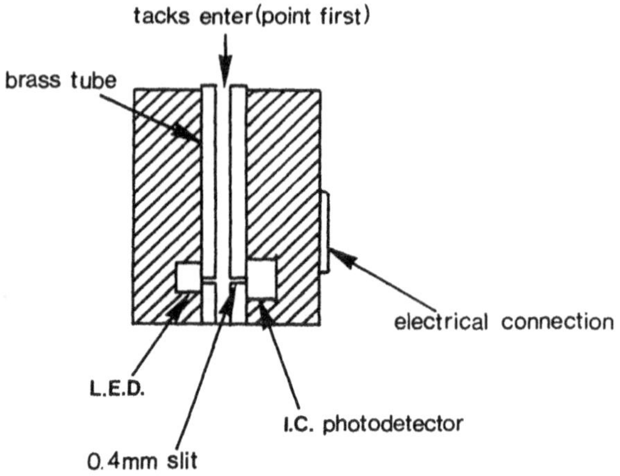

Figure 11.9. Drop-through sensor.

transmitted is proportional to the width of the tack; the point of the tack starts a time check, at the end of which the tack width is checked again. The time elapsing corresponds to a fall through a distance of about 1 mm; if the measured width exceeds a threshold at this distance from the tack point, a feather is indicated. This approach works irrespective of the length of the tack. Bench trials showed that this detected feathers well; the next stage was to see what would happen if tacks were fed automatically. This is a "drop through" method, in which tacks are fed from a bowl feeder, which incidentally ensures that the tacks always fall point first; unfortunately, tacks tended to jam when falling past the sensor slit, and it was not possible to cure this. The maximum time obtained between jams (half an hour) was too small, and the approach had to be abandoned. This is a pity, since the cost per installation for sensing is only about £150, which is low enough to contemplate installing a sensor on each insertion machine. The low throughput of 800 tacks/min is acceptable for screening the product before delivery since many systems can be used together, in parallel. Tacks indicated as being bad are blown out of the stream by a jet of air, initiated by a relay driven by the detector. It is impossible to make this short enough to avoid also rejecting a few good tacks, but operationally this does not matter; the objective is to remove every single bad tack.

An alternative "drop through" method using magnetic sensing

was also investigated.[3] This involved dropping the tack through a pot core which formed part of a tuned circuit. Passage of a tack decreased the reluctance of the air gap, causing the frequency of oscillation to fall in proportion to the decrease in reluctance (and hence to the quantity of material in the gap). An FM discriminator was used to convert the variation in oscillation frequency to a signature showing the variaιion of material within the gap with time. Figure 11.10 shows two such signatures, obtained from tacks with and without feathers; the bump due to the feathers is clearly visible. The signature was sampled, digitized, and processed by a Horizon microcomputer, and a considerable quantity of statistical material obtained to determine the best position on the signature for setting a threshold, and to establish the performance available from the system quantitatively (Figure 11.11). The magnetic sensor is cylindrically symmetrical and hence the form of the signature is unaffected by change in tack orientation. The system will inspect ferromagnetic tacks only, but the fraction of nonferromagnetic (brass) tacks manufactured is small. Unfortunately, it is still a "drop through" method, and hence cannot be used because of jamming.

Clearly, it is necessary to use a method of presenting the tacks that will not cause jamming; the raceway approach, shown in Figure 11.12, was most promising; it is a simple modification of the then

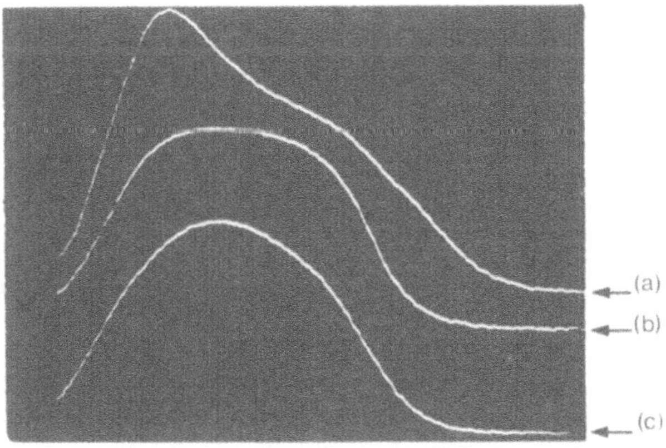

Figure 11.10. Signatures from an electromagnetic sensor. Trace A shows the signature from a good tack; trace B shows the signature from a tack with one feather (see Figure 11.7.), and trace C is from a tack with two feathers.

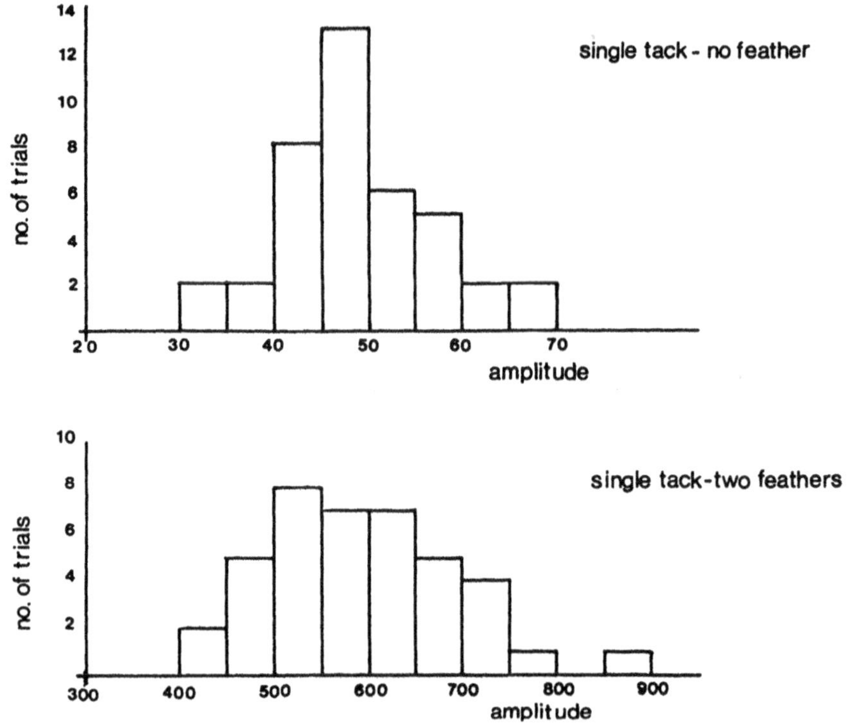

Figure 11.11. Histogram showing performance for tack detector described in ref. 11.3. Comparison of the amplitude scales shows that the distributions are well separated, indicating excellent potential classification performance.

existing feed, with the troublesome drop eliminated. However, a different method of sensing is now required that can tolerate variation in the speed of tack movement.

The solution here is to pass the tacks in front of a low-resolution CCD linescan camera. A packaged Fairchild I-SCAN system with 256 elements was used, with a bellows mounting for the lens to facilitate the close focusing required. This measures the silhouette of the tack as it passes. The silhouette is obtained by counting the number of pixels occluded by the tack. This computation is performed by a simple hardware preprocessor comprised of standard logic components. Thus, a simple microcomputer (a PET was used in the initial development) suffices to control the instrument. More than 98% of feathered tacks were detectable. Why the remaining 2% did not succumb is still un-

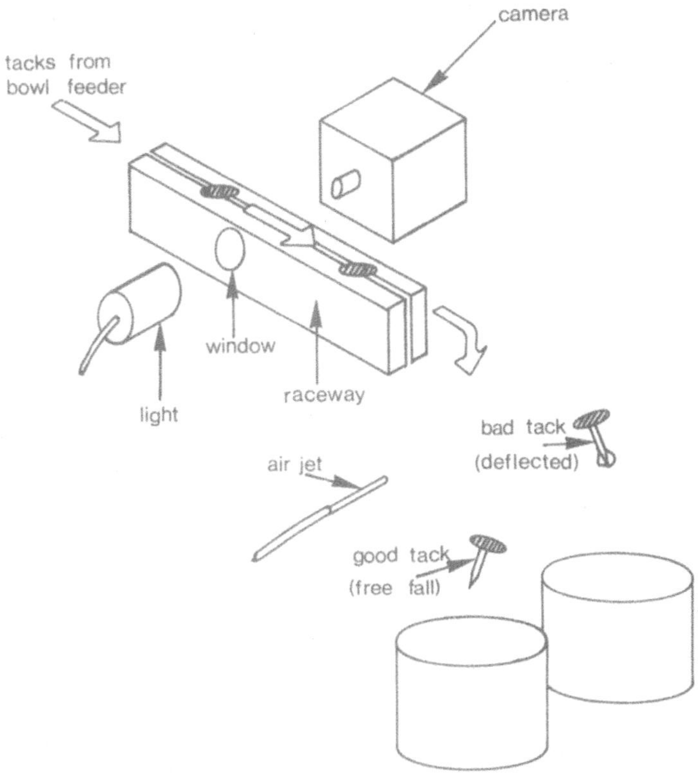

Figure 11.12. Raceway system for tack inspection.

known. Slow-motion cinephotography suggests that this is due to tacks rocking as they pass the sensing point, but other causes (such as length variation within the tolerance allowed for the tacks) cannot be excluded. This configuration is now being perfected by an instrument manufacturer (Airmatic Ltd. of Sileby, Leicester) for use on-line and for manufacture in batch quantities.

It is frequently necessary to measure the performance of the tack inspection instrumentation experimentally. This involves examining several thousands of tacks to ensure results which are reliable statistically, since the error rates required are a fraction of 1%. One method is to seed the batch inspected with a known number of defective tacks and count the number rejected; if, however, this batch already contains some defective tacks, the test may yield an optimistic result. Another method used is to examine a sample containing several thousand tacks, of which an unknown number are defective, using the instrumentation repeatedly. Suppose the first trial yields x bad tacks, which are removed from the batch, and y bad tacks are then revealed by a second trial immediately afterwards. When the system is working more or less correctly, y will be much less than x, and we can conclude that the probability of missed detection (per bad tack) is then y/x. This approach avoids the need to sort the tacks by hand (likely for aesthetic reasons to be no more accurate than the inspection instrument), but (like the first test) reveals only missed detection due to random errors, such as random orientation and oscillation of the tacks as they pass the camera. Missed detections due to systematic errors (such as tacks which are too long being missed by the camera) are not revealed, and must be detected by modifying the sensing to expose specific suspected causes. False alarm rejection of good tacks is measured by counting the number of good tacks deposited in the reject bin, bearing in mind that their cause may lie in the system which deflects bad tacks from the stream, rather than in the sensing device. The total number of tacks in the sample is determined either by weighing or by using the instrumentation to count the tacks directly.

Early in the project, difficulties were experienced in making the raceway transport system reliable, and before these were overcome, a further configuration had been considered which eliminates these completely. This is illustrated in Figure 11.13; the tacks are scattered

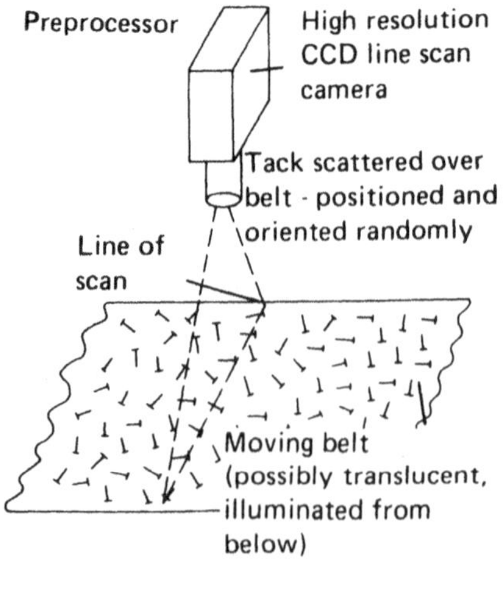

Figure 11.13. Arrangement for inspecting tacks scattered on a moving belt.

randomly over a moving belt (hardware for this is used in the existing visual inspection) and viewed from above by a linescan camera having a high resolution, at least 2000 points. Diffuse illumination from below, provided possibly by using a white belt with a rough surface, ensures that the tacks can be sensed as silhouettes. A preprocessor is then used to "window out" the tacks, so that each tack is presented in a window 32 pixels^2, and this is processed by a fast processor to detect feathers, etc. If more than one tack is present at any time in the viewing window, recognition is frustrated. Though it is impossible to guarantee that this will not occur, it is possible to control the mean density of the tacks to ensure that (with random positioning) the probability of interference is kept to a level which is acceptably low (in which case, both interfering tacks are rejected, causing some waste). The intention was to use the commercial version of the CLIP4 processor for this; it is very fast, and some preliminary experiments (Figure 11.14 shows the skeleton of a tack obtained using the prototype CLIP4 at University College) suggest it can do the job. The chief difficulty with this approach is its cost; the preprocessor indicated in Figure 11.15 would have to be developed specially to perform the windowing, and a

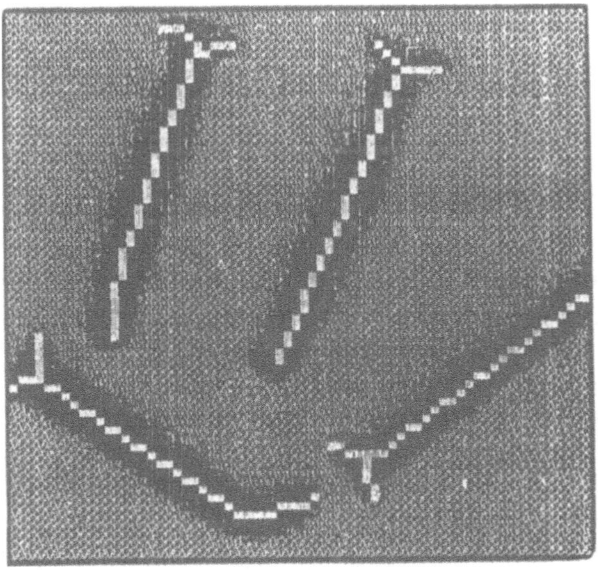

Figure 11.14. Tack skeletons obtained using CLIP4.

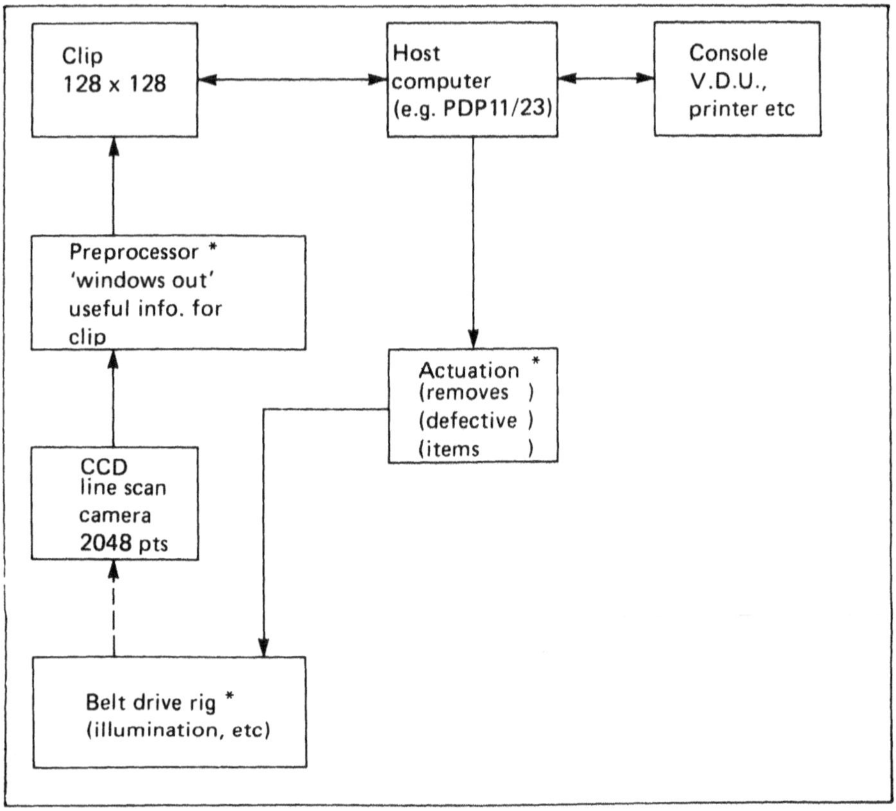

Figure 11.15. Processor system for inspecting tacks on a belt.

commercial CLIP with a 32-pixel square costs about £30,000, which would make the complete instrument extremely expensive! Also, means must be developed for removing defective tacks from the belt. Hence, it was decided to persevere with the raceway system.

11.4. Inspection of Bare Printed Circuit Boards

The inspection of printed circuit boards for visually perceivable defects is currently performed using human operatives; the operation is costly and inefficient since the task is so boring that the inspectors cannot concentrate for very long. This is alleviated by allowing them to rest at intervals. Purely electrical inspection of printed circuit boards is inadequate since it merely determines whether electrical paths are correct at the time of the inspection; latent defects (which will cause trouble later, possibly after components have been inserted into the board and the whole system has been delivered to the end user) are not detected. Some examples of the defects that must be detected are shown in Figure 11.16.

Thus, a need exists for instrumentation to automate the visual inspection of the boards. The goal of the project to be described was

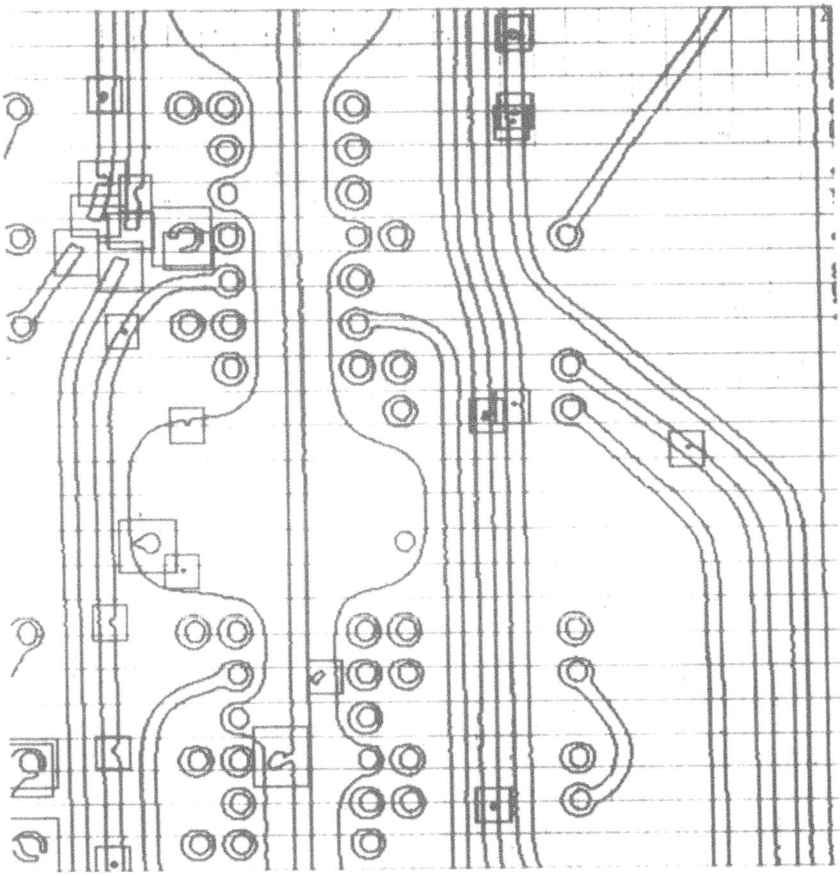

Figure 11.16. Section of a printed wiring board with defects. Squares have been inserted by computer processing around each defect. A CCD linescan camera was used as sensor; the roughness in the track boundaries is due to spatial quantization error.

to produce a general purpose instrument able to handle the widest possible range of tasks and to be self-trained by scanning a perfect prototype. This project began as an unsponsored academic study[4,5]. Discussion with potential users suggested the following specifications:

Size of board—5 × 5 cm–50 × 50 cm
Resolution—50 μ
Inspection time—20 sec per board (will vary according to board size)
Trainability—Can learn any desired track pattern in about 1 min by scanning a perfect specimen
Performance—100% of defects must be detected, though possibly with some "false alarms" to be corrected later by a human inspector
Indication—VDU or paper output showing nature and location of defects
Cost—£30,000 per instrument, in batch production at 1981 prices

The chief difficulty in inspecting pc boards is the high resolutions required in sensing because many significant defects are very small. Thus, the quantity of information to be sensed and processed is very large. With a board 10 cm^2 and a 50 μ resolution, 4,000,000 pixels are required to specify the board completely. The obvious approach, to have two scanners operating in synchronism, one examining the inspected sample, the other a perfect prototype and signaling a fault when the two disagree, is undesirable because of the high cost of two carefully synchronized high-quality camera systems, and because sample and prototype boards may well differ slightly due, for example, to slight differences in etching that do not constitute a fault. The Philips company is, however, known to have produced such a system for use in-house. Thus, a configuration having only one scanner, with the characteristics necessary for the inspection stored inside the system, was adopted.

Some pc board inspection systems have been investigated that eliminate any requirement to store the prototype pattern by, for example, using the sequence shrinkage–expansion (to original size) comparison with unprocessed pattern to detect small defects such as hairline breaks and whisker bridges. This processing does not detect gross defects such as portions of missing track or pad, and thus is not acceptable.

Since defects are effectively modifications to the track pattern, only data close to the track boundaries need be examined in great detail. This decreases the processing requirement considerably. The approach adopted exploits this property. It uses two methods simultaneously. The first detects large defects; the board is divided into 2-mm squares, and the area of track and length of track boundary within each square, learned by scanning a perfect prototype, is stored in the inspection system. The second detects small defects; it comprises a filter that is moved along the track boundaries to detect defects such as hairline cracks. The two methods operating together should detect significant defects of all sizes.

The chief task in the research investigation was to devise a suitable form of filter, and to demonstrate that it worked effectively. Operationally it is necessary to detect all defects (albeit with some false alarms); this is the "inverse Neymann-Pearson" approach discussed in Section 2.6. An initial form of the instrument would probably generate so many such indications that a human operative would still be required, but since the system would draw attention directly to areas of putative defect, the inspection task would be much quicker and less boring, and accordingly more efficient. Later developments would include additional processing, more complex and more powerful than the basic edge filter, but needing to process only data indicated by this filter as "probably" containing a defect, and therefore operationally acceptable.

Two forms of edge filter (shown in Figure 11.17) were tried as follows:

1. The sum S of seven consecutive segments of contour code is computed within a "window" that is moved along the track boundary. Given that the nth contour component $C(n)$ is the sum modulo 8 of the nth and $(n + 1)$th Freeman code segments $F(n)$ and $F(n + 1)$, then

$$S = \sum_{n=1}^{n=7} C(\mathrm{n})$$

2. The distance $D1$ measured along the boundary between pairs of points that are not adjacent is compared with their separation $D2$ as measured directly, and a threshold is applied to their

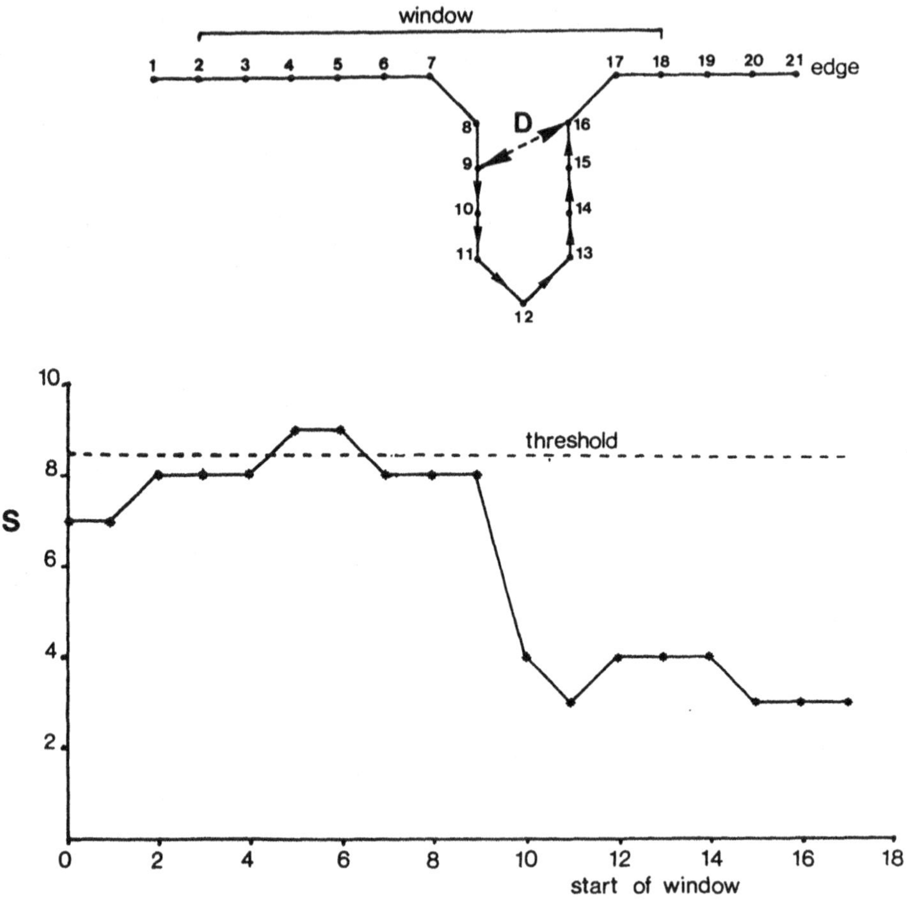

Figure 11.17. Operation of filters for the detection of defects in track boundaries. The lower picture shows how the sum S defined above for method 1 varies as the window shown in the upper picture is moved along the track boundary. When S exceeds a threshold, a defect is indicated. The upper picture also shows how the distance D measured directly between points 9 and 16 is much less than that measured along the boundary (method 2).

difference ($D1 - D2$). The distance $D1$ along the boundary is calculated by summing the number of edge-points between the two sample points, and weighing horizontal and vertical links (see Fig. 2.4) with 1.0 and diagonal links with 1.414. The direct distance is calculated from the coordinates of $D1$ and $D2$ using Pythagoras' theorem. Figure 11.18 shows a receiver

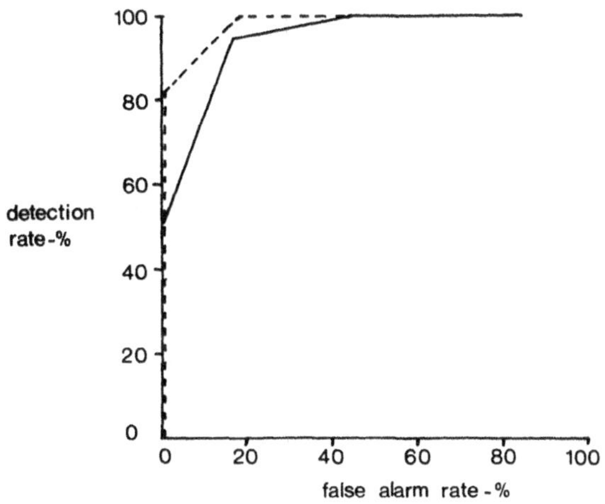

Figure 11.18. Receiver operating characteristic comparing filters for detecting defects in printed circuit track boundaries. The solid line is for the type "1" filter; the broken line for type "2." The superiority of the type "2" filter is evident after reference to Figure 2.16 (b).

> operating characteristic (ROC) in which the performance of the type (1) filter (solid line) is compared with that for the type (2) filter (dotted line). The type (2) filter is clearly superior, though it is costlier to compute.

The chief difficulty with the instrument is in producing the hardware within allowable cost, since the mounting and transport of the board must be done with high precision (±2/1000 inch). Covering the full width of the board in a single swathe would require a camera of very high resolution, or several more modest cameras. The project is now being examined by commercial organizations, following 7 years of intensive investigation at the City University in London.

11.5. Steel Strip Inspection

Next, we describe a project to produce an instrument for the automated visual inspection of cold-rolled steel strip. This task is particularly difficult since the data processing rate required is extremely high, the contrast of many defects is small, and the properties of the various defects exhibit wide statistical variations. However, since a vast investment is required to produce a modern high-speed steel strip production line incorporating automatic control, justified by the throughput obtained, it is necessary to provide compatible means for inspecting the strip, and to bear the cost of developing suitable instrumentation. The on-line inspection problem exists for steel producers worldwide, and collaboration within the European Coal and Steel Community provided partial support for this project, and enabled the results of investigators in several European countries to be shared.

Cold-rolled steel strip is produced on highly automated lines at only a few locations worldwide. It is used for a variety of purposes, each of which requires different quality specifications. Currently, production capacity exceeds demand, and selling is highly competitive, with customers able to exert pressure to ensure high quality.

Steel to be used for the outer parts of car bodies is required to have a surface which is completely defect-free; however, the surface is generated by a roller made rough by shotblasting to give the surface sufficient texture for paint to stick, and for the steel to flow in the press. Steel for the inside of cars may have defects which spoil appearance but do not degrade structural strength. The requirement for other uses, such as oil drum manufacture, is even less stringent. The poorest quality strip has to be scrapped and remelted.

Defects are caused by malfunctions at various stages in the manufacturing process; it is important therefore to identify defects by type so that the cause can be identified and corrected. The steel starts as an ingot that is rolled into a slab about 10 inches thick. Bubbles in the ingot give rise to laminations (thin detached wafers of steel) in the strip. The ingot is then hot rolled until it is about 0.1 inch thick; hot rolling produces a greater reduction in thickness per pass than cold rolling, but yields a poor surface finish. The surface is in fact covered by oxide scale (which may get rolled into the surface, causing yet another defect), and the final stage in hot rolling is to remove this by pickling in an acid bath. Inadequate control of this latter process causes rust spots.

To restore the metallurgical properties of the strip, the coils after hot rolling are annealed (a process taking several days); slight variations in the properties of the steel cause the laps of strip to stick together; upon unrolling they must be wrenched apart, causing a particularly insidious defect termed sticker wrench. This seems to comprise local variation in surface texture in lines across the coil

Figure 11.19. Sticker wrench photograph. The sticker lines appear bright against a dark background. Reproduced by kind permission of Dr. Campbell Watts of the British Steel Corp. Welsh Laboratory.

(Figure 11.19), which are of low contrast under most conditions of viewing, but stand out strongly in a few.

The strip is then cold rolled until its thickness is correct; a malfunction at this stage, e.g., two thicknesses of material passing through the rollers, damages the roller surface, leaving an imprint ("roll mark") on subsequent material. A final skin pass reduces the thickness by only 0.001 inch, but tempers the surface and imprints a shot-blasted texture.

Inspection for surface defects is a final and indispensible part of the manufacturing process. This has been performed visually, but becomes ineffective when the strip speed exceeds 1.5 m/sec. The introduction of on-line automatic control to the rolling process has raised linespeeds to 25 m/sec, and it is desirable to perform the final inspection at this speed; the only solution is an automatic system using machine vision.

The first stage in this (as in all) problem was to define the task. Following discussion with the Welsh Laboratory of the British Steel Corporation, it emerged as follows.

The objective of the inspection was to detect, delineate (i.e., determine the extent of), and identify defects visible on the surface of the strip that were larger than 1 mm in diameter. At least 95% of defect occurrences were to be detected, and 85% were to be correctly identified. Further, not more than 0.2% of good material could be wasted

E ROLL MARKS

Type		Cause
16A	Tail mark and corner mark	Bad tail end of coil passing through mill - sometimes occurring on break caused by bad edges or welds etc.
16B	Draw-out	Strip squeezing out of rolls at tail end, with rolls hard down - (sometimes nips the T/end)
16C	Bruise	Proud inclusion in or on strip contacting roll
16D	High Spot	Indentation in roll due to edge scrap
16E	Pinch Mark	Extra thickness of strip due to crimping or doubling on passing through rolls leaves corresponding mark on the roll
16F	Drag Mark	Caused by rolls making contact with each other during a roll
16G	Fracture Mark & Firecrack	Crazing of roll due to stresses in the roll
16H	Bite Mark	Usually caused when entering strip into rolls which sometimes snatch the leading edge and drags small bits of metal onto the roll
16J	Fen Mark	Usually caused by rolled in side scrap from pickle line
16K	Chain Mark	Line of high spots-equi-distant, around circumference of roll
16L	Shot Blast Marks	Marked rolls leaving roll shop

Figure 11.20. Schedule showing some defects occurring on steel strip. Reproduced by kind permission of the British Steel Corp. Welsh Laboratory.

due to false alarms. The material could be as wide as 1.5 m moving at 25 m/sec. A schedule was prepared, naming 35 types of defects to be considered and describing their properties. Some examples from this are shown in Figure 11.20.

The work was divided among three organizations. The British

Description and/or Diagram	Size Range	S, A or D	Remarks
	Tail mark - any length Corner mark - several cm	A	
	Up to full width	A	
Very light shiny area 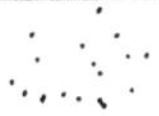	2-4 cm	A, S	
Small spots in relief on strip surface 	2-3 mm	D	
	Any length usually in 20-40 cm range. 1-2 cm wide	A	
	5 mm x 20 cm approx.		
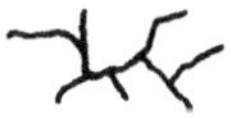 Thin marks proud of strip surface	Up to several cm normally but can be larger	S & D	Does not occur very often
	Any size		
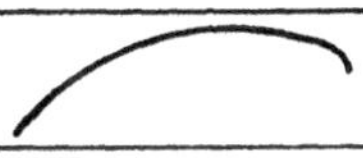	Any length		
	3 mm spots	S, D	
Small pits	Less than 1 mm	A, S	

Figure 11.20. (*Continued*)

Steel Corporation Welsh Laboratory provided sample data and coordinated the project overall. The SIRA Institute (now SIRA Ltd.) provided instrumentation for scanning, and converted the sample data into computer compatible form; it would eventually engineer, build, and install a prototype system. The Department of Systems Science at The City University, London, provided and evaluated methods for signal processing capable of implementing the inspection. Alternative approaches were investigated and comparison analysis of performance provided, so that the end user (British Steel) could exercise commercial and technical judgement and select the best configuration to be built and operated on-line. An analysis of likely cost savings amortized over a reasonable period indicated that a cost per installation of about £400,000 (in 1974 values) would be acceptable.

The ways in which a surface defect can interact with incident light[6] are shown in Figure 11.21; all possibilities of absorption, reflection, and scattering can occur, providing independent information; it is desirable that all are sensed. Some defects give greatest contrast when viewed with specularly reflected light, whereas others are best sensed just off-specular, or a long way off. Thus, a laser scanner was selected to interrogate the surface. Its advantages in this application lie in high operating speed and good signal-to-noise ratio. Further, a design (SIRA type 1500, Figure 11.22) was available which could sense

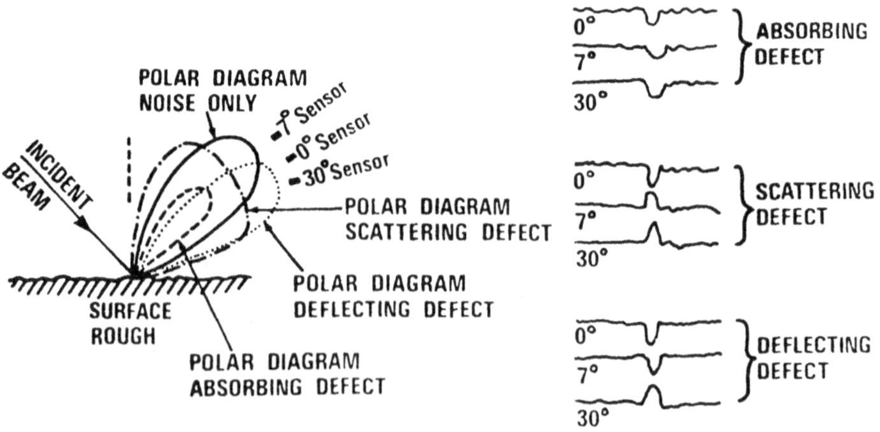

Figure 11.21. Interaction with incident light of various kinds of defect on a rough surface.

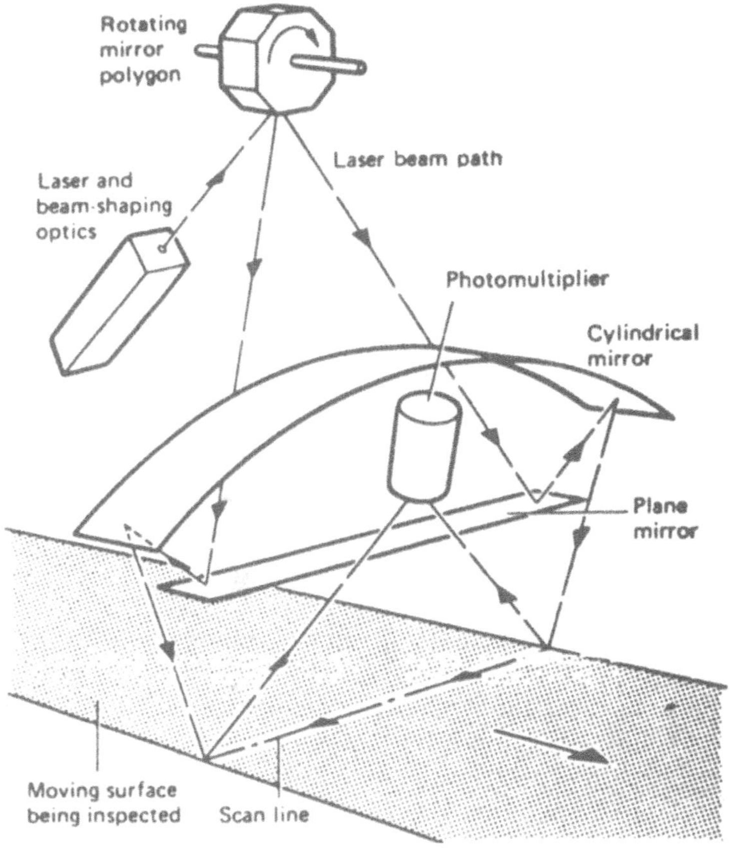

Figure 11.22. SIRA type 1500 laser scanner.

at three angles (specular, 10 degrees off-specular and 30 degrees off-specular) simultaneously, generating a vector signal (Figure 11.23), and this was used. This uses a 5-mw Helium–neon laser deflected by a 12-sided mirror drum rotated on air bearings at 24,000 rpm and can move a 1 mm-diameter spot over the surface perpendicular to the direction of strip movement at 1.5 m in 200 μsec. Since three photomultiplier tubes and associated circuitry were needed to sense at the three angles, the device was complex and expensive. The monochromaticity of the laser scanner (which throws color information away) does not seem to be a disadvantage, but the coherence of the beam adds noise due to laser speckle, and hence reduces defect con-

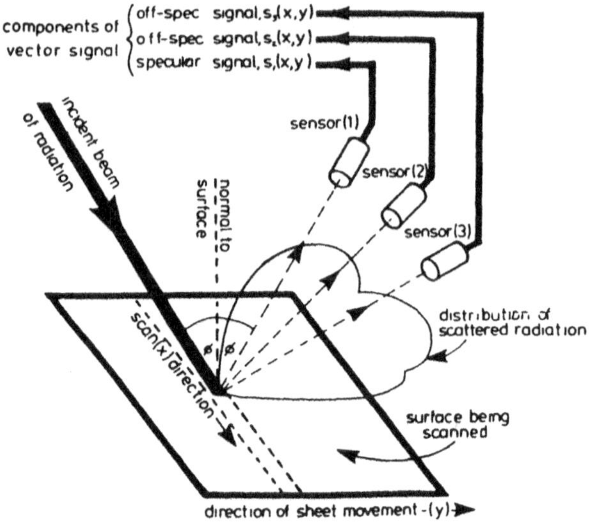

Figure 11.23. Vector signal as obtained from a multisensor arrangement.

trast. The rms amplitude of the noise was roughly doubled due to the speckle.

The only competitive alternative would have been a high-resolution CCD linescan camera, but doubt existed as to whether sufficient light could be projected onto a line of surface to give adequate contrast at the required operating speed. Also, three synchronized cameras would have been needed to sense simultaneously at the three angles. Rasterscan cameras would require a shutter to freeze the motion of the sheet, and again, need three instruments for the three views; an expensive version such as a Chalnicon would be needed to obtain adequate sensitivity.

Selection of the signal processing methodology is critical because information must be processed at upwards of 30 Mbits/sec in each of three channels. To enable alternatives to be evaluated quantitatively and compared objectively as quickly as possible and with minimum cost, it was decided to investigate this aspect using computer simulation. A CDC7600 computer at the University of London Computer Center was used; although a batchmode machine with programs submitted on punched cards, this was the fastest scientific computer then available to the research community. A microfilm plotter was available that offered fast, high-resolution (16,000 plotted points in each direction), hardcopy graphical output, and proved indispensible. The NAG and other libraries of subroutines for implementing important algorithms, such as Fast Fourier transformation and matrix manipulations, were an added attraction. The procedure was to develop a suite of mutually compatible processing subroutines written in Fortran, apply these in various combinations to sample data obtained using a laser scanner examining real sample material, and display the results on microfilm to assess the effectiveness of the processing. This approach has the additional merit that, once gathered, the data does not (unlike the sheet steel it was obtained from) deteriorate if reasonable precautions are taken to protect the database against, for example, accident erasure.

About 470 sheets of cold-rolled strip containing samples of 35 types of defects were gathered by British Steel at its plants at Port Talbot and Llanwern, as shown in Table 11.1. Originally it had been intended to gather 50 samples of defects showing wide "within class" variation and 10 samples of those showing little "within class" variation, obtained from different lines at widely varying times, to include the full statistical spread of property. However, some defects occur so rarely that during the 2-year period occupied by this stage of the project, this was not possible. The sheets were 30 cm^2 for ease of handling, and were coated with palm oil and housed individually in plastic envelopes to prevent rust.

The sheets were scanned at the SIRA Institute using a fully automatic rig controlled by a PDP11 minicomputer. A previous investigation on tinplate inspection had shown that the data acquisition task was so tedious that automation was essential, or the data would be useless due to operator error. The scanning process for sample data acquisition comprises the following stages:

1. The sheet is placed on a table which can be advanced by a stepper motor under computer control. Sheet number, defect type, etc. are keyed in, and the acquisition process is initiated.
2. One line of signal from the specular sensor is sensed, digitized to 8 bits, and recorded on industry-standard magnetic tape together with the output from a simple threshold crossing detector for validation purposes.
3. Signal from 10° off-specular sensor is digitized and recorded.
4. Signal from 30° off-specular sensor is digitized and recorded.

Table 11.1
Data Base for Steel Strip Inspection

	Defect type	*Number of samples*
01	— Skin Lams	37
02	— Seams	7
03	— Slivers	13
04	— Fleck Scale	35
05	— Jet Scale	10
06	— Holes	3
07	— Hot Mill Scrp.	9
08	— Scale Pits	3
09	— Serr. Edge	3
10	— Steel Pits & Digs	4
11	— Scrap Mks.	15
12	— Sticker Wrench	25
13	— Point Wrench	9
14	— Oxid Edge	7
15	— Sand Pits	15
16A	— Tail & Cnr. Mk.	3
16C	— Bruise	10
16D	— High Spot	8
16E	— Pinch Mk.	17
16F	— Drag Mk.	4
16G	— Fract. Mk. & Fk.	1
16J	— Fewmark	2
17	— Chatter Mark	3
18	— Cold Mill Scr.	4
19	— Pick-Up	26
20	— Carbon P.V.	14
21	— Coil Digs	32
22	— Coil Break	9
23	— Edge Strain	21
24	— Indents	2
25	— Rust Spots	38
26	— Feather Mks.	5
27	— Water Stain	20
28	— Rolled-In Scale	1
29	— Brkr. R1 Scale	3
30	— Burnt Scale	3
32	— Pickle Line Digs.	0
33	— Frictional P.U.	0
34	— Grease and Dirt Pits	4

Table 11.1
(Continued)

	Defect type	Number of samples
35	— Rolled-In Scrap	3
36	— Dirty Surface	1
37	— Scuff Marks	8
38	— Oil Contam.	2
39	— Transfer Stain	3
42	— Herringbone	2
44	— Unknown	0
46	— Bar Scale	1
49	— Pickle Line Stain	1
53	— Snaky Edge	0
54	— Wiper Board Scratch	0
55	— Rolled in Scratches	1
57	— Surface Jam	1
61	— Skid Mark	1
62	— Alum. Lines	0
99	— Prime Sheet	10

5. The sheet is advanced by 1 mm, and steps 2–4 repeated.
6. Data gathering continues until 300 scans have been recorded, or until the edge of the sheet is sensed by a fall in reflectivity.

The next stage was to validate the data, i.e., to confirm that it has been stored correctly and that identification and orientation were correct. The data from each sheet comprised three files (representing different angles of view), each 300 × 700 8-bit samples, in addition to binary data. This is so vast that the only practical way to examine it was to plot it on microfilm to give a graphical display. The 8-digit sampled and quantized analog data was displayed using the perspective plot (Figure 11.24), in which each successive line of scan is plotted above and a little to the right of the previous one with hidden lines eliminated, giving an effect of relief corresponding to signal amplitude. The videoprint (Figure 11.25) shows binary data (output triggers from the threshold detector) plotted on a plan view of the sheet. The squares (inserted manually after scanning) show where the defect marks are

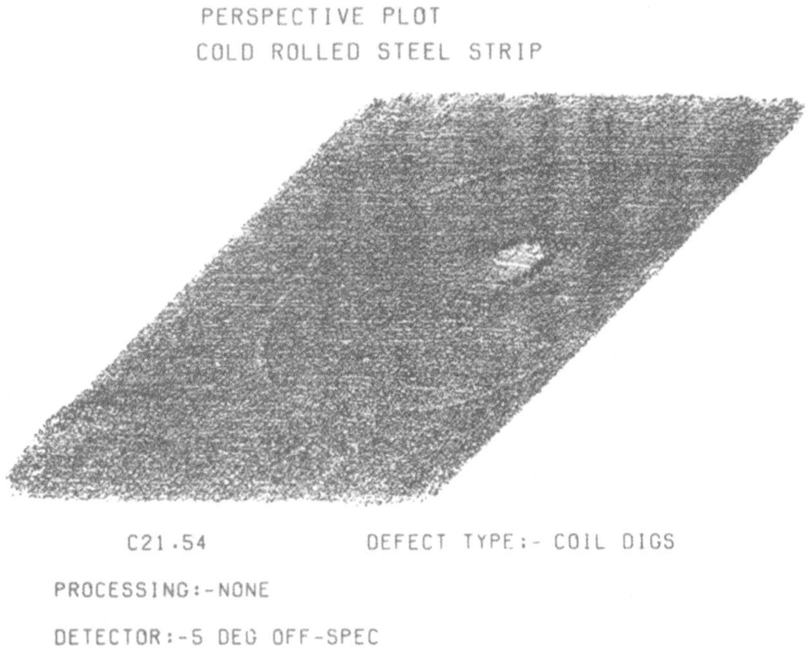

Figure 11.24. Perspective plot (computer onto microfilm).

known to be present; a good detector should therefore generate triggers inside all squares, but none outside. The serial number stamped on the top right-hand corner of each sheet should generate a cluster of triggers in this position, giving another check for correct orientation.

Two kinds of detector were tried.[7] The first (the SIRA detector) compared the signal with an attenuated, low-pass filtered version of itself (to eliminate the effect of slow variation in surface reflectivity); it responded to both positive and negative peaks in local reflectivity (Figure 11.26). The second (the "matched filter bank") detector passed the signal through a number of filters whose shapes corresponded to typical (triangular) signals of various sizes. If the output signal from any filter exceeded a (normalized) threshold, then a signal corresponding to the filter giving the largest response was indicated (Figure 11.27). They were about equally effective, but the Matched Filter Bank (MFB) detector gives more useful information about the extent of the defect. Both detected adequately all but six types of defects, namely, sticker wrench, point wrench, chatter marks, pinch marks, edge strain, and seams.

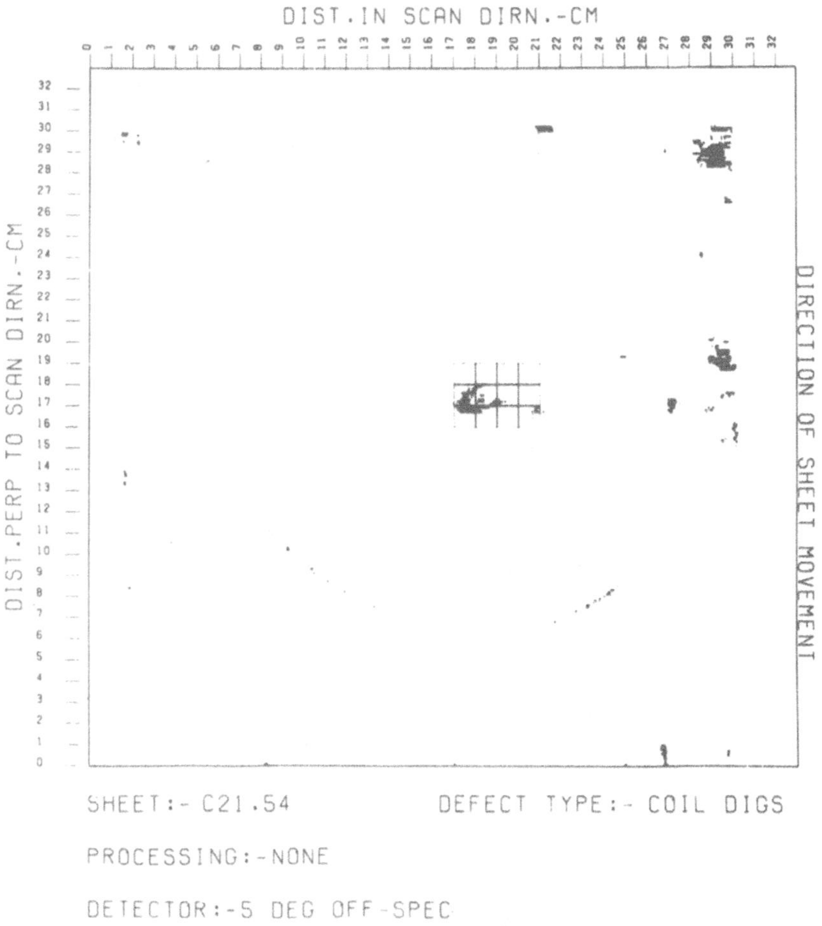

Figure 11.25. Videoprint, cold rolled steel strip (computer onto microfilm).

These "difficult" defects were dealt with as follows:

1. Point wrench—The detection threshold was lowered far enough to sense the defect; the excessive number of false alarm triggers were then eliminated using a two-dimensional binary filter (described in Section 2.6.3). The sequence of results is shown in Figure 11.31.
2. Pinch marks—Same solution as for point wrench.
3. Seams (a defect elongated along the rolling direction)—De-

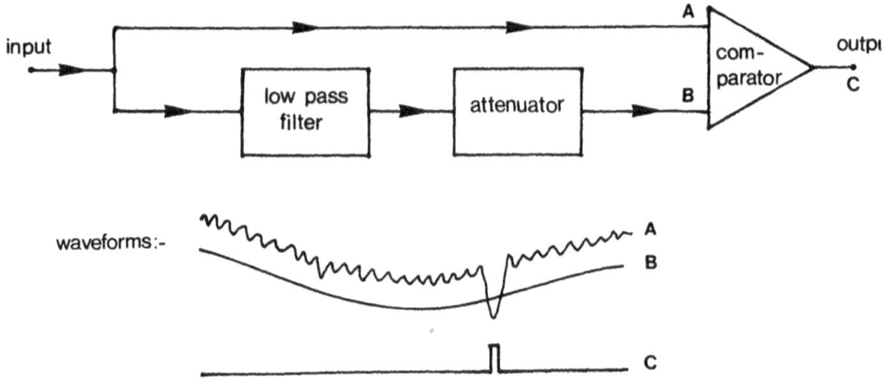

Figure 11.26. Detector adaptive to variation in signal level.

tected by increasing contrast using a two-dimensional matched filter, then using a normal SIRA detector to generate triggers.

4. Edge strain (that appears as lines TRANSVERSE TO the rolling direction)—Detected by turning the SIRA detector around to operate in the rolling direction instead of the normal scan direction.
5. Chatter marks (that consist of bands of low contrast, alternately dark and bright, running along the scan direction and with period 0.5 cm—Detected by averaging the amplitude of each scan (to reduce noise), then computing the Fourier transform of 256 successive averaged scans. The power spectrum showed a pronounced peak (see Figure 11.28) at 2 cycles/cm which is not present on sheets without chatter marks. This peak is detectable using a simple threshold.
6. Sticker wrench—Despite all efforts, there were not really detectable using the output signal from a laser scanner. The most successful of many methods tried was to treat the output signal from the laser scanner as a vector, and find the threshold (as a surface in a three-dimensional space) whose crossings most faithfully depicted a known pattern of sticker lines. A result is shown in Figure 11.29. This should be compared to Figure 11.30, which shows the videoprint obtained from a longitudinal SIRA detector operating on the signal produced by a CCD linescan camera. This uses monochromatic incoherent illumination so there is no noise due to speckle, and the contrast

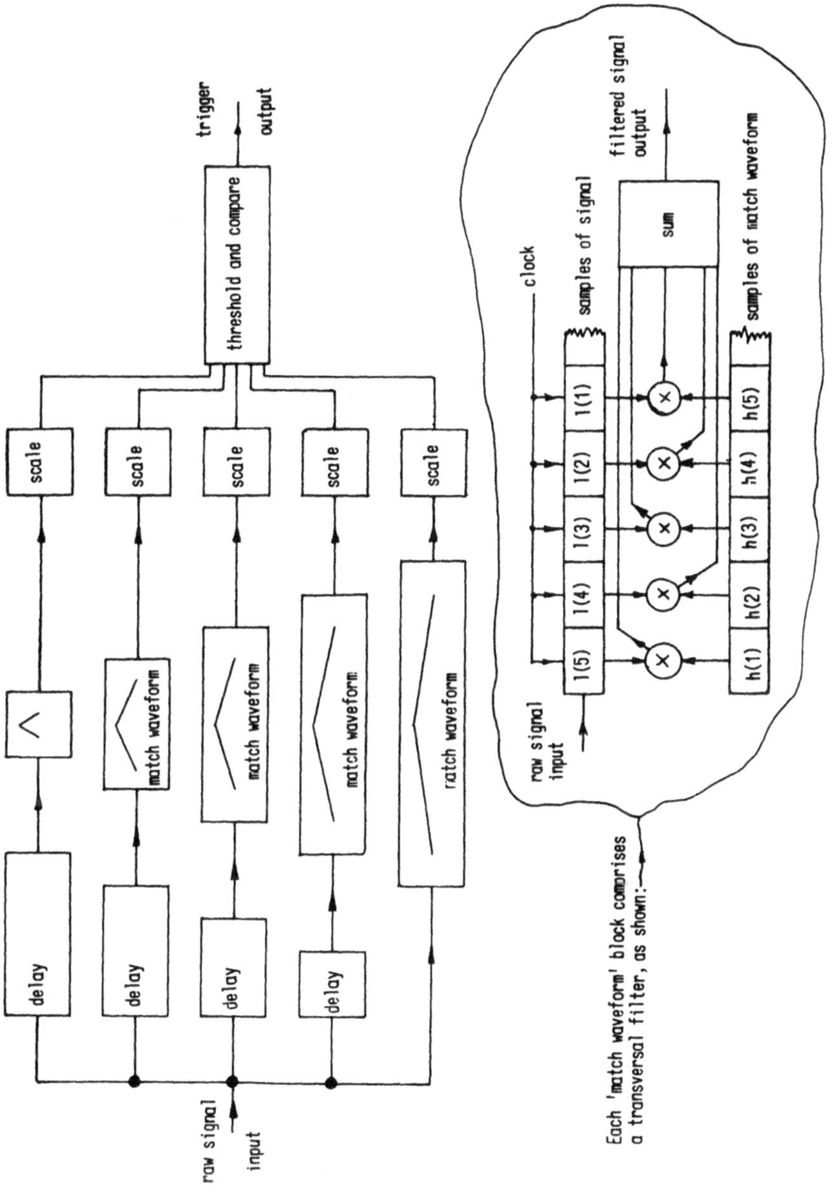

Figure 11.27. Matched filter detector.

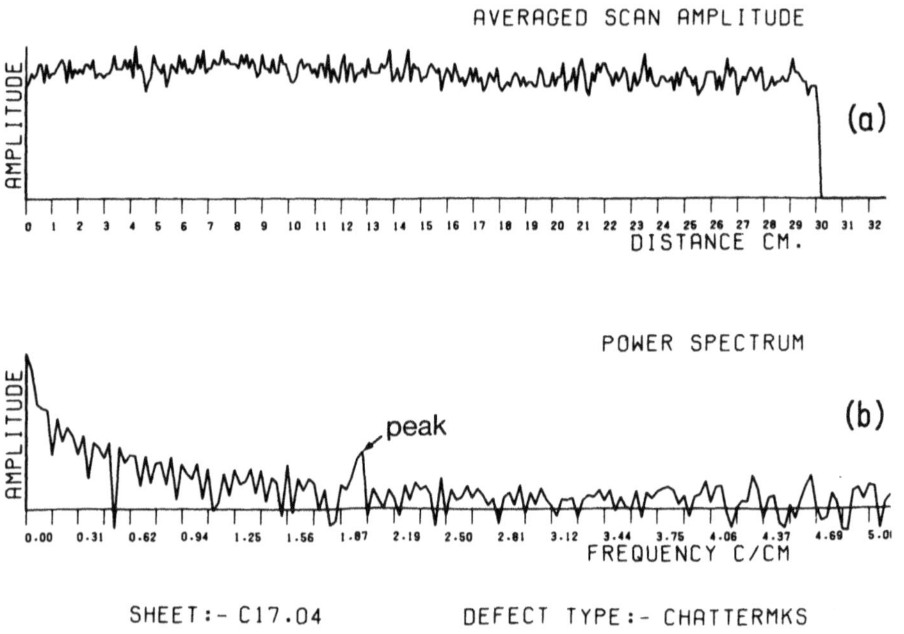

Figure 11.28. Detection of chatter marks using Fourier analysis. The plot of the averaged sample amplitude for each scan down the sheet in trace (*a*) does not show the periodicity indicative of the defect despite the improvement in message/noise ratio brought about by the averaging. On Fourier transformation, however, a peak at the frequency of the periodicity is evident. The contrast of the peak is considerable since the amplitude scale for the transform is logarithmic.

of the defect marks is about twice that for the laser scanner. Note that Figures 11.29 and 11.30 were both obtained from the same sheet of steel.

The detection processing exploited an analogy between the detection of defect signals in the midst of noise produced by surface roughness and the detection of target signals in the presence of noise in radar. Defense needs have justified extensive and profound analyses of detection for radar that are well represented in scientific publications, even in textbooks.

With the detection problem solved, it was possible to proceed to classification. But before defects can be classified, they must be delineated,[8] i.e., the boundaries of the defect must be determined. This defines the information which must be provided to the defect classifier;

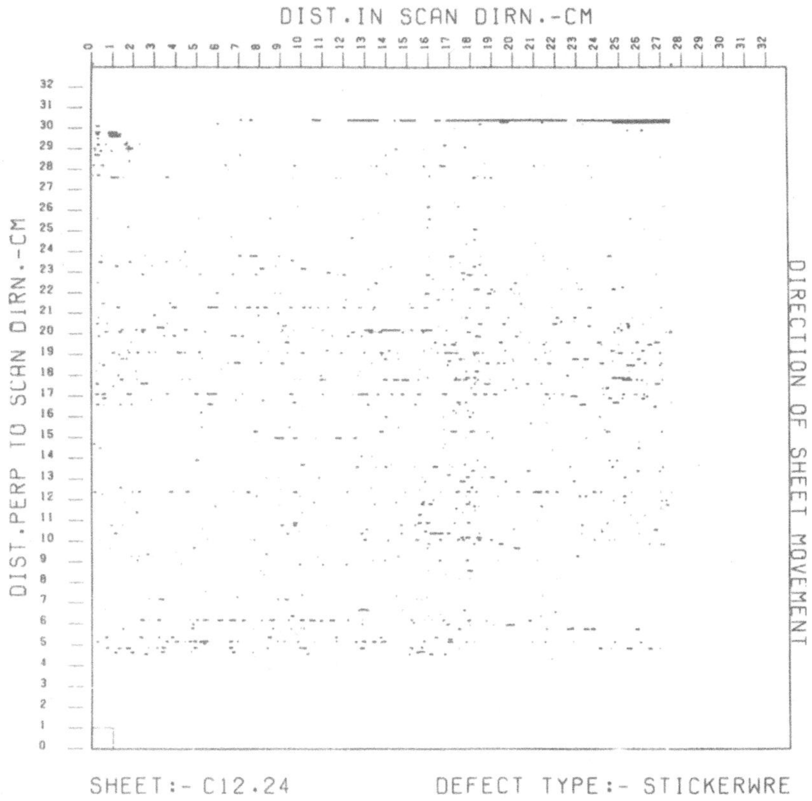

Figure 11.29. Videoprint from sticker wrench.

it is important also in ensuring efficiency in processing, since most of the surface of the steel is by hypothesis free of defects, and this can be ignored during recognition to keep the data rate down. The delineation adopted was a two-stage process. In the first, clusters of triggers were generated from each scan, using a MFB detector. In the second, triggers from successive scans considered by the processing to arise from the same defect were associated using a "capture rectangle" approach. This involved (in principle) including all triggers within a certain distance of an original trigger as belonging to the same defect.

The pattern recognition task in identifying defects is particularly difficult because of the large "within class" variation in defect appearance, so that defect characteristics are hard to define, and because

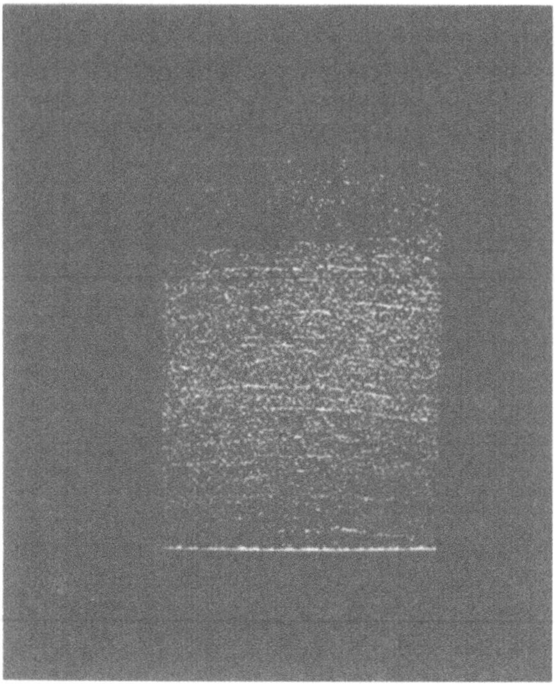

Figure 11.30. Sticker detected using a CCD camera and coherent illumination.

suitable candidate features are hard to identify intuitively. These far outweigh aspects in which the task is reasonably simple; for example, in that the number of classes to be recognized is reasonably small, and that self-training is not required, since it is not envisaged that any "new" types of defects will emerge during the lifetime of the instrumentation. The approach adopted was therefore as follows. Information useful for characterizing defects is contained both in single scans (in properties such as the area, perimeter, maximum width, and maximum depth of the waveforms due to defects, and in the plan view provided by many successive scans. The latter may be considered to comprise a silhouette as a first step. Many possible features are available, and the extent to which they are useful, or even mutually independent, must be determined computationally by simulation analysis. Thus, it was decided[(8)] to use feature space analysis for recognition, applied first to successive scans, then to combinations of scans. Kernel discriminant and linguistic classifiers were considered but eliminated owing to various shortcomings, and a linear classifier was selected as the best approach.

Performance is clearly not adequate, but in view of the difficulty of the task, this is not surprising; much work remains to be done. The repertoire of approaches available is far from exhausted. Instrumentation benefitting from this work is already operational, for detecting pinholes in the strip.

11.6. Glass Tubes on a Mesh Belt

11.6.1. Scanning

The goal of the project was to find the location of glass tubes after they had been carried through an annealing oven on a mesh belt so that they could be removed by a robot.[10] The tubes were U-shaped with an exhaust tube, for evacuating, as shown in Figure 11.31. The photograph shows that the contrast between the glass and the belt, under normal lighting conditions, is very poor. Other existing problems included the high temperature, about 60°C, above the belt and the presence of broken tubes on the belt. Tubes could break during the annealing process. All tubes that were complete in the U-section were to be removed, including those that had lost their exhaust tubes. The latter would be placed apart from the complete tubes for repair. Broken glass could fall from the end of the belt.

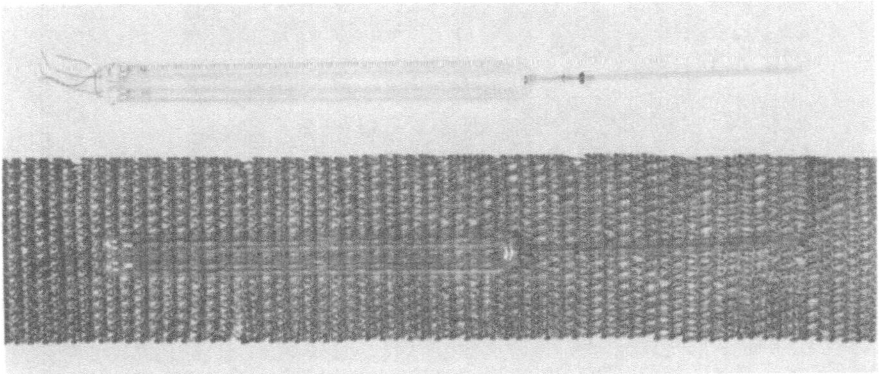

Figure 11.31. The photograph shows the high patterning of the belt and the low contrast of the tube in normal lighting.

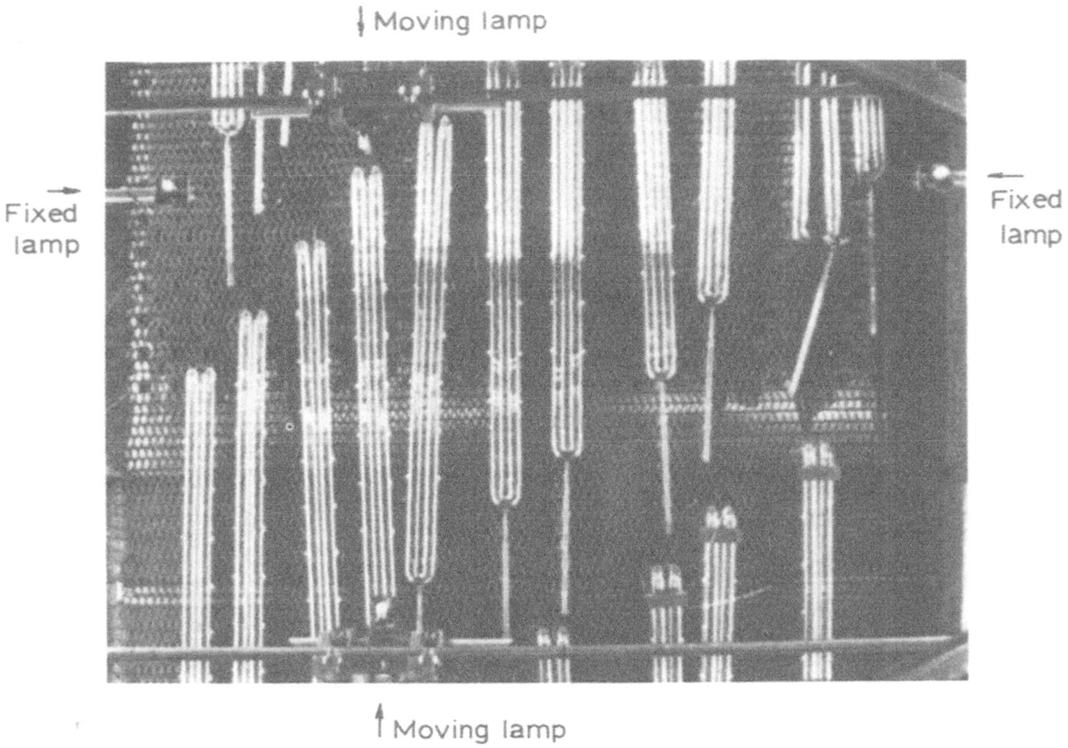

Figure 11.32. A frame from the film of the photographic camera showing the four marker lamps indicating the location found for a tube. One tube in the picture is broken.

The tubes were placed on the belt by hand in a staggered pattern as shown in Figure 11.32, with their center lines within 10° of the run of the belt. The tubes were made in batches, not mixed, in four sizes, the longer tubes having the larger of two diameters. In terms of the scanning pitch, the diameters were 13 and 17 pixels and lengths of the U-sections were from 170 to 570 pixels. The center lines of the legs of the U were 19 or 24 pixels apart near the bend but could be up to 2 more or 4 less at the other end. On the return run of the belt below the conveyor, it was supported on rollers that produced shiny flat strips on the belt. The velocity of the belt corresponded to 15 pixels/sec.

Since motion, provided by the belt, was continuous, the use of television cameras was inappropriate. Flash lighting would have been required to freeze the motion. Also, several cameras would have been needed to give the required resolution. A 1024-diode linear array camera was used. The scan line was set to be across the belt with the camera above the center of the belt. The distance from the belt was such that the angle subtended by the scan line was about 50°. In consequence, no problems arose in selecting a lens and the $\cos^4$ effect on the light pickup was not excessive.

The first problem was to produce a lighting system that would illuminate the glass without illuminating the belt. It was necessary to avoid specular reflection from the flats worn on the belt and the support plate beneath the belt, but specular reflection from the glass

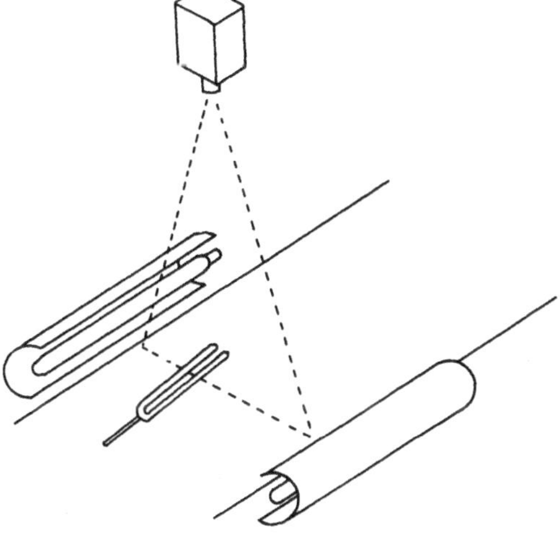

Figure 11.33. The arrangement of the camera and lighting for scanning the tubes on the belt.

was the best method to reveal the tubes. The only positions for the lamps that satisfy these requirements are by the sides of the belt, as shown in Figure 11.33. The lights cannot be too low or light to a tube may be blocked by its neighbor. It is necessary to consider only the lighting along the scan line. For light to be reflected to the camera as the angle of the tubes on the belt changes, it is necessary that it come from points ahead of, or behind, the scan line. For this reason, long light sources are required. A small source spread by a cylindrical mirror or lens is a possibility. Linear discharge lamps, including fluorescent, were considered and the low pressure sodium lamp chosen because of its efficiency and output. Reflectors, roughly elliptical but shaped empirically, were added to improve the uniformity of the reflections as displayed in the video output. As shown in Figure 11.34, each section of the tube is revealed as two bright lines.

The integration time of the camera was set at 40 msec, corresponding roughly to an integral number of cycles from the lamp. The motion of the belt was sensed by an optical incremental encoder and, after counting down, produced 15 interrupt pulses per sec. The next complete scan after an interrupt was accepted by the processing system. The slight variation in the scan spacing resulting from this method is not important and is better than using scan timing phase locked to the belt movement, since this method would have required more complex circuits and the integration time would have been variable and not related to the lamp supply frequency. A count was kept of the number of scans.

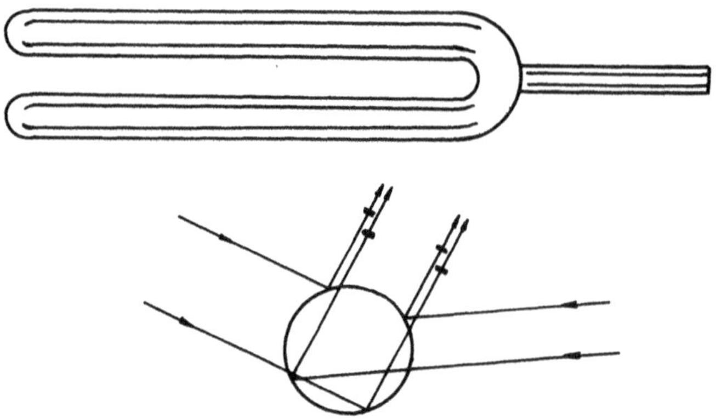

Figure 11.34. The use of reflection from the tube walls to reveal the position of the tube.

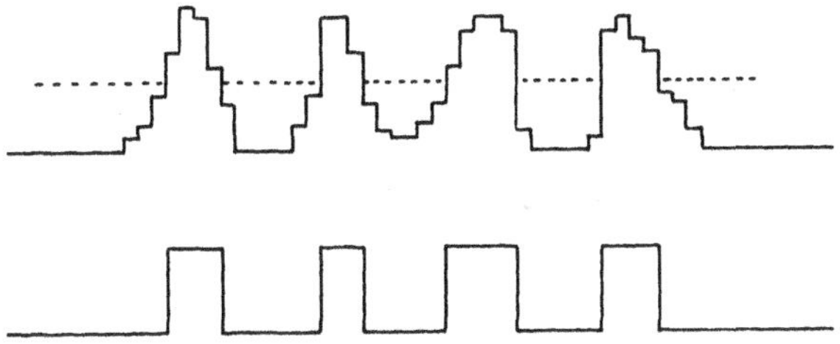

Figure 11.35. The analog video signal thresholded to give a binary output. Note the effect of light scatter from the side walls.

The video signal in the camera is shown in Figure 11.35. In the first system, a fixed threshold was used to convert this to a binary signal. It can be seen in Figure 11.35 that the signal falls gradually outside of the lines from each leg, due to some light being scattered from the vertical sections of the walls of each tube. Where the legs are close together, these extra signals merge. Consequently, the threshold level must not be too low if the bright lines are to be separated. The signal height was not constant across the belt, being lower near the sides, and this required a low threshold. The resulting compromise rendered the system intolerant to the additional variations in the light reflected from individual tubes. This can be overcome by using a threshold generated from values stored in the camera so that it follows the effective variation in the light level across the belt.

11.6.2. Processing

The binary stream from the camera, 1024 pixels at 160 kHz, spaced at intervals of 40 or 80 ms, were fed to a small computer, a Philips P851. The computer was programmed in the real time language RTL/2. The computer was provided with information concerning the diameter of the glass tubes, in terms of the separation of the bright lines, i.e., 9 or 12 pixels, the separation of the legs, and the length of the tube. Since this type of computer is not designed for picture processing, it was appropriate to get away from the picture format as soon as possible. First, the center of each bright line was found and its distance from adjacent lines calculated. If a pair matched the expected separation within a given tolerance, they were nominated as a candidate leg and the position stored. In subsequent scans, the positions of new candidates were compared with the candidate list. Eventually, the list consisted of a start position, in terms of the scan count and pixel position, and similar coordinates for the point at which it was last seen. When it had not been seen for a preset number of scans, e.g., 25, it was deemed to have ended. A gap less than this was allowed before a candidate was ended, to allow for the varying quality of the reflections from the tubes.

As each candidate ended, its length was compared with the stored data to check whether it was complete, i.e., not broken. If complete, it was compared with adjacent candidates to determine whether either were at the correct separation for the U, and whether they had started and finished at about the same time. In most cases this led to the establishment of the position of a U-tube. It could fail for several reasons. When the U-tube had broken at the bend, the legs might not be in the correct relative positions. It was also possible that two U-tubes had their legs equally spaced or that a single broken leg was at the appropriate spacing from a good U-tube. In the first case, the existence of the four legs facilitated a straightforward decision that two tubes were present. The second case is more difficult. Either a guess must be made—fortunately such an occurrence is rare—or the presence of the exhaust tube may reveal the correct pairing. The existence of the exhaust tube has to be noted to determine whether the tube is good or in need of repair.

Thus it is seen that after the initial stage of detecting the lines, processing becomes a matter of manipulating numbers and applying simple logical tests. These are processes for which a standard computer is most suited. With about six tubes in view, the computer could complete its processing within the scan interval. There could be up to 14 tubes; in this case some scans would be ignored. The scan count was correctly maintained and the only effect was some loss of accuracy in the positions of the tubes along the belt.

11.6.3. Testing

The system posed an interesting problem in determining its performance. It could be set up over the conveyor during production, but a robot could not be used to demonstrate the accuracy of locating the tubes. Production could not be interrupted. Clearly, the best method of recording the positions of the tubes on the belt was by photography. The computed positions could have been printed but, with the intention of using 5000–10,000 tubes, correlation with the photographs would have been a major task. Instead, the computed results were used to control the camera and the positions of some small lamps that were visible in the photograph. An area downstream from the scanner was illuminated by fluorescent lamps placed at the side of the conveyor, similar to the positions of the sodium lamps. A 16-mm cine camera, operated in single-shot mode, was placed above the belt to take pictures, of which Figure 11.32 is one. Two small lamps in fixed positions above the side of the belt defined a line across the belt. The computer calculated the time at which the end of a U-tube should cross this line and caused the camera to take a photograph. Prior to this, it had computed the positions for two small lamps that could be moved across the belt under servo control. These lamps then defined a line through the center of the U-tube. This can be seen in the photograph. As a result, each photograph, giving the results for one tube, could be examined in a few seconds.

Several runs were made using two tube sizes, with about 4000 tubes of each size. In all, about 150 tubes were incorrectly located, giving a success rate of 98%. Many of the failures were believed to be due to poor quality of the video signal. At this time, the camera was using the single fixed threshold and could not cope with very low signals or merging of the bright lines between the legs of the tubes. A better performance was expected from the modified hardware and improved software. From measurements of many film frames, using a microscope, the r.m.s. errors in the positions across the belt were 0.6 pixels and along the belt, 4.8 pixels. The extreme errors were 3.5 pixels and 13 pixels respectively, for these directions. The position across the belt is the most important of the two in guiding the gripper of the robot.

11.7. Unpacking and Mounting of Deflection Units

The systems for this task incorporated robots and vision systems, each unit with its own computer.[11,12] Each system contained a control program which specified the task to be performed. The programs invoked vision system commands to determine identity, or to extract location information, that was then converted to the reference frame of the manipulator that performed the required actions. There were also facilities for sequencing the operation of the various parts of the system, and for coping with foreseeable errors and failures. The success of such systems depends significantly on the use of standard mechanical and software modules, and the ease with which these modules can be reconfigured to perform a different task when required. When a single manufacturer's equipment is used, some system building aids and programming tools may be provided. In this case, the two robots and the vision systems were based on different computers. To cope with such a situation, ROBOS, a ROBot Operating System, was developed (see Section 7.7). ROBOS coordinated the activities of the various machines, maintained information about the relationships between the frames of reference, performed transforms between frames, supervised a system of process interlocks, and generally dealt with much of the "housekeeping" necessary in a robot application.

11.7.1. Unpacking

The deflection units (DUs) for television receivers are packed in layers; in each layer are 14 boxes containing 4 DUs (Figure 11.36). Alternate DUs face opposite directions. Play between each DU and its box, and between the boxes and the carton, resulted in a misplacement of each DU, with respect to a nominal grid, of ± 15 mm and $\pm 10°$ about a vertical axis. This error is too great to permit a blind robot to unpack the DUs, so a vision system was used to find the position and orientation of each DU with sufficient accuracy for it to be removed by a gripper.

A Cartesian gantry robot was constructed with an X, Y, θ carriage and an arm that could move vertically. A vidicon camera was mounted on the carriage so that it maintained the same relative position with respect to the gripper. The robot placed the gripper above the nominal position of a DU (Figure 11.36) and from the picture obtained, the exact position was determined. The arm lowered the gripper onto the DU which was then withdrawn and transferred to a conveyor belt. The gripper was fairly compliant and had two thin fingers that slid around the ferrite yoke and were locked. As the DU was pulled from its box, a pneumatically operated "pusher" held down the bottom of the box. The DU was placed cone downwards on the conveyor by having a horizontal joint in the wrist which rotated as the DU dragged on the belt (Figure 11.37). All of the major functions of the gripper were checked by simple sensors.

Figure 11.36. Deflection units in their packing case and the robot gripper poised above one unit.

The ease with which a picture can be processed is largely dependent upon the degree of control that can be exercised on the appearance of the object and its background. If the object can be provided with readily visible reference marks of suitable geometry, or if the object can be set against a contrasting background, then simple methods can often be used. This was not possible in this case. Also, the camera-to-DU distance varied, depending upon the layer being unpacked, and would have changed the image size. This problem was resolved by the use of a parallel projection optical system (see Section 5.1.5) that produces an image of constant size. The scene could not be lit from behind; therefore, only the Fresnel lens placed in front of

Figure 11.37. The first robot placing a deflection unit onto the conveyor.

the camera was used. This collected the paraxial light scattered from the scene which was lit by lights near the camera. The resulting image is shown in Figure 11.38.

A multipass picture processing method was used in which a number of features were identified in sequence. This method relies on the fact that the camera was calibrated, as described later, and that the image size remained constant as the object distance varied. It was then possible to adjust the threshold to isolate a feature of known size from its background, and to find its position. With this knowledge, a search could be made for the next feature as both its approximate position and required threshold were then known. This process could be re-

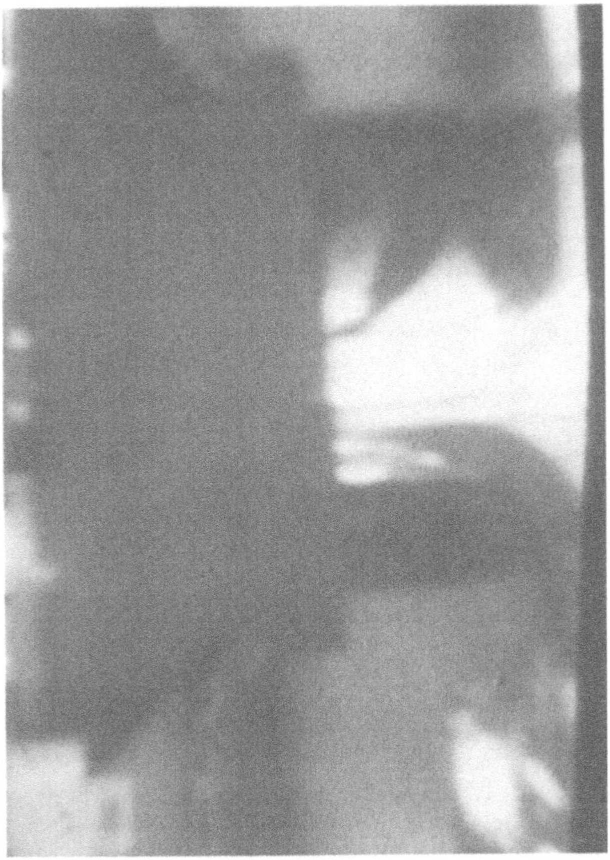

Figure 11.38. The view of the deflection unit in its case seen by the camera through the parallel projection optical system.

peated until the required feature was found. The value of this technique is that a progression can be made from features which, although reliably present and easily isolated, are not accurately positioned on the object, to features that are difficult to find but that can be used to locate the object accurately. For the DU the feature sequence was:

1. A high light on the frame coil winding (Figure 11.39a).
2. The edge of the ferrite yoke next to the winding (Figure 11.39b).
3. The edges of the coil against the background of the box (Figure 11.39c).

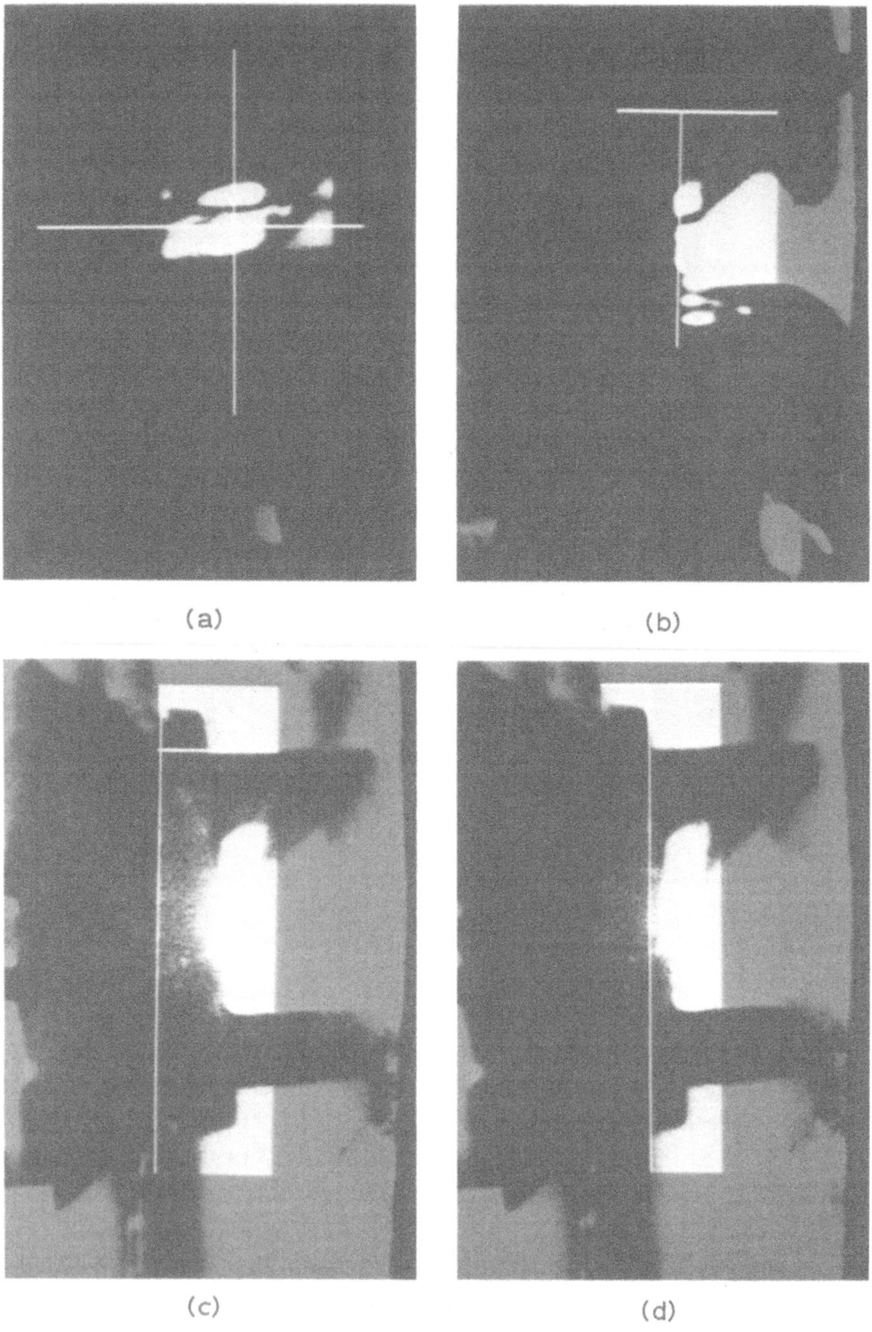

Figure 11.39. Four thresholded pictures derived from the analog view. The bright rectangles mark the processed areas and the lines indicate the positions found.

4. The edges of the yoke against the background of the box (Figure 11.39d).

Thus, the sequence started with the location of an easily detected highlight, and ended with the accurate measurement of the positions of visually obscure edges. The position and orientation of the DU could then be calculated from this measurement.

The system could remove a DU from the box in about 10 sec, with the required gripper position and orientation being determined in about 0.5 sec. The absolute errors were less than 1 mm in position and 1° in orientation. These errors were acceptably small, especially considering the compliance of the gripper.

11.7.2. Mounting

The DUs travelled on the conveyor to a barrier (Figure 11.40) but at that point their positions and orientations were not accurately known. Each had to be transferred to the neck of the television tube in a specific orientation so that three "landing areas" on the DU rested on three ceramic studs on the cone of the tube. When correctly placed, the DU had to be pushed down, with a specified force, while a screw clamp onto the tube neck was tightened to the specified torque.

The DU was scanned by an upward facing vidicon camera using a parallel projection system (see Section 5.1.5), with lighting both from above and below the DU. The parallel projection system was used to avoid the parallax errors that would have occurred had the DU been off of the axis of the camera. The DUs were located by a feature-sequence technique similar to that used for their location in the box. The features used were the central hole, the edges of the windings where they lay on the plastic former, and three small holes in the plastic, the holes having been added to the design for this purpose. The provision of such modifications to assist the vision system is a useful technique, but with the risk that a designer may remove them from new designs as having no function in the operation of the unit.

The scans, set to cover the area of the central hole and surrounding windings, were processed to check the range of gray levels (Figure 11.41). When correct, this revealed the presence of a DU. This was confirmed by selecting a threshold which separated the hole from the surrounding DU and by checking the value of this threshold and the

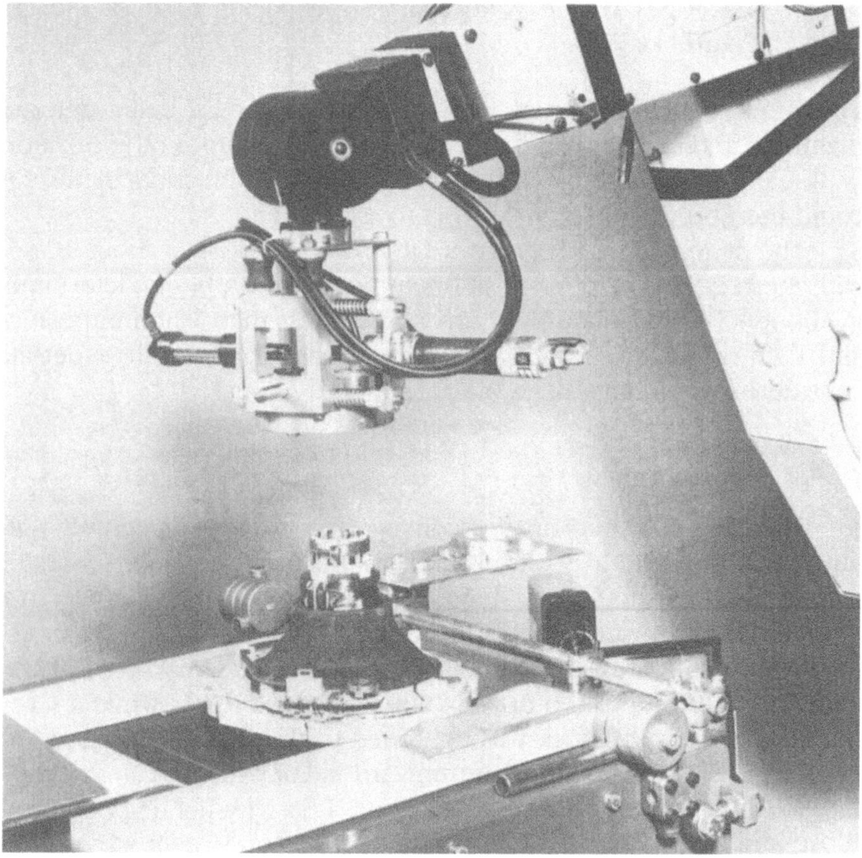

Figure 11.40. The second robot preparing to remove the deflection unit from the end of the conveyor.

position of the center of gravity of the hole. The scan was repeated until this center position ceased to change, at which time the DU was known to be stationary.

The orientation of the DU was found in two steps. First, a circular annulus of pixels, centered on the center of the hold but of a larger diameter, was examined. Simple thresholding separated the windings from the plastic and revealed the positions of the edges of the windings (Figure 11.41). After checking that the winding width was within an allowed tolerance, the edges were used to determine the orientation. The 180° ambiguity arising from the symmetry of the windings was

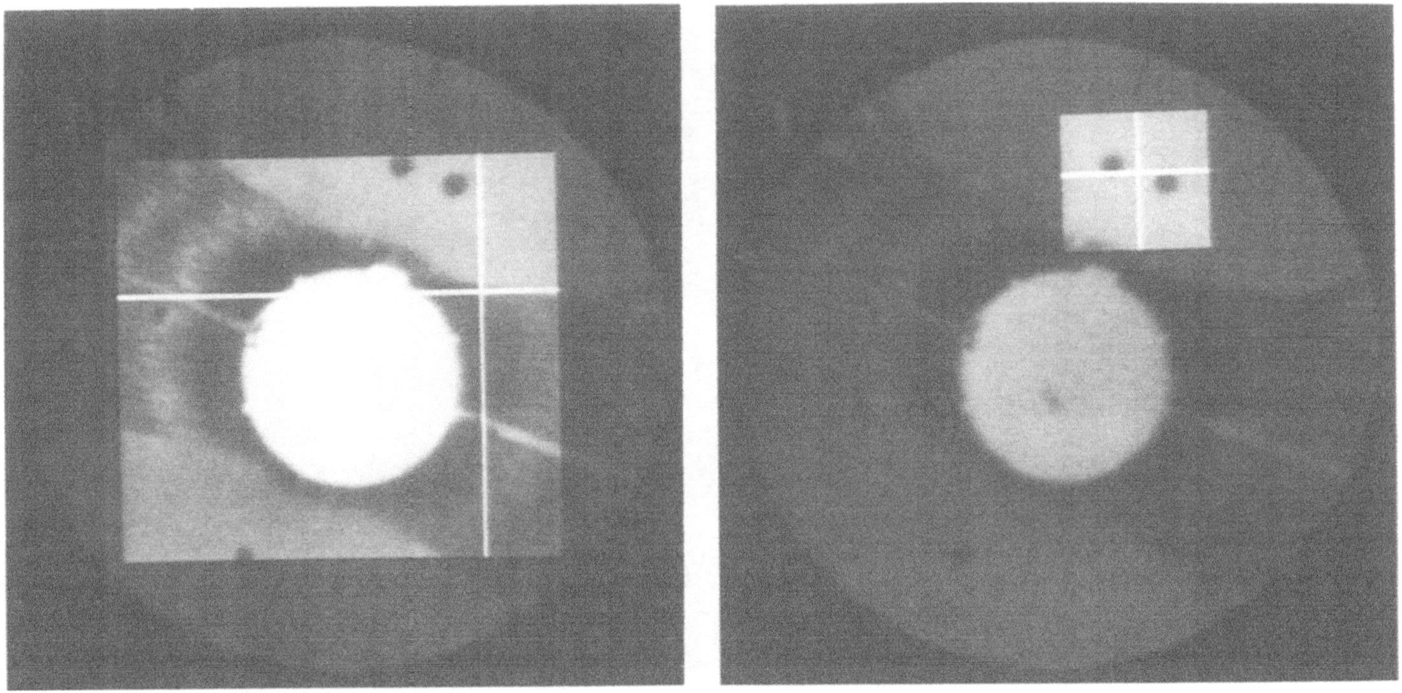

Figure 11.41. The rectangles show the areas scanned to locate the windings and the two holes which indicate the orientation.

resolved by scanning the areas containing the small holes (Figure 11.41), one having a single hole and the other, a pair. The position of the center of the neck hole and the orientation were now known within the television coordinate frame.

The DUs were transferred from the conveyor to the television tubes by a PUMA 560 robot equipped with a specially designed gripper (Figure 11.42). The gripper had pneumatically operated clamping pads to grasp the top of the ferrite yoke, and a pneumatic screwdriver to fasten the clamp. The required vertical, lateral, and torsional compli-

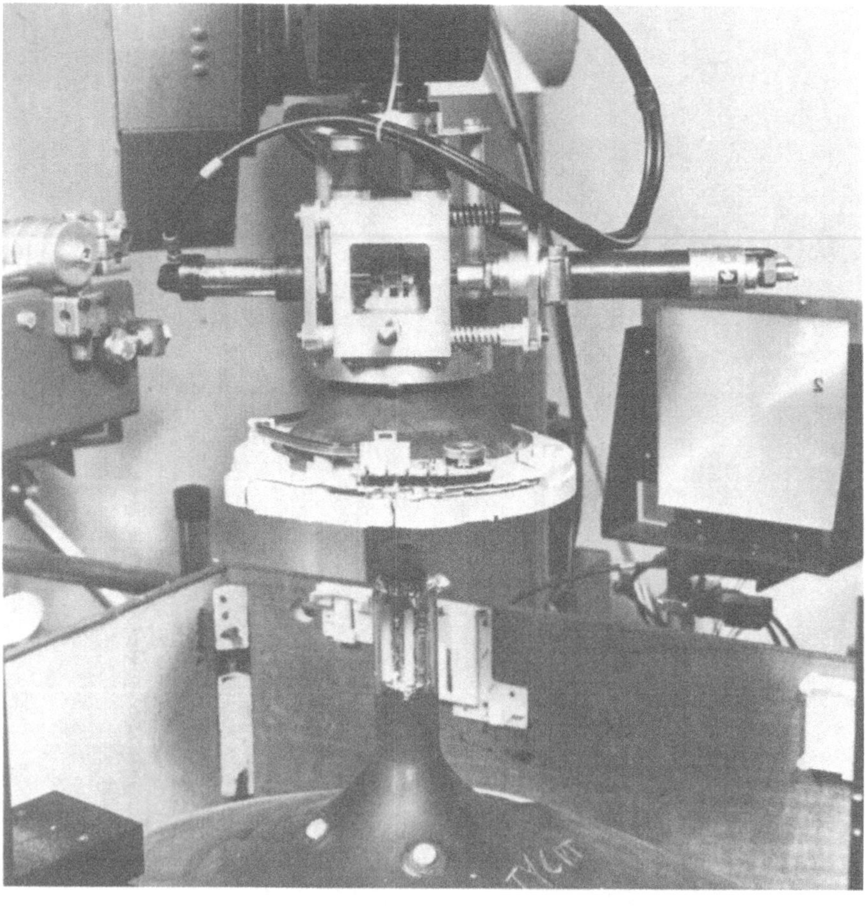

Figure 11.42. The robot placing the deflection unit onto the neck of the tube. The gripper carries the screwdriver to tighten the clamp.

ance was provided by a combination of a compression spring module between the gripper and the PUMA wrist flange, and "metalastic" rubber mountings. The location coordinates of the DU were converted to the coordinate frame of the robot, and the robot instructed to pick up the DU. The orientation of the gripper resulted in the screwdriver being aligned with the clamping screw. Correct grasping of the DU was confirmed by a photodetector.

While the robot was picking up the DU, the vision system used two cameras to determine the position of the neck of the tube. The orientation of the tube was adequately defined by virtue of its position in the television cabinet, and the clamping of the cabinet. Each camera used the backlit parallel projection system. They viewed horizontally but with their axes at an angle (Figure 11.43). The images were thresholded at a level to give the correct width to the top of the neck, and the center of the top found. This gave the displacement from the two optic axes and hence the location of the neck in the coordinate frame of the cameras. This was converted to the frame of the robot and the robot placed the DU on the neck. The DU was lowered until the compliant mountings were sensed to be slightly compressed. It was placed with the orientation deliberately slightly out of position and rotated through an angle sufficient to move it through the correct angle. The shock of the engagement of the studs with the recesses

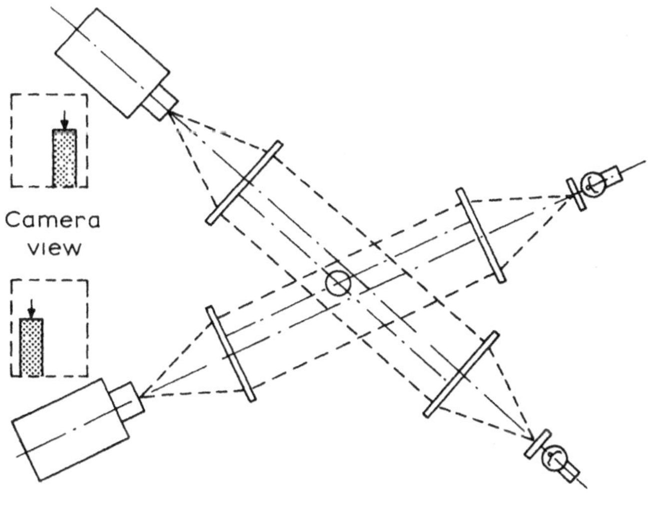

Figure 11.43. The twin parallel optical system used to locate the neck of the tube.

was detected by an accelerometer to prove correct engagement. The over rotation was accommodated by the compliant mountings. The screw was then tightened to the correct torque while the DU was held down by the required force.

The mounting cycle took about 10 seconds. The location of the DU and the neck each took about 1 second, but this occurs in parallel with the operation of the robot.

11.7.3. Calibration and Control

Calibrating a television camera equipped with conventional perspective optics usually means determining its position and orientation in three-dimensional space. Using parallel projection, it is sufficient to determine the camera orientation relative to "world coordinates," and the magnification of the lens.

The camera in the unpacking machine moved in *X, Y,* and θ with the gripper. It was calibrated by using a special plate, carried by the gripper, which could be swung into the field of view of the camera by a pneumatic actuator. The plate was black with four white spots, chosen for ease of picture processing. They allowed the position of the gripper in the television image to be found, and the *X* and *Y* spacings of the spots permitted the scaling factors to be calculated.

In the mounting system, each camera was calibrated using a pointed rod that was carried by the robot. The spatial offset between the gripper and the rod tip was known. The tip was moved into the field of view of each of the three cameras in turn. The position of the tip in the television image was determined using a simple recognition algorithm. The tip was then moved through known distances, in *X, Y,* and *Z,* in the coordinate frame of the robot, and its locations in the coordinate frames of the cameras found by the same algorithm. The transformation matrix was then calculated.

Both systems were controlled by software based on the robot operating system ROBOS (see Section 7.7). In this system there were the two robots, with their own controllers, and a vision system operating the four cameras. Although part of a single vision machine, the cameras and the two robots each had their own coordinate frames. Each of these machines was controlled by its own software module called a machine controller. The machines were connected by standard ROBOS communication channels that were implemented by shifting data in memory, or by physical connection if the machine controller were installed in a separate computer. At the center was the system supervisor, and the data structure that was connected to the task description program and the individual machines.

11.8. Other Systems

Following are brief descriptions of a few examples of sensory systems and signal processing. Further information can be found in the associated references. Other than the first, these systems have not been designed by the authors but serve to demonstrate a wider range of techniques in this field.

11.8.1. A Trainable Component Feeder

Although it tends to take a subordinate role in robotics, the presentation of components to an inspection or assembly system is an important consideration. It is not always possible to retain the orientation of a component from one stage of its production to the next

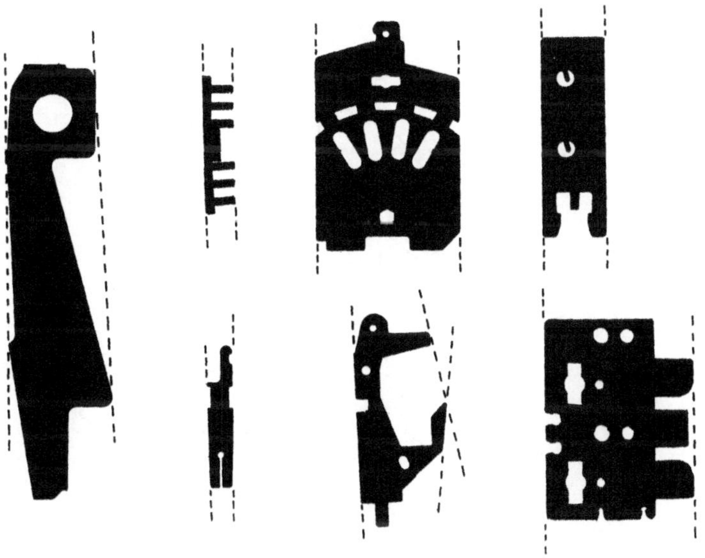

Figure 11.44. Typical components having a limited number of orientations when moving on a track. The surfaces which will run against the side of the track are marked.

and reorientation is often necessary (see Section 9.3.4). Many forms of feeders are available for this purpose; many select components having the right orientation, or reorientate other components, by feeding the components through tooling, an assault course of carefully shaped metal. Some complex components will cause jamming to occur in the tooling. Also, it is often expensive to design the tooling even when a solution is possible. This feeder was designed to avoid these problems; it uses a simple vision system in place of tooling and can be "taught" a new component very easily.[13,14] It makes use of the fact that many components move on a track in a limited number of orientations (Figure 11.44).

A linear array camera is used and the signal is processed by a single board computer (Figure 11.45). A high contrast is obtained by using backlighting through a slit in the track or by using a contrasting track. The output from the camera is a binary signal. Many components can be fed along a track in a limited number of orientations; the vision

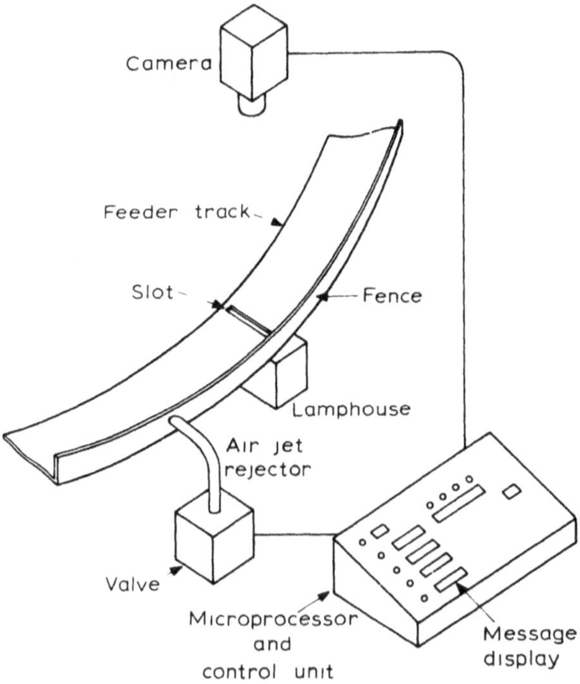

Figure 11.45. A vision system used to determine the orientation of a component on a feeder.

system discriminates between these orientations. The system was intended for components which are almost flat, but it can also be used with any components which take up a limited number of orientations. The camera could view horizontally, if that provided a view of the significant features, or the system could be extended to two cameras. The velocity of the components need not be known and some variation in the speed of each component as it is scanned can be tolerated. Speed changes from component to component of 5 to 1 during training and 30 to 1 during sorting are acceptable.

The camera uses 128 detectors but only 64 of these are used in the recognition process. To avoid the necessity of maintaining accurate positioning of the camera with respect to the track, the processor monitors the position of the fence beside the track and the output from the next 64 sensors are used for processing. When a component is not present, the fence/slit edge is visible as a black-to-white transition. The 64 binary values are condensed into two 16-bit words as shown in Figure 11.46.

The recognition system does not, as such, use concepts of holes, etc., but uses the one-dimensional view of the camera as its basis. In the learning phase, the equipment indicates that a few components should be fed first in the required and then in other orientations. For each batch, it records the black and white word pairs that occur, and

64 bits from camera	BBBW	WWBW	WWBB	BBBB	BWBW	WBWW	WBBB	WBBW	....
16 bit Black word	1	0	0	1	0	0	1	0	.
16 bit White word	0	1	0	0	0	1	0	0	.

	Black word	White word	Identity
Part of reference table	1111 1111 1100 0000	0000 0000 0001 1111	D
	1111 1111 1100 0000	0000 0000 0000 1111	E
	1111 0000 0010 0000	0000 0011 0000 1111	F
	1110 0000 0010 0000	0000 0111 1000 1111	G

			Matches
Condensed scans	1111 1111 1100 0000	0000 0000 0001 1111	D (also E)
	1111 1111 1100 0000	0000 0000 0010 1111	E
	1111 1111 1000 0000	0000 0000 0010 1111	none
	1100 0000 0010 0000	0010 0011 0000 1111	none
	1111 1000 0010 0000	0000 0111 1001 1111	F (also G)

Figure 11.46. Condensation of the scan into the black and white words, and matching the words with the reference table.

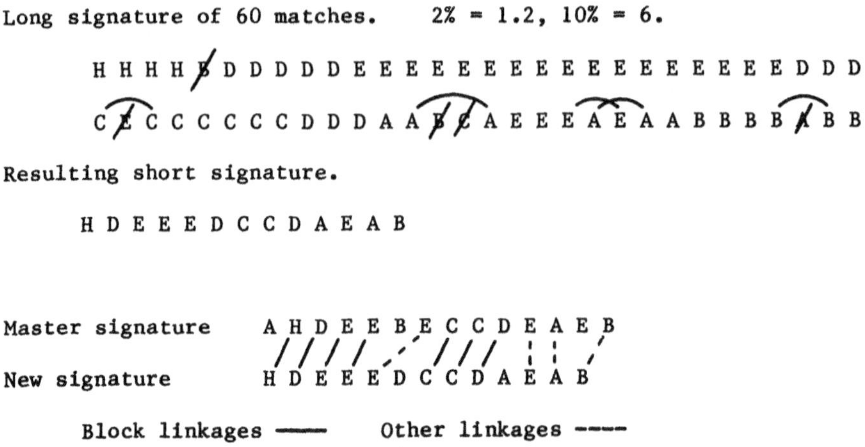

Figure 11.47. Condensing the long signature into a short signature and comparing this with the master signature.

keeps those most frequently occurring. These are placed in a reference table, typically 20 from the required orientation and 12 from the others, and each is given an identity (Figure 11.46). The identities for those from the wrong orientations are marked so that during scoring in the sorting phase they will carry a negative score.

In the final part of the learning process, the system asks to see components in the required orientation. Each new word pair from the scan is compared with the reference table and, if a match is found, the identity is recorded (Figures 11.46 and 11.47). To allow for some variation in the edge of the component and the thresholding in the camera, the words are slightly modified before they are placed in the

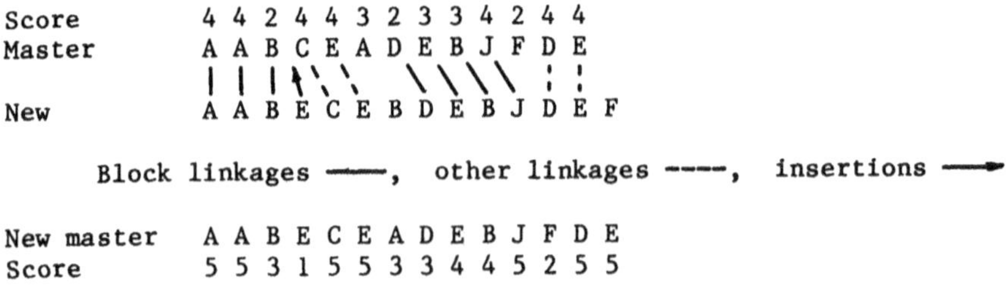

Figure 11.48. Combining the new signature with the temporary master signature to form a new master during the learning process.

reference table to improve the chances of a match. As the camera has a constant scan rate, the number of identities in the long signature so formed will depend upon the speed of the component. The long signature is reduced to an almost standard length as shown in Figure 11.47. After some regrouping, small groups that are shorter than 2% of the total are deleted. One entry is made in the short signature for each group less than 10%, two for 10% to 20% and so on. The length of the short signature is independent of the component velocity.

The short signature from the first component is taken as a temporary master and is modified after comparison with the signatures from the following components (Figure 11.48). Each is matched with the master, first by finding similar blocks of three or more identities, then single matches between these blocks. When there is only one position where an extra identity appearing in the new identity can be inserted, then this insertion is made. Scores of the occurrences of each identity are kept. When one score reaches seven, then all identities with a score of four or less are deleted and the remainder form the master signature. The reference table and the master signature are the data for the component. To avoid retraining after each switch on, this data could be stored in RAM with battery support or in another semi-permanent memory. This memory could retain data for several components that could be selected by a control switch as required.

During sorting, a similar process occurs, the new black/white words are compared with the reference table, the long signature is reduced to the short, and this is compared with the master (Figure 11.47). Of course, the master is not modified. The score for the match is the percentage of the identities of the new signature that match. This is reduced if any word matches have been made with word pairs in the reference table derived from a wrong orientation. The component is deemed to be in the required orientation if the score exceeds a preset threshold.

The system selects one orientation, but by a simple extension it could be extended to recognize others. For example, the reversed orientation results in a reversed signature. Provided that the differences from the training components are not too small, it can reject damaged or alien components. The cost of the system is equal to that of only a few retoolings of a feeder for difficult components, yet it can be set up by low-skill labor.

11.8.2. Fitting Windows to Cars

Windscreens and back lights are fixed directly to the metal of the body by adhesive placed as a band near the edge of the glass.[15] For a clean result it is necessary that the window be placed in the right position the first time. The adhesive is placed on the windscreen by a robot and passed to a second robot that places the glass into the window aperture with an accuracy of 0.5 mm. Another robot places the back light. The car is stationary while the windows are fitted and the operation is completed within 10 seconds, of which about half is used by the vision system.

The vision system consists of four 256-diode linear array cameras with associated lighting. The lamps are mounted near their respective cameras and each produces a parallel beam. The cameras scan orthogonally across the edge of the screen, two cameras being at the two sides and the others scanning at widely separated positions at the top of the screen. The lights are highly reflected from the metal, but as the glass is brought close to the aperture the adhesive casts a shadow on the metal. The cameras monitor the width of the remaining metal. The sensors at the top ensure that the widths are equal and of the required size, while those at the sides equalize their measurements.

The accuracy required is at the limit of a robot of the size necessary to handle the glass, and therefore it is not appropriate to use the robot for the fine adjustments. Instead, the suction cups that hold the glass and the vision system are mounted on a plate which can be moved with respect to the end of the robot arm by stepping motors. Compliance in the suction cups allows the glass to be brought against the metal, and springs supporting the plate control the force applied when the glass is in position. The system is stationary while the scans are made, and the processor switches between the cameras as required.

This is an example of a specialized system, justifiable for such an application, based on a relatively simple technique. The principal problems arise from defects on the metal surfaces that result in the video signals having to be processed to remove noise.

11.8.3. A Miscellany

The use of color in vision systems is demonstrated by a system for automatically picking oranges from a tree.[16] It is necessary that the oranges be ripe when picked; this condition is revealed by their color. By measuring the parameters, e.g., hue and saturation, of the colors in the scan, areas having the correct values may be found. A complication is that some oranges may receive a significant quantity of light reflected from leaves. This will have a high green content and make the fruit look less ripe than it is, resulting in the fruit being left on the tree.

Color has been used in the inspection of car brake shoe assemblies.(17) Several cameras view the scene, which is illuminated by different colored lights. Each camera has a filter appropriate to the color of the light that it is to use. In this way, the cameras can operate simultaneously and have the lighting angles that they require without being affected by the lighting for the other cameras. The system checks for the correct positions of the components, the thickness of the pads, the surface condition of the pads, and the use of the correct spring types.

Carding of the soles of shoes, i.e., grinding a strip around the rim of a sole to aid the adhesion of rubber soles,etc.,(18) and the fettling of castings, i.e., removing the molding lines,(19,20) can be performed by robots, but it helps if the force applied can be measured. The output from a force sensor in the end effector can be related to the motion of the tool and the grinding rate controlled.

An example of the use of a robot to assist a sensor and to acquire a "sense of smell" is illustrated by the system used by British Leyland for the detection of leaks in car bodies.(21) A small quantity of helium is released within the car and the robot moves a helium gas sensor along the areas where leakages are likely to occur.

Many other applications are described elsewhere.(22–30)

References

1. L. Norton-Wayne, CLIP offers superfast image processing, *Sensor Review,* 133–135 (July 1984).
2. L. Norton-Wayne and G. Hutton, Intelligent instrumentation for the automated inspection of tacks, *Proc. 2nd Intl. Conf. of Products and Services for the Automated Factory,* 155–164 (Nov. 1985).
3. S. Holland, Instrumentation for the automated inspection of tacks, B. Sc. project report, Leicester Polytechnic, School of Electronic Engineering (1982).
4. G.A.W. West and L. Norton Wayne, British Patent App. No. 8122618 (July 1981).
5. G.A.W. West, Ph.D. thesis, The City University, London (1983).

6. L. Norton-Wayne, W.J. Hill, and C. Watts, The automated visual inspection of steel strip, *Proc. 5th Intl. Conf. Automated Inspection and Product Control,* Stuttgart (1980).
7. L. Norton-Wayne, The Detection of defects in automated visual inspection, Ph.D. thesis, The City University, London (1982).
8. L. Norton-Wayne, W.J. Hill, and R.A. Brook, Automated visual inspection of moving steel surfaces, *Brit. Journal of NDT 19*(5), 242–248 (1977).
9. W.J. Hill, L. Norton-Wayne, and L. Finkelstein, Signal processing for automatic optical surface inspection of steel strip, *Trans. Inst. Meas. Control 5* (3), 137–154 (1983).
10. A. Browne, Location of glass tubes on a mesh belt conveyor, *Sensor Review 4*(1), 11–14 (Jan. 1984).
11. P. Saraga, C.V. Newcomb, P.R. Lloyd, D.R. Humphreys, and D.J. Burnett, Visually controlled robots for unpacking and mounting television deflection units, *Optical Engineering 23*(5), 512–517 (Sept.–Oct. 1984).
12. P. Saraga, C.V. Newcomb, P.R. Lloyd, D.R. Humphreys, and D.J. Burnett, Vision guides robots in unpacking and assembly, *Sensor Review 4*(2), 78–82 (April 1984).
13. A. Browne, A trainable component feeder, *Proc. 5th Intl. Conf. on Assembly Automation,* 85–93, Paris (May 1984).
14. A. Browne, Vision-based component feeder teaches itself, *Sensor Review 4*(3), 143–146 (July 1984).
15. Anon., Automatic direct glazing of cars, *Sensor Review 4*(3), 125–128 (July, 1984).
16. C. Loughlin, Inspectrum: high speed colour inspection, *Proc. 2nd European Conf. on Automated Manufacturing,* 109–116, Birmingham (May 1983).
17. Anon., Knowledge-based vision software for car industry, *Sensor Review 4*(4), 163–166 (Oct. 1984).
18. M.Y. Chirouze, P. Coiffet, and A. Vacant, Modern shoe machining needs a robot with a force controlled tool, *Proc. 12th Intl Symp. on Industrial Robots and 6th Intl. Conf. on Industrial Robot Technology,* 155–161, Paris (June 1982).
19. H.J. Warnecke, M. Schweizer, and E. Abele, Cleaning of castings with sensor-controlled industrial robots, *Proc. 10th Intl. Symp. on Industrial Robots and 5th Intl. Conf. on Industrial Robot Technology,* 535–544, Milan (March 1980).
20. E. Abele and W. Sturz, Sensors for the adaptive control of fettling tasks with industrial robots, *Proc. 2nd Intl. Conf. on Robot Vision and Sensory Controls,* 133–145, Stuttgart (Nov. 1982).
21. Anon., Robots to 'sniff out' those water leaks, *The Industrial Robot 9,* 150–152 (Sept. 1982).
22. A. Kochan, Case studies, in: *Machine Vision, the Eyes of Automation,* J. Hollingham (ed.), IFS (Pub.) Ltd., Bedford, and Springer-Verlag, Berlin (1984).
23. K.H. Bitter, Application of opto-electronic image sensors, *Proc. 2nd Intl. Conf. on Robot Vision and Sensory Controls,* 93–116, Stuttgart (Nov. 1982).
24. U. Ahrens, W. Friedrich, and S. Deliev, Sensor application in the loading and unloading of pallets with industrial robots, *Proc. 2nd Intl. Conf. on Robot Vision and Sensory Controls,* 177–188, Stuttgart (Nov. 1982).
25. B.G. Batchelor and S.M. Cotter, Recent advances in automated visual inspection, *Proc. 2nd Intl. Conf. on Robot Vision and Sensory Controls,* 307–326, Stuttgart (Nov. 1982).

26. R.N. Kay, Gray scale processing for small batch manufacture, *Proc. 2nd Intl. Conf. on Robot Vision and Sensory Controls,* 339–355, Stuttgart (Nov. 1982).
27. F. Maali, C.B. Besant, and C.D. Backhouse, Visual recognition of engine bearing caps, *Proc. 4th Intl. Conf. on Robot Vision and Sensory Controls,* 427–435, London (Oct. 1984).
28. E.R. Davies, Design of cost-effective systems for the inspection of certain food products during manufacture, *Proc. 4th Intl. Conf. on Robot Vision and Sensory Controls,* 437–446, London (Oct. 1984).
29. B.G. Batchelor and S.M. Cotter, Inspecting complex parts and assemblies, *Proc. 4th Intl. Conf. on Robot Vision and Sensory Controls,* 447–476, London (Oct. 1984).
30. J. Wilder, Automatic optical inspection of keyboards, *IEEE Computer Society Conf. on Industrial Applications of Machine Vision,* 133–138 (May 1982).

Future Directions

12.1. Introduction

In this chapter, we consider the directions in which intelligent automation is likely to move in the future, and highlight shortcomings that will have to be overcome to ensure continued progress. Developments in the tools of automation, both hardware and software, are considered. Areas in which the application of intelligent automation is likely to expand are suggested. The chapter concludes by speculating as to whether there exists any fundamental limitation to what intelligent automation might ultimately achieve.

It must be emphasized at the outset that the factor most significant in determining the rate at which intelligence is applied to automation is, simply, money. There is no shortage of good ideas or of applications. Although imagination, self-confidence, optimism, and technical competence are all important, progress is determined ultimately by the provision of resources regulated in turn by the state of the economy, the health of the market for particular products, and government policy influenced by the expectations of the community. The emergence of new ideas and methodologies is generally most significant because of the increase in cost-effectiveness it brings about—what has hitherto been merely possible becomes viable—though a desire to impress by having the latest innovations on one's production lines may also be a prominent motivation. A tendency for the costs of labor to rise has increased the relative competitiveness of developing countries; even Japan now finds it expedient to import textiles from, e.g., India, and semiconductor chips designed and diffused in the USA are packaged abroad since this labor-intensive activity would otherwise be too expensive. The introduction of automation might bring the work back home, and eliminate the possibility of an indispensable external supplier becoming unfriendly, thus exerting pressure on manufacturers.

More recently, an increasing concern for health and safety in fields such as nuclear power and mining, and underwater operations, such as pipeline inspection, has provided a stimulus to automation which would not be justifed on economic grounds alone.

World markets are bitterly competitive, and the winner is often he who gets his blow in first. Timely application of the techniques and methodology of intelligent automation depends critically on relevant knowledge being disseminated, with the minimum of delay, to those who can use it. The education system is responsible for spreading understanding of the principles in-depth; more immediate distribution of information regarding state-of-the-art advance is the province of specialist technical societies. Much of the theory behind intelligent automation is too specialized to be included within the already overcrowded batchelor's degree, hence provision is needed for covering it at postgraduate level. But it is often inconvenient to force the would-be automation specialist to quit his post in industry, or to commit himself to the arduous study program required for a higher degree. Short courses (of a duration of a week or so) are one solution here, but evening courses provided by local educational institutions can provide deeper and more quantitative expertise than the intensive but necessarily brief encounter provided by the short course.

Many organizations are active worldwide in disseminating technical expertise. In the UK, the British Pattern Recognition Association (BPRA), the British Robotics Association (BRA), and the Institution of Electrical Engineers (IEE) are prominent in organizing relevant technical meetings and conferences. The two latter organizations also publish journals containing technical articles describing advances in the state of the art; BRA tends to be more practical in its coverage than the IEE or BPRA. The various Proceedings of the US Institution of Electrical and Electronic Engineers (IEEE) are respected internationally as a prime source of information regarding innovation in all branches of electronics relevant to industrial automation. The International Association for Pattern Recognition (IAPR), of which the British Pattern Recognition Association is the UK member, coordinates activities of pattern recognition societies in its member countries (which span the developed world), and organizes biannual conferences held in member countries in rotation. Developments in theory applicable to intelligent automation and progress in its application are described in a large number of journals; an up-to-date list is given elsewhere.[1]

12.2. Factors Limiting Progress

In the past, progress has followed two technical developments. The first has been in the hardware available for sensing and processing; electron tube rasterscan cameras such as the vidicon that have been available for many years have been too slow and unstable, too expensive and fragile, and too incompatible with digital processing to be applicable widely in industrial automation. The advent of linescan cameras based on scanned laser beams and solid-state sensors, the former offering speed, sensitivity, and resolution; the latter, value, simplicity, and stability, has been a boon. The second key development has been the dramatic improvement in the cost-effectiveness of computation brought about by the development of the microprocessor. Although the tasks they perform could have been performed by minicomputers that had been available for several years previously, minicomputers were simply too expensive. Development of methodology or software has on the other hand not proved to be a significant limitation—much can be achieved with processing which is conceptually very simple. The state of the art in machine vision, for example, is very much in advance of application to automation. The technical background (and the optimism and self-confidence) necessary to apply methodology which has been available since the mid 1960s has, however, often been lacking. Particular technical areas that are conspicuously in need of concentration of effort to generate innovation in order to eliminate known bottlenecks will be discussed in Section 12.3.

12.3. Areas of Innovation

12.3.1. Mechanical Hardware

Although the development of robots has encouraged developments such as improved drive mechanisms and position sensors, concentration is needed in the mechanical aspects of intelligent automation if the capabilities of existing processing hardware and software are to be fully exploited. For example, manufacturing operations in the garment industry involve pieces of soft, flexible material such as textiles, and a need exists to orient and position these with great accuracy. Image sensing devices which can identify soft pieceparts, orient them to within 1°, and position them to within 0.005 inch are readily available, but cannot be exploited since the capability in mechanical handling needed to complement them does not exist. Although there is little difficulty in devising machine vision capable of inspecting and identifying the mechanical pieceparts of which mechanisms are constructed, actually moving the pieceparts past a sensor at a realistic speed without jamming remains a serious obstacle. Imitation of tasks that a human operative finds easy, such as inserting a screw into a threaded hole, firmly grasping an egg, or climbing a staircase, remain beyond the capability of all but a few experimental prototypes operating (slowly and unreliably) in research laboratories. These problems are currently the subject of vigorous research.

12.3.2. Electronic Hardware

There *will* be continued progress in this area—although the capacity of what exists already far outstrips application! The electronics industry is so geared to research and development (essential for continued profit in the face of intense competition) that it innovates spontaneously. The solid-state sensors used in robot vision are used for many purposes besides automation (such as TV for entertainment and surveillance, facsimile transmission, weapon guidance, and airplane and satellite electronic photography) which generate a demand that preceded automation and, in effect, subsidizes it.

Regarding sensors for machine vision, the most urgent need is for a reduction in the cost of two-dimensional solid-state (e.g., CCD and CID) chips. The current high price is a result of the difficulty encountered in fabricating without defect the large chips that are required; yields are low. Two-dimensional solid-state sensor chips are being made from CCD dynamic memory chips with the opaque cover replaced by a transparent quartz window; these cost an order of magnitude less than chips produced for the purpose, but exhibit inferior performance. Looking further ahead, there is a need for sensors to have parallel outputs; the time taken to clock the output from a large serially-addressed array prevents the obtainment of full benefits from developments in fast digital processors. Ideally, some preprocessing (for example edge detection) should actually take place on the sensor chip. This could be implemented by producing chips comprising several parallel layers of processing; in fact, the production of such multilayer chips has just been announced.[2] The recent appearance of solid-state sensors possessing both line and area scans, with several sites per pixel, each sensitive to a different primary color, thus enabling color imaging with a single camera, will prove useful in many applications.

Laser scanners continue to improve, giving increasingly better resolution and data throughput. Use of fiber optics (or lightguides) rather than mirrors to gather returned light makes design compact and economical. Replacement of the photomultiplier sensor by a solid-state sensor can reduce cost and improve reliability. In a few applications, solid-state lasers can replace the gas lasers currently used. There is merit also in sensing outside the visible wavelength range; light in the infrared at about 10 μ can often penetrate surface dirt and produce reduced noise due to surface roughness, in addition to being less susceptible to disturbances from ambient lighting. The chief difficulty here is that sensors that are both fast and sensitive must be cooled to operate efficiently.

The greatest scope for progress lies, however, with nonoptical sensors. There exists, for example, a need for sensors that can measure the spatial distribution of pressure with a resolution of 1 mm, as does the human fingertip. But there is much less scope for application outside automation here, and so correspondingly less drive for development by the electronics industry; the bulk of work is consequently performed by academic research laboratories, and progress is slow.

12.3.3. Hardware for Processing

The microprocessors on which automation currently depends are used so widely that industry-funded development, fueled by aggressive competition, ensures rapid progress and good value for money. Within little more than a decade, 4-bit processors have developed into 8-bit, 16-bit, and even 32-bit devices. Second-sourcing arrangements ensure dependable delivery, and widespread collaboration provides bus systems, peripheral chips, etc. Special purpose chips incorporating a degree of parallelism are beginning to appear, often as optional extras in general purpose systems, to give a tenfold increase in processing speed.

Pipeline processors for speeding operations on one-dimensional data vectors are well established, and again, are good value since their application is not confined to automation. At least one general purpose convolutional processor is now being marketed. More general systolic processors are being developed. Parallel proessors for two-dimensional vector signals such as CLIP and WISARD are now available commercially, but constitute an expensive option which is not always cost-effective despite vastly increased speed. WISARD has an advantage over CLIP in using standard random access memory rather than chips which must be specially made. Economically viable intelligent automation is possible in almost all applications without *any* of these special purpose processors; at present they are mainly of academic interest, and for use in research rather than on the production line. They will, however, find steadily increasing application in the future, as their costs decline due to quantity of production, and problems due to limited resolution are overcome. A tendency to process ever more complex images, such as those involved in picking randomly arranged components out of bins, in interpreting stereo images, and in processing in real time the views from moving vehicles, is likely to render array processors viable eventually.

The potential superiority of parallel computing even with the SIMD architecture is beyond doubt; the challenge is to make parallel computers cheaply. The solution here may be to make parallel computers using arrays of standard microprocessors, which are made commercially in vast quantitites and are hence inexpensive. We note that CLIP4 achieves economy in chip count by including eight processors on one chip, though each processor handles an image only 1 bit deep. A standard microprocessor such as a Z80 contains only one processor, but this is 8 bits deep. Thus, the number of chips required to handle an image with 8 bits per pixel is the same whether CLIP4 or an array of standard microprocessors is used. But CLIP4 does have a real advantage in its multiplicity of data pathways, which enable it to communicate directly with its eight nearest neighbours. Standard micoprocessors have only one data pathway, and providing a multiplexing capability using extra chips would increase the chip count sufficiently to destroy any cost advantage. Some ingenious methods are however being researched to overcome this shortcoming.[(3)]

The Transputer[(4)] manufactured by INMOS is a revolutionary 32-bit microcomputer whose architecture includes four data pathways; it is thus ideally suited to forming arrays for parallel processing. Similar processors such as GAPP from Martin Marietta and NCR have 72 processors plus memory on a single chip, with a throughput of 28 mips (million instructions per second); many more will undoubtedly appear. But if they are not cost-effective, their future in automation applications is doubtful.

The WISARD processor is noteworthy for being able to distinguish patterns, even though the attributes have not been selected as features and used explicitly by an experimenter. WISARD may be trained for ANY pattern merely by being shown; no other action is required of its human operator. Its tolerance of noise is outstanding. It may be regarded as possessing intuition, and is in this sense unique. Its operation is essentially nonnumeric; of all electronic computers, it is the form probably most similar in its operation to the human brain. In the very long term, WISARD-like processors may well become of dominant importance in intelligent instrumentation.

The sole nonelectronic method of processing still viable is optical processing, particularly that using coherent optics. Though competitive for high data rate operations in defense applications such as synthetic aperture radar, coherent optical processing is as yet little used in automation, and little growth is foreseeable without a major breakthrough. Optical fibers provide an ideal solution to the problem of high-rate data communication in the noisy environments encountered in factory automation.

12.3.4. *Software and Methodology*

The proliferation of existing general purpose languages (Fortran, Pascal, Basic, etc.) is more than adequate. Languages such as "C," that combines assembler and high level instructions, represent a welcome advance. Well-used languages such as Fortran tend to have efficient compilers which generate object code giving fast execution. Special purpose, very high-level languages for specific applications such as robot control and image processing are also appearing; an important objective here is standardization. Perhaps the greatest merit of the dominant general purpose languages is their standardization; dialects are a nuisance in processing. Despite the appearance of books dedicated to high-level languages in image processing [(5)] it appears that no one has specified what such a language should contain, which instructions must be included, and how best instructions should be configured for brevity and clarity.

The chief shortcoming of existing software is a lack of standard, user-friendly packages to implement existing methodology such as machine pattern recognition, and to enable this methodology to be more widely applied. The packages (used for designing classifiers rather than performing classification) should run on the widest range of machines, from micros to mainframes. Processing utilizing the algorithmic approach of traditional machine pattern recognition is underused, partially because of lack of software packages, but also because it is too sophisticated for most engineers to understand. A "law of diminishing returns" effect has taken hold in research, as academics expend increasing effort on trivial details. This is a great pity, since further developments in the imitation by electronic processing of the logical reasoning used by human beings would be more effective.

So far as methodology itself is concerned, that for processing images is healthy. Because of its applications in medicine, defense and other well-funded fields, the state of the art in computer image processing far exceeds current application in automation. The development of methodologies is, however, far from complete, and steady progress continues towards the analysis of ever more general kinds of scenes. However, the approach is changing. Most methods of image analysis currently used depend on a "bottom-up," or inductive approach, in which information initially at the lowest level (i.e., code information in the form of a pixel array) is interpreted to extract higher-level information such as semantic understanding. Current research is inclining towards a "top-down" or deductive methodology, in which general truths are used to interpret information at a lower level. This imitates the way in which the human brain interprets a scene; most of the information used is actually taken from a store in the brain built-up over many years, rather than extracted immediately from the scene.

Machine synthesis of speech is well developed; understanding of speech has reached commercial exploitation at least for isolated words. Use of speech facilitates noncontact communication between man and machine; it leaves hands free for other work, and will probably be used with increasing frequency for issuing instructions to robotic systems. An outstanding characteristic of human information processing is the ability to extract messages from noise, even with message/noise ratios well below unity. This uses the enormous redundancy contained in natural messages to correct errors. Thus, noisy speech can be understood well although individual phonemes are recognized rather poorly; in view of this principle, effort should be concentrated on exploiting context to infer high-level information, rather than on improving phoneme recognition. This misdirection of research effort, towards improving phoneme recognition, is inhibiting progress in the development of speech communication systems.

The most promising current development in methodology for intelligent automation is the application of logical processing (reasoning using high-level information directly) to automation tasks, such as inspection, using special languages such as Prolog and Lisp. Both *expert* systems, and *intelligent knowledge-based* systems are involved; the former incorporate the special skills and knowledge of an expert in the narrow field considered. Application to intelligent automation can only follow publication by research workers of basic results; not much is available as yet, though a useful publication does exist.[6]

The availability of expert systems will have an impact in two ways. The development of a processing system with sensors and actuators and possibly several interacting machines is a complex task involving a knowledge of many technologies. There are many pitfalls to watch for. It is not easy to find experts in many of these areas, let alone all, yet it is necessary to develop and maintain equipment in situations where it is uneconomical to employ such experts. As an alternative, an expert system can be incorporated into a development system which will propose solutions to multidisciplinary problems, guide the engineer, and give warning of pitfalls. This will reduce the cost and timescale required to produce good automation systems.

Automation systems currently used usually incorporated a simple cause-and-effect structure; something is sensed at one place, and a set of simple laws causes something to happen elsewhere. Humans are capable of more complex behavior, since they can learn during operation and adapt to changing circumstances; eventually the same capability will have to be incorporated into machines. Thus, an inspection system looking originally for certain types of faults may itself realize that a new kind of fault has appeared and train itself to identify it. The "unsupervised learning" briefly mentioned in Section 3.3.1 may be required to produce instrumentation which is self-training.

In Section 8.8.2 we pointed out the need for the provision and acceptance of standardized tests for the evaluation and comparison of both instrumentation and methodology. Though the best way of benchmarking an instrument for a particular task is to evaluate it actually on that task, much instrumentation is devised and marketed without a specific application in mind. It is, further, difficult to judge the extent and significance of innovation in methodology without standards for comparison; such standards must be internationally agreed upon and administered by an authoritative body (such as the National Physical Laboratory in the UK) if they are to be effective.

12.3.5. Systems Approach in Automation

A production process (such as the manufacture of motor vehicles or garments) generally comprises a long sequence of operations performed in succession. It is rarely possible to automate a complete line simultaneously; individual operations are automated piecemeal as viable instrumentation becomes available to perform them. To enable new instruments to be added as technology progresses, there is a need for standardized bus systems with which all instrumentation is compatible. To meet this need, General Motors is developing the manufacturing automation protocol (MAP) network, which uses high bandwidth coaxial links and a token passing protocol to prevent congestion due to contention when the traffic rate is high. The objective is to produce a transparent network, to which all compatible instrumentation can "plug in and go," with a standardized modem available in Large Scale Integration (LSI) for incorporation by Original Equipment Manufacturers (OEMs).

Another requirement is for standardization of interfaces so that the component units of an instrumentation system can be interchanged easily. Interface standards such as RS232 are too general to be efficient in most automation tasks, but special purpose interfaces are emerging. For example, the most recent general purpose systems for machine vision use cameras conforming to the RS170 standard, giving the user a wide choice with plug-in compatibility.

Automation systems must be planned to ensure the best possible efficiency with due consideration of the capabilities of instrumentation available now and in the foreseeable future. One obvious requirement is that expensive operations should not be repeated unnecessarily—the classic illustration here is the automated sorting of mail, in which machine vision is used to read postal codes (a difficult and expensive task) only at the initial sorting stage. An identifier comprising dots of ink readable with ultraviolet light is then imprinted and is used in all subsequent stages of automated sorting, since it is much easier and less expensive to read. In other manufacturing operations it is less easy to spot opportunities for avoiding duplication. In apparel manufacture, for example, expensive machine vision can be used to identify pieces of material presented as a random selection, sort them into kits, and orient and position them before an ultimate joining operation. But it is essential that the machine vision be used as sparingly as possible and that less expensive machines be provided to hold pieces in a specified position, bundle, and release them as necessary to meet kit demand, etc. To maximize the efficiency with which such systems are configured, it is beneficial to apply computer simulation and modeling techniques.

12.4. Automation outside the Factory

Intelligent automation is becoming widely accepted inside the factories; though opportunities to exploit it are often greater outside, these are very slow to develop. For example, robots to perform chores in the home such as cleaning, fetching, and carrying are still a long way off, despite the obvious need for them. The required versatility combined with safety (home robots would be operated by the technically illiterate) is beyond the current state of the art. But the economics of high volume production possible in a product for consumer use could (as has happened for washing machines, color television, and cars) result in a complex product being excellent value for money.

The labors of gardening (such as mowing lawns) are quite easy to automate, and it is surprising that robotic or at least telechiric lawnmowers are not available commercially. The guidance involved in covering a grass plot, remaining within its boundaries, and avoiding obstacles is not very complex, though it may be difficult to ensure the necessary degree of safety in unskilled hands.

Perhaps the greatest scope for the spread of intelligent automation in the medium-term future lies in agriculture. There have been some intriguing experiments on robot shearing of sheep, though with limited success as yet. Ploughing fields should be simple using a robotically guided tractor, and could, for example, operate 24 hours a day to take advantage of brief spells of fine weather during a critical time of year. Mechanization of the milking of cows (which involves attaching suckers to numerous udders and then removing them) would speed up a laborious task. Computer image analysis could be used to identify individual cows and enable milk yield to be recorded. Automated crop picking (with individual items such as cabbages tested to ensure they were ready for harvest) is also worthwhile; papers reporting successful progress in this are already being published. The automated sorting, grading, and packing of harvested produce such as apples and kiwi fruit for size and quality is well within the capability of existing know-how. In the past, problems such as pest control, selective breeding, and marketing have preoccupied agriculturalists, and appreciation of the potential for intelligent automation in ensuring efficient and profitable operation has been slow to develop.

The potential for intelligent automation in the fishing industry is much smaller. It cannot easily be applied to actually catch the fish, although commercial fishing is the most dangerous of occupations. There may be some application in sorting catches by species and by size.

Intelligence applied to automation in mining (already well developed, e.g., for cutting coal), to minimize the requirement for human operatives to spend time underground, would improve the safety of a dangerous industry, and facilitate exploitation of reserves in inaccessible places such as under the North Sea. Automation could similarly be applied to oil production, e.g., for inspecting offshore platforms and pipelines deep under waters.

Many other areas remain largely untapped. Light industries, such as footwear and clothing, traditionally dependent on skilled human labor, have exploited little intelligent automation; introducing automation would make the industries in developed countries such as the UK more competitive with countries whose labor cost is low. A large investment is needed to develop the necessary instrumentation, and conditions do not yet seem opportune for providing the necessary capital. In transportation, the navigation and control of ships and aircraft is highly automated in the interests of safety. The safety record of scheduled commercial airlines is consequently outstanding; it is far safer to travel by air than in one's own car. Introducing some automation to the guidance of the family car (for example, to avoid obstacles, or to identify a desired turn-off on a motorway), would equally aid safety. The costs involved could be reasonable because of the vast number of installations involved.

Education in schools and colleges consumes a sizable fraction of the national budget, but still depends on techniques (teacher and class, talk and chalk) which have remained largely unchanged for two thousand years. The potential benefits of information technology have been to a large extent ignored. It is, however, simple to provide computerized teaching in which each pupil has a terminal to himself, through which a dialogue is maintained with a teaching program. The student gets constant individual attention from his "teacher"; attention and interest cannot flag. Instruction can proceed at the optimum rate for each individual, and records of progress can be maintained continuously, automatically, and objectively. Unlike teachers, educational computers cannot suffer "off days"; an outstanding program (unlike a gifted teacher) can be duplicated and enjoyed by millions of students; it need never die. Undoubtedly, certain functions of the human teacher, particularly with younger pupils, can never be automated since they depend on relationships between teacher and pupil which can never exist between a human being and a machine, such as those required to promote discipline and positive social behavior. Could a machine ever convincingly exhibit sympathy or anger, approval or distaste? The main obstacles to the increased and effective application of intelligent automation in education would seem to lie in reactionary and unimaginative human attitudes rather than in the costs involved, though these are undeniably formidable.

In complete contrast, entertainment has made the fullest use of machine intelligence. Computer games whose intellectual level covers the whole gamut of human interest, from chess to space invaders, are now well established; they can be found in any home. The huge scale of production has kept costs down. In a society with ever increasing leisure time, the computer game may be fulfilling a serious social need.

12.5. Long-Term Prospects

Finally, let us consider what direction automation is likely to take in the distant future, whether there are any activities which may never be automated and why, and whether there exists any fundamental limitation to what automation might be made to do.

Some people hold what amounts to a religious belief that the capability for thought displayed by human beings (and to a lesser extent, animals) is unique, and can never be equalled by a machine; it exists only because the living being possesses a soul. A parallel controversy existed in the early 19th century, when the theory of Vitalism asserted that organic chemicals possessed a special life force, and differed from ordinary chemicals in that they could be produced only by a biological process. Wohler's synthesis of urea (unquestionably organic) from cyanate (unquestionably inorganic) in 1828 debunked this theory, and highly complex proteins are now synthesized routinely. As yet, the thought capability of machine systems does not approach that of man because of the latter's complexity, but progress in miniaturization is steadily closing the gap. There is no fundamental difference between the electrochemical processing used in animal nerve systems and the purely electronic processing used in computers. Thus, there appears to be no reason why equality in capability cannot be achieved. Intelligent machines do not have souls because building them in would detract from the machines' usefulness. The nervous system of the animal has evolved to ensure survival; that of the robot is engineered to do a useful job of work. Why complicate matters by building in a capability to suffer?

Continued investment in the engineering of intelligence into machinery is thus very sound. No limit has yet been encountered to the capability of software; though the mathematician Turing showed that calculations can be devised which a conventional computer cannot perform, these are of no practical significance.[7] Although limits to the capability of electronic hardware do exist, set by basic physics, these are hardly a cause for worry. The speed at which information can flow is limited by the speed of light to about 1 nsec per foot, but this is countered by microminiaturization. Use of very large-scale integration (VLSI) keeps signal paths very short. A finite quantity of energy is required to store each bit of information (otherwise there will be corruption by thermal noise), which in turn limits the minimum size permissible for a memory cell. But this is far from being reached; semiconductor processing technology imposes a restriction first.

Thus, we can foresee no *physical* limitation to what intelligent automation might be made to do; a *practical* limit is set by cost-effectiveness, but the limit of economic viability is ever extending as the cost of processing hardware decreases. There is a limitation of another kind, since many activities depend at least partially upon emotional relationships which can exist only between human beings—like respect, loyalty, and sexual attraction. Will a man obey an order to expose himself to danger, knowing this has been given by a machine? And, consider the nursing of the sick; this indispensible activity consumes a substantial fraction of the national budget. In hospitals, machines which can automatically monitor the heartbeat, blood pressure, breathing, etc. of a patient are commonplace. If any abnormality is sensed, and if automated corrective action (like electronic pulses to restart the heart) does not suffice a doctor can be summoned. But there is much more to nursing than merely monitoring the vital signs of patients. It is most improbable that in the foreseeable future there will emerge a machine for comforting the distressed, positioning a bedpan, administering an injection, or ejecting pestering visitors. Where the participation of a human personality is paramount, automation cannot take over, but wherever it is not, automation is always a possibility.

References

1. J. Hollingum, *Machine Vision,* IFS (pubs) Ltd., Bedford, U.K. (1984).
2. G.R. Nudd, The Impact of VLSI on Image Analysis, Proceedings of the 10th European Solid State Circuits Conference, 19–21 September, 1984, Edinburgh.
3. H.J. Siegel, L.J. Siegel, F. Kemmerer, P.T. Mueller, Jr., H.E. Smalley, Jr. and S.D. Smith, PASM—a partitionable multimicroprocessor system for image processing and pattern recognition, *Tech. Rept. TR-EE-79-40*, School of Electr. Eng., Purdue University (1979).
4. D. May, T. Kingsmith and I. Pearson, The T414 Transputer—the End of the Beginning, *Electr Eng. 57*(707):51–59 (1985).
5. M.J.B. Duff and S. Levialdi (eds.), *Languages and Architectures for Image Processing,* Academic Press, London (1981).
6. Applications of artificial intelligence, *Proc. SPIE 485* (1984).
7. M. Minsky, *Computation-Finite and Infinite Machines,* Chap. 8, Prentice-Hall International, London, 1972.

Index

GPSR Compliance

The European Union's (EU) General Product Safety Regulation (GPSR) is a set of rules that requires consumer products to be safe and our obligations to ensure this.

If you have any concerns about our products, you can contact us on ProductSafety@springernature.com

In case Publisher is established outside the EU, the EU authorized representative is:

Springer Nature Customer Service Center GmbH
Europaplatz 3
69115 Heidelberg, Germany

Batch number: 10761599

Printed by Printforce, the Netherlands